AF901026

BIBLIOTHEK
Zentrum für
Ultrastrukturforschung
Inv. Nr.
H630-B1522

Universitätsbibliothek
Bodenkultur
Ausgeschieden
mit Zl. 2020'0376

UB BOKU
+H23877905
HB 05

Methods and Tools in Biosciences and Medicine

Microinjection,
edited by Juan Carlos Lacal, Rosario Perona and James Feramisco, 1999
DNA Profiling and DNA Fingerprinting,
edited by Jörg Epplen and Thomas Lubjuhn, 1999
Animal Toxins – Facts and Protocols,
edited by Hervé Rochat and Marie-France Martin-Eauclaire, 2000
Methods in Non-Aqueous Enzymology,
edited by Munishwar Nath Gupta, 2000
Techniques in Molecular Systematics and Evolution,
edited by Rob DeSalle, Gonzalo Giribet and Ward Wheeler, 2002
Methods for Affinity-Based Separations of Enzymes and Proteins,
edited by Munishwar Nath Gupta, 2002
Analytical Biotechnology,
edited by Thomas G.M. Schalkhammer, 2002

Analytical Biotechnology

Edited by
Thomas G.M. Schalkhammer

Birkhäuser Verlag
Basel · Boston · Berlin

Editor
Prof. Dr. Thomas G.M. Schalkhammer
TUDELFT
Kluyver Lab. for Biotechnology
Julianalaan 67
2628 BC DELFT
The Netherlands

I
605 231/6
UNIVERSITÄTSBIBLIOTHEK
BODENKULTUR
Wien
Peter-Jordanstraße 82
A-1190 Wien

Library of Congress Cataloging-in-Publication Data
Analytical biotechnology / edited by Thomas G.M. Schalkhammer.
p. cm. – (Methods and tools in biosciences and medicine)
Includes bibliographical references and index.
ISBN 3764365897 (soft cover : alk. paper)
ISBN 3764365900 (hbk. : alk. paper)
1. Analytical biotechnlogy. 2. Analytical biochemistry – Methodology. I. Schalkhammer, Thomas G.M. 1961–. II. Series.

Deutsche Bibliothek Cataloging-in-Publication Data
Analytical biotechnology / ed. by Thomas G. M. Schalkhammer. – Basel; Boston; Berlin: Birkhäuser, 2002
(Methods and tools in biosciences and medicine)
ISBN 3-7643-6589-7
ISBN 3-7643-6590-0

ISBN 3-7643-6589-7 Birkhäuser Verlag, Basel – Boston – Berlin
ISBN 3-7643-6590-0 Birkhäuser Verlag, Basel – Boston – Berlin

The publisher and editor can give no guarantee for the information on drug dosage and administration contained in this publication. The respective user must check its accuracy by consulting other sources of reference in each individual case.

The use of registered names, trademarks etc. in this publication, even if not identified as such, does not imply that they are exempt from the relevant protective laws and regulations or free for general use.

This work is subject to copyright. All rights are reserved, whether the whole or part of the material is concerned, specifically the rights of translation, reprinting, re-use of illustrations, recitation, broadcasting, reproduction on microfilms or in other ways, and storage in data banks. For any kind of use permission of the copyright owner must be obtained.

© 2002 Birkhäuser Verlag, PO Box 133, CH-4010 Basel, Switzerland
Member of the BertelsmannSpringer Publishing Group
Printed on acid-free paper produced from chlorine-free pulp. TCF∞
Cover illustration: Color assay using assemblies of DNA-conjugated nano-clusters (see p. 264)
Printed in Germany
ISBN 3-7643-6589-7
ISBN 3-7643-6590-0
9 8 7 6 5 4 3 2 1

www.birkhauser.ch

Contents

List of Contributors

ALGUEL, YILMAZ, Institut für Biochemie und Molekulare Zellbiologie, Universität Wien, Dr. Bohrgasse 9, A-1030 Wien, Austria

BAUER, GEORG, NovemberAG, Ulrich-Schalk-Str. 3, 91056 Erlangen, Germany; e-mail: bauer@november.de

GABOR, FRANZ, Institute of Pharmaceutical Technology and Biopharmaceutics, Althanstraβe 14, A-1090 Wien, Austria; e-mail: franz.gabor@univie.ac.at

HOFFMANN, OSKAR, Institute of Pharmacology and Toxicology, Althanstraβe 14, A-1090 Wien, Austria; e-mail: oskar.hoffmann@univie.ac.at

MAYER, CHRISTIAN, Analytical Biotechnology, TU-Delft, Julianalaan 67, 2628 BC Delft, The Netherlands; e-mail: c.mayer@tnw.tudelft.nl

PALKOVITS, ROLAND, Hämosan Life Science Services, Dr. Bohrgasse 7B, A 1030 Wien, Austria; e-mail: roland.palkovits@haemosan.com

PITTNER, FRITZ, Vienna Biocenter and Ludwig Boltzmann Forschungsstelle für Biochemie, Dr. Bohrgasse 9, A-1030 Wien, Austria; e-mail: fp@abc.univie.ac.at

STICH, NORBERT, Analytical Biotechnology, TU-Delft, Julianalaan 67, 2628 BC Delft, The Netherlands; e-mail: n.stich@tnw.tudelft.nl

VERHEIJEN, RON, Euro-Diagnostica B.V., Beijerinckweg 18, P.O. Box 5005, NL-6802 EA Arnhem, The Netherlands; e-mail: ron.verheijen@eurodiagnostica.nl

WIRTH, MICHAEL, Institute of Pharmaceutical Technology and Biopharmaceutics, Althanstraβe 14, A-1090 Wien, Austria; e-mail: michael.wirth@univie.ac.at

Preface

Modern analytical biotechnology is focused on the use of a set of enabling platform technologies that provide contemporary, state-of-the-art tools for genomics, proteomics, metabolomics, drug discovery, screening, and analysis of natural product molecules. Thus, analytical biotechnology covers all areas of bioanalysis from biochips and nano-chemistry to biology and high throughput screening. Moreover, it aims to apply advanced automation and micro fabrication technology to the development of robotic and fluidic devices as well as integrated systems.

This book focuses on enhancement technology development by promoting cross-disciplinary approaches directed toward solving key problems in biology and medicine. The scope thus brings under one umbrella many different techniques in allied areas. The purpose is to support and teach the fundamental principles and practical uses of major instrumental techniques. Major platforms are the use of immobilized molecules in biotechnology and bioanalysis, immunological techniques, immunological strip tests, fluorescence detection and confocal techniques, optical and electrochemical biosensors, biochips, micro dotting, novel transducers such as nano clusters, atomic force microscopy based techniques and analysis in complex media such as fermentation broth, plasma and serum. Techniques related to HPLC, capillary electrophoresis, gel electrophoresis, and mass spectrometry have not been included in this book but will be covered by further publications.

Fundamentals in analytical biotechnology include basic and practical aspects of characterizing and analyzing DNA, proteins, and small metabolites. The structure of this book should provide a clear sense of where each technology is used and how to implement most effectively each technology in developing and using effective, efficient analytical methods for characterizing and analyzing biomolecules. The protocols included as a workshop will be of value to all students, scientists, or regulatory, technical, and quality-insurance personnel.

Additional modules are included that address basic experimental procedures, acquisition of transferable skills, security concerns, and awareness of the commercial and ethical considerations of scientific research. It is clear that not all readers will have practical experience in biotechnology, although a basic understanding of chemistry and physics (optics and electro) is assumed. Coupled with this knowledge the book starts with familiar techniques providing the chemical basis and understanding of how to bind molecules to surfaces and how to analyze chemical surface groups as a basis for further reading. Novices in the field should read chapter by chapter, whereas experts can use any chapter without needing information from others (except where cited). Student readers' initial exposure is simply one of orientation, such that they are able to recognize the equipment and become familiar with some of the basic concepts and further on to operate them routinely. Thereafter, specialized chapters such as AFM or nano-particles introduce novel techniques in a "hands-on" way. A variety of techniques, which may be used for the identification and characterization of cells and biomolecules down to single-molecule detection, provide a tool-kit for modern bioanalysis. Aside from the instruction to a variety of practical techniques, the add-on-modules also ensure that the reader is equipped with the skills that are essential to the training of good scientists. Risk management and experimental planning are all dealt with if hazardous chemicals are to be used. Further more specialized guidance is offered with respect to focus of the chapters, resources available for the literature review, and practical hurdles that must be overcome.

The central idea is that the book should not be prescriptive but should offer the greatest flexibility. All chapters offer the opportunity to undertake related experiments. The practical information at the application stage is outlined in protocols, associated with helpful information. A wide span of the "level of choice" is offered, starting from simple readout of fluorescence and moving up to single-molecule handling with atomic force microscopes. Detailed protocols ensure that a number of selected experiments may be used as a basis for practicals and courses. Our experience is that basic knowledge combined with protocols permits a greater depth of investigation and a more fruitful application.

Thomas G. M. Schalkhammer — März 2002

1 Immobilized Biomolecules in Bioanalysis

Fritz Pittner

Contents

Methods and Tools in Biosciences and Medicine
Analytical Biotechnology, ed. by Thomas G.M. Schalkhammer
© 2002 Birkhäuser Verlag Basel/Switzerland

1 Introduction

For analytical purpose, immobilized biomolecules may be employed in various areas. Because of their specificity, enzymes have been used for several decades for the assay of analytes in complex samples to avoid laborious purification procedures. Beyond that, immobilized enzymes offer further advantages because they are more stable, reusable, and may be easily attached to microtiter plates, chips, test strips, electrodes, and bioreactors connected to bioanalytical devices.

To isolate or enrich analytes followed by qualitative or quantitative determination, affinity chromatography techniques employing specific and selective biorecognition may be used. For this purpose, antigens, antibodies, receptors, hormones, or other biomolecules exhibiting strong biorecognitive affinities may be immobilized to – in most cases – solid supports. Also, such bio-components may be immobilized to construct new types of ELISAs, gene probes, etc., connected to analytical biochips, microelectrodes, and so on.

In order to reach these aims, it is of vital interest to have good and reliable immobilization techniques at hand, which will be reported in this chapter. Techniques to immobilize biomolecules comprise micro-encapsulation, entrapment in polymeric networks, adsorption, covalent binding, and cross-linking with bi-functional reagents. For our purposes, in most cases these last two techniques turned out to be the most practical and will be treated preferentially, to provide you with recommended, simple, and efficient immobilization techniques to reach your aims.

To attach a biomolecule to a matrix, either reactive groups of this biomolecule or groups of the respective matrix have to be employed. If there are no reactive groups available, they must be introduced. Sometimes cross-linking reagents are useful for coupling, and sometimes spacers are necessary to make the biological ligand more accessible, especially if it is a small molecule.

However, most coupling reactions can be carried out with comparatively few reactive groups, which will be introduced now, together with an appropriate choice of coupling techniques.

Reactive residues of biomolecules recommended for coupling are:

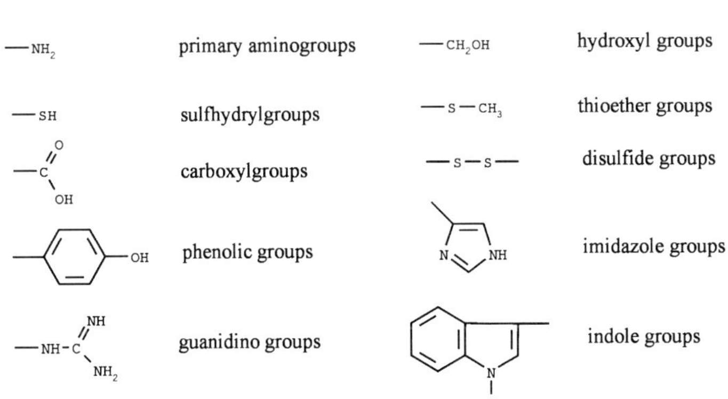

Figure 1

As carrier or support for the biomolecules nearly every nontoxic substance may be used that possesses reactive groups or is capable of introducing them.

2 Methods

2.1 Basic activating and coupling techniques:

Zero cross-linking
The smallest reagents available for bioconjugation are the so-called zero-cross-linkers. They act as activators, mediating the conjugation of two reactive groups of different molecules by forming a bond containing no additional atoms. In this case no intervening linker or spacer will be incorporated.

This is necessary, e. g., in the preparation of hapten-carrier conjugates. The intention here is to generate only an immune response to the attached hapten. Therefore, additional epitopes created by cross-linkers or spacers must be avoided. In this case zero-length cross-linking agents eliminate the potential for this type of cross-activity, since the only mediate is a direct linkage between two substances.

The reactions presented in the following may be performed in aqueous or nonaqueous environments, depending on the desired application.

The types of bonds involved here are:
- amide linkages (i. e. condensation of a primary amine with a carboxyl group)
- phosphoramidate linkages (i. e., reaction of an organic phosphate group with a primary amine)
- secondary or tertiary amine linkage (i. e., reductive amination of primary or secondary amines with aldehyde groups)

Carbodiimides
These compounds are employed to mediate the formation of amide linkages between an organophosphate and an amine [1, 2].

Protocol 1 EDC [or EDAC; 1-ethyl-3-(3-dimethylaminopropyl)carbodiimide hydrochloride]

This reagent is very soluble in water as well as the isourea formed as a byproduct of the cross-linking reaction. Both may be removed easily from the desired products by dialysis or gel filtration [3–11]. Caution, the presence of both carboxyl groups and amines on one of the molecules to be conjugated with EDC may lead to partial self-polymerization.

1. The protein to be modified may be dissolved (about 10 mg/ml if possible) in one of the following media:
   * water
   * 0.1 M MES, pH 4.7–6.0
   * 0.1 M sodium phosphate pH 7.3
2. Dissolve the substance to be coupled in the same solution used in step 1. Small molecules should be added in more than 10 molar excess to the protein amount present.
3. Add the solution prepared in step 2 to the protein solution
4. Add EDC to the above solution to obtain a 10-fold molar excess – or more, if possible – of EDC to the protein. Avoid precipitation and scale back to added amount until a soluble incubation mixture is obtained.
5. React for 2 h at room temperature.
6. Purify the resulting conjugate by gel filtration or dialysis. Recommended buffer for this procedure: 0.01 M sodium phosphate, 0.15 M NaCl pH 7.4 works in most cases. Unwanted turbidity may be removed by centrifugation. In case of immunogen conjugates this is not necessary, since precipitated immunogens are often more immunogenic than the soluble ones.

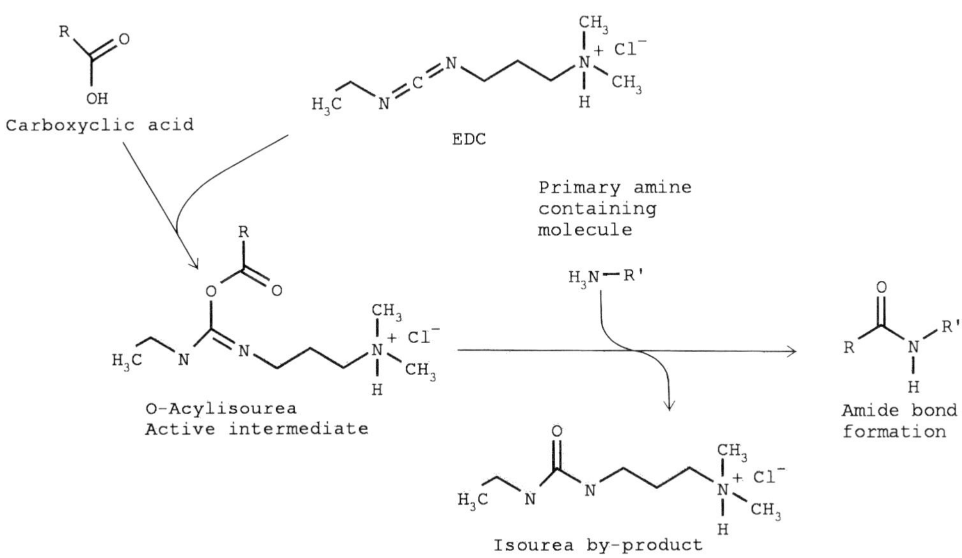

Figure 2

Protocol 2 EDC combined with sulfo-NHS

Sulfo-NHS is able to increase the efficiency of EDC-mediated reactions by forming sulfo-NHS ester intermediates. These intermediates are more stable in aqueous solutions than are those formed from the reaction of EDC alone with carboxylate which results in higher yields of desired amide bond formation [12]. This technique is very useful to create activated proteins [13]. Problems to solve: In addition to potential side reactions of EDC, the high efficiency of the sulfo-NHS mediated reaction may result in insoluble conjugates, particularly when coupling peptides to carrier proteins. In such cases scale back the amount of EDC/sulfo-NHS added to the reaction. Sometimes sulfo-NHS has to be omitted completely.

Protocol

1. Dissolve the protein to be modified (1–10 mg/ml) in 0.1 M sodium phosphate buffer pH 7.4. (To increase the solubility of some proteins; addition of an appropriate amount of NaCl is recommended).
2. Dissolve substance to be coupled in the same buffer in at least a 10-fold molar excess, if solubility allows increase the excess even further.
3. Combine the two solutions to obtain an at least 10-fold molar excess of small molecule to protein if possible.

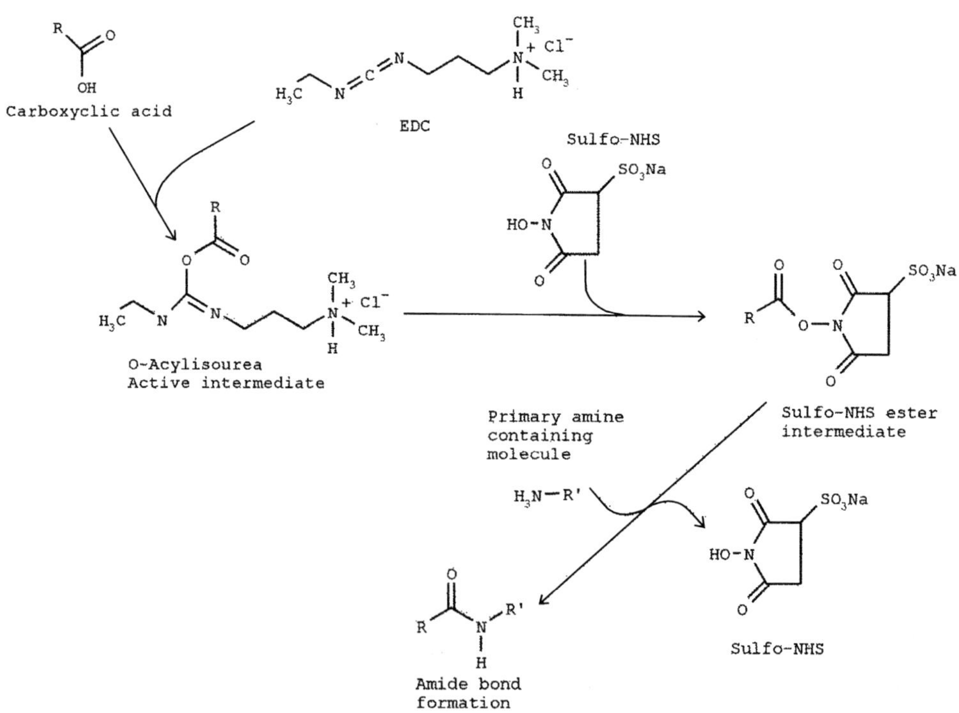

Figure 3

4. Add EDC (at least 10-fold molar excess to the protein) and sulfo-NHS (final concentration 5 mM). Use of highly concentrated stock solutions is recommended. Mix well. If undesired precipitation occurs, scale back the amount of addition until conjugate remains soluble.
5. Incubate for 2 h at room temperature.
6. Purify the product by dialysis or gel filtration.

Schiff base formation followed by reductive amination

Primary and secondary amines can react with aldehydes or ketones to form Schiff bases. Formation of Schiff bases is enhanced at alkaline pH values, but they are not completely stable unless reduced to secondary or tertiary amine linkages by $NaBH_4$ or $NaCNBH_3$[14]. $NaBH_4$ also converts the remaining aldehyde groups into nonreactive hydroxyls, whereas $NaCNBH_3$ will not affect the original aldehyde groups or the activity of some sensitive proteins.

For the following procedure it is assumed that the requisite groups are present either on the two molecules to be conjugated or on the carrier and the biomolecule to be immobilized.

Protocol 3 Schiff base formation followed by reductive amination

1. Dissolve the amine containing compound to be conjugated at a concentration of 1–10 mg/ml in a buffer having a pH between 7 and 10. (The higher the pH, the more conjugate will be formed.) Buffers to be recommended: 0.1 M sodium phosphate buffer or 0.1 M sodium citrate buffer. Do not use amine-containing buffers.
2. Add an amount of the aldehyde-containing compound to the mixture in step 1 to obtain the desired molar ratio for conjugation.
3. Add about 10 ml 5 M cyanoborohydride in 1 M NaOH/ml of the conjugation mixture. Cyanoborohydride is highly toxic: Avoid skin contact, do not inhale, and use a fume hood when carrying out this procedure!!
4. Incubate for 2 h at room temperature.
5. Add 20 ml 3 M ethanolamine (pH adjusted to the desired value with HCl) per ml of the volume of the reaction mixture. Allow to react for 15 min. at r. t.
6. The conjugate may be purified by gel filtration or dialysis using an appropriate buffer. In the case of immobilized biocompounds, the immobilisate may be washed with buffer, 0.1 M NaCl solution, ice cold distilled water followed by buffer.

Homobifunctional cross-linkers

These cross-linkers act as a molecular rope, tying amino group-containing biomolecules together by reacting covalently with the same common groups on both molecules. Also, carriers containing amino groups may be activated with excess of cross-linker to immobilize biomolecules containing primary amino or hydrazide groups.

Activation of hydrazide carriers with glutaraldehyde
This is a good method to immobilize biomolecules to inorganic or organic insoluble carriers, providing spacers long enough to make immobilized ligands or proteins sterically accessible to the bulk solution [15].

Protocol 4 The cyanogen bromide method of coupling

Activation with cyanogen bromide is a good means to activate carriers with plenty of –OH groups for immobilization of biomolecules with sterically accessible aminogroups.
CAUTION: Cyanogen bromide is highly toxic and explosive in the solid state upon friction. It must be kept under alkaline conditions; otherwise, HCN or dicyan will be formed.
Various procedures for cyanogen bromide activation are given in the literature [15]. One of the most convenient is the following one.
Protocol

1. Wash wet packed polysaccaride carriers (10 g) (e. g., Sepharose 4B or Cl-4B (Pharmacia)) with water.
2. Suspend it in 20 ml 2 M K_2CO_3 solution, cooled to 0 °C under gentle stirring or shaking.
3. Activate with 1 ml cyanogen bromide solution in N,N-dimethylformamide (DMF) (1 g/ml) for 90 s.
4. Filter and wash with 50% aqueous DMF followed by ice water and 0.2 M aqueous $NaHCO_3$ solution.
5. Add the substance to be coupled (50–500 mg) dissolved in about 20 ml of appropriate buffer pH 7.6–9.0 (avoid buffers containing primary amino groups) and incubate the mixture by rotating slowly in the cold for 24 h. Decrease in pH may result in a decrease of the amount of ligand coupled but sometimes may be compensated for by a better survival of biological activity. Below pH 6, the coupling efficiency is usually too low to be of practical use. Higher pH may increase the coupling yield, but also may damage the ligand and cause unwanted cross-linking.
6. Remove unbound ligand by washing with buffer.
7. Store wet in the cold before use.

Bisoxirane coupling
This technique is useful for introducing low molecular weight ligands through amino or hydroxyl groups. The reaction proceeds in two steps: (1) activation in which one oxirane group of the reagent reacts with hydroxyl in the polymer matrix, leaving the other group free; (2) the coupling itself, where the remaining oxirane ether is allowed to react with the ligand forming substance containing either amino groups (or other nucleophils) or OH groups. Agarose is one of the best carriers to be activated with this technique. It is recommended to synthesize such a highly activated gel on a small scale to obtain optimal results [16].

Protocol 5 Bisoxirane coupling

1. Wash about 7 ml wet, packed Agarose gel (Pharmacia) with distilled water on a glass filter; remove water by gentle suction.
2. Suspend the gel in 5 ml 1 M NaOH containing 2 mg $NaBH_4$/ml and 5 ml 1,4-butanediol-diglycidylether and incubate the mixture at r. t. for 10 h under slow stirring or gentle agitation.
3. Wash the activated gel thoroughly with distilled water and store until used in distilled water at 5 °C. The gel may be kept about 1 week without loosing its coupling properties.
4. For coupling of the ligand, suspend the gel in 5 ml 0.2 M Na_2CO_3 solution at pH 9, containing up to 600 mg of the ligand. Keep in suspension by stirring gently for 2 days at room temperature. (If the stability of the ligand allows, keep the mixture at higher temperature.)
5. Wash the products sequentially with water, 1 M NaCl solution and the buffer to be used.

Divinylsulfone coupling

This coupling procedure is useful for the activation of hydroxylic polymer carriers for coupling of a ligand:

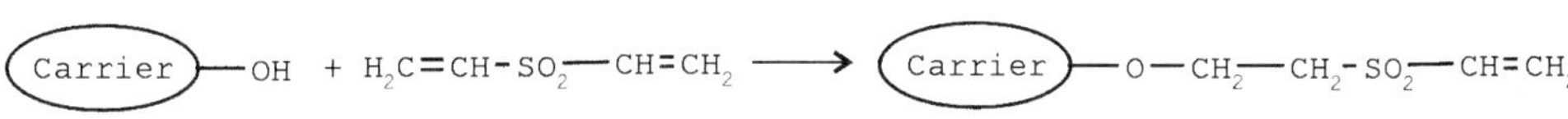

Figure 4

The use of an excess of the reagent is recommended to minimize unwanted cross-linking side reactions. The vinyl groups thus introduced into the matrix are more reactive than the oxirane groups described above. They are able to couple amines alcohols and phenols under very mild conditions [16].

Caution: The products are unstable under alkaline conditions (pH 8–10). Only alkali-stable proteins and other biocomponents may be immobilized by this method.

Protocol 6 Divinylsulfone coupling

1. For activation, 10 ml of packed hydroxyl-containing carrier (e. g., Sepharose) is suspended in 10 ml 1 M Na_2CO_3 solution (pH 11); divinylsulfone (about 2 ml) is added.
2. The suspension is kept for 70 min. on a shaker at r.t.
3. Stop the reaction by adjusting the pH to 7.0 and wash the gel on a glass filter funnel with excess water (about 500 ml) in small portions.
4. For coupling, the activated gel is suspended in 10 ml 0.3 M Na_2CO_3 solution (pH 10) containing about 200 mg of the biocomponent to be immobilized. The suspension is agitated gently on a shaker for 2 h at r.t.

5. Wash the product excessively with the following solutions:
   a) 0.3 M Na_2CO_3 solution (pH 10) containing 1 M NaCl
   b) 0.3 M glycine (pH 3) containing 1 M NaCl
   c) 0.05 M Tris/Cl buffer (pH 7.5) containing 0.5 M NaCl
6. Store the gel in buffer (5 c) at 5 °C
7. Treatment with 1 M glycine (pH 7–8) is recommended to ensure complete blocking.

Immobilization of biomolecules on aldehyde-containing gels
A very good means of obtaining various carriers containing reactive aldehyde groups for coupling of biomolecules is use of spacers containing hydrazide groups followed by glutaraldehyde [15]. Such spacers are useful with low-molecular-weight ligands to increase steric accessibility to the bulk solution. On the other hand, covalent multiattachment of macromolecules (like proteins) with the help of spacers offers the opportunity to stabilize the ligand, but keeps it flexible enough to maintain the biorecognitive properties to a high extent. The following recipes are given to synthesize various types of hydrazido-carriers.

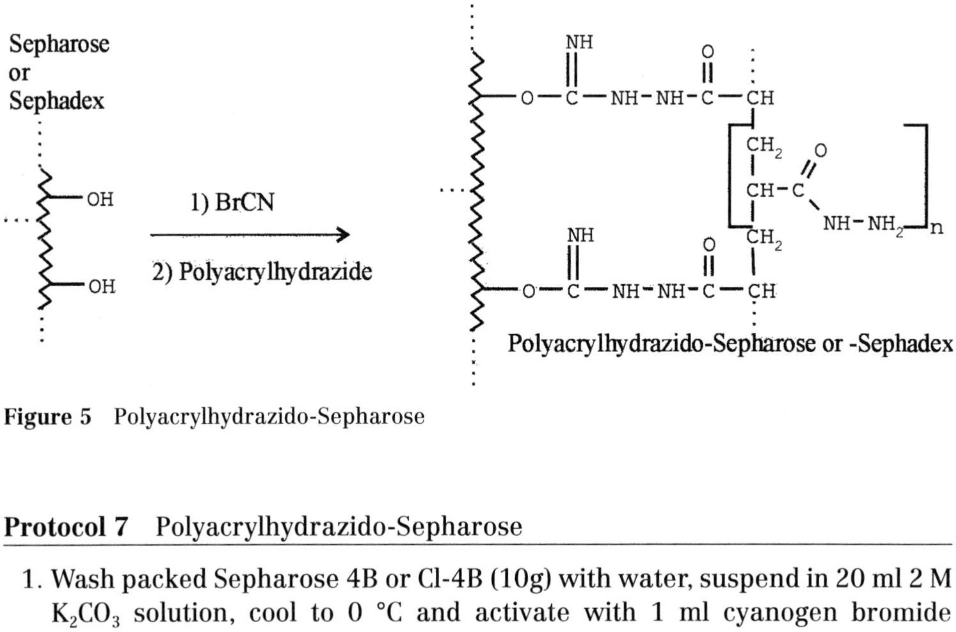

Figure 5 Polyacrylhydrazido-Sepharose

Protocol 7 Polyacrylhydrazido-Sepharose

1. Wash packed Sepharose 4B or Cl-4B (10g) with water, suspend in 20 ml 2 M K_2CO_3 solution, cool to 0 °C and activate with 1 ml cyanogen bromide solution in DMF (N,N-dimethylformamide) (1 g/ml) for 90 s.
2. Filter and wash with 50% aqueous DMF, followed by ice water.
3. Supend the activated gel immediately in 30 ml 0.2 M aqueous $NaHCO_3$ solution containing 0.1–0.6 g water-soluble polyacrylhydrazide and agitate the mixture on a shaker for 16 h at 4 °C.
4. Filter off the gel and wash with 0.1 M NaCl solution until samples of the solution show no color when tested with 2,4,6-trinitrobenzene-sulfonic acid (TNBS).

5. Store at 4 °C as wet filter cakes.
6. The quantity of available hydrazide groups on the matrix may be tested by reaction with an excess of TNBS. The quantity of unreacted TNBS can be measured and subtracted from its initial concentration.

Protocol 8 Polyacrylhydrazido-Sephadex

1. Allow Sephadex G25 coarse (Pharmacia) or Sephadex G200 (Pharmacia) (10 g) to swell in hot water for several hours.
2. Filter and suspend in 2 M aqueous K_2CO_3 solution.
3. Activate with cyanogen bromide as described above for 90 s.
4. Couple activated Sephadex immediately to 0.1–0.6 g polyacrylhydrazide dissolved in 30 ml of 0.2 M $NaHCO_3$.
5. Continue the procedure as described above.

In order to synthesize these carriers, polyacrylhydrazide is needed, which is not available commercially and has to be prepared first.

Protocol 9 Preparation of polyacrylhydrazide

1. Suspend polymethacrylate (5 g) in 70 ml hydrazine-hydrate (98%) and reflux at 100 °C for 3 h (or until you obtain a clear viscous solution).
2. Allow to cool to r.t. and filter off insoluble particles.
3. Pour the filtrate under stirring into 500 ml ice cold methanol containing 1 ml concentrated acetic acid.
4. Filter off the precipitate formed and wash with a cold mixture of methanol/acetic acid 500/1 (v/v).
5. Resuspend the precipitate in 150 ml water and filter off insoluble particles.
6. The polyacrylhydrazide solution thus obtained may be used directly for coupling or can be freeze-dried for storage over prolonged periods.
7. Caution: The freeze-dried polymer has to be completely free of water to avoid cross-linking leading to insolubility.

Protocol 10 Synthesis of polymethylacrylate

1. Add consecutively under stirring Na-laurylsulfate (0.25 g), 5 ml freshly redistilled methylacrylate, 1 ml 1% thioglycolic acid, and 0.125 g ammonium persulfate to 50 ml water.
2. Caution: This procedure must be carried out under a hood to avoid inhalation!!
3. Reflux at 80 °C for at least 2.5 h.
4. Pour the emulsion under stirring into a mixture of 100 ml ice-cold water and 100 ml ice-cold 2 M HCl and keep the white, cloudy precipitate in the cold for 30 min.
5. Wash by decantation with excess cold water.

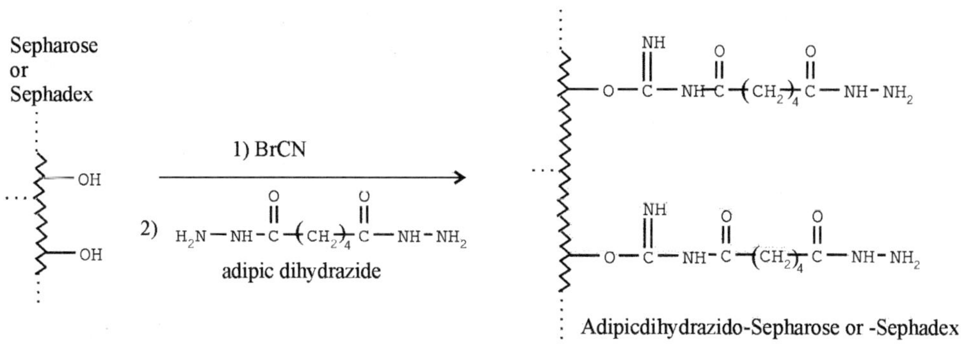

Figure 6 Adipicdihydrazido-Sepharose or Sephadex

Protocol 11 Adipicdihydrazido-Sepharose or Sephadex

1. Activate the carrier with cyanogen bromide as described above.
2. Treat the activated carrier with a 2 M solution of adipic-dihydrazide in 0.5 M $NaHCO_3$.
3. Filter off and wash with 0.1 M NaCl solution until samples of washing show no color when tested with TNBS.
4. Store at 4 °C as wet filter cakes.

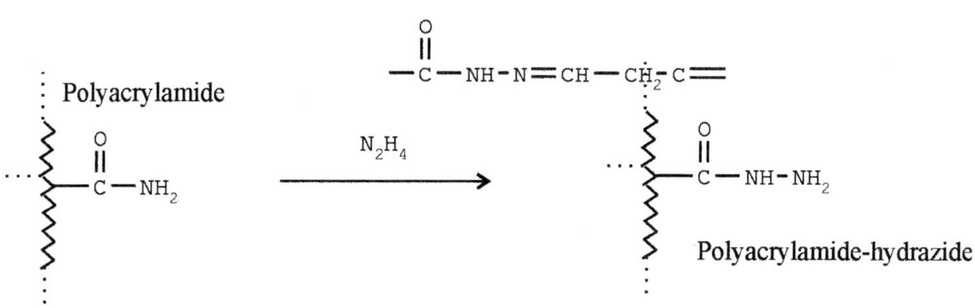

Figure 7 Polyacrylamide-hydrazide

Protocol 12 Polyacrylamide-hydrazide

1. Allow 10 g dry acrylamide beads (e. g., Biogel P60) to swell overnight in water at r.t.
2. Filter and suspend in 70 ml water, heated to 47 °C.
3. Treat with 68 ml 90% hydrazine-hydrate and keep at 47 °C for 1 to 72 h, depending on the degree of hydolysis desired.
4. Filter and wash with 0.1 M NaCl.
5. The amount of hydrazide per g packed gel may be determined with the aid of TNBS

Since hydrazinolysis is a rather slow reaction, this procedure can be used to obtain controlled amounts of reactive groups for the coupling of biocomponents.

Caution: Do not inhale hydrazine; it is very toxic and allergenic!

For the coupling of biocomponents containing amino groups the hydrazide carriers must be activated with glutaraldehyde. Excess is recommended to avoid cross-linking.

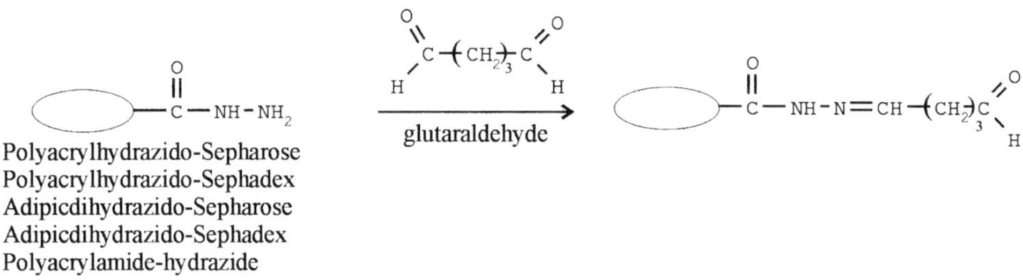

Figure 8 Glutaraldehyde activation

Protocol 13 Glutaraldehyde activation

1. Suspend hydrazide group containing carriers in excess 10% aqueous glutaraldehyde solution (about 3–5 volumes per volume packed gel) and keep under shaking for 4 h at r.t.
2. Filter and wash with ice-cold water until the filtrate shows no reaction with 2,4-dinitrophenylhydrazine. (The completion of the reaction is indicated by a negative TNBS test.)

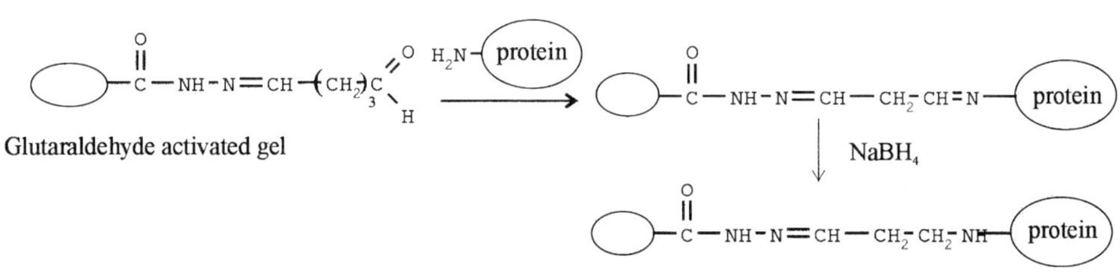

Figure 9 Coupling of protein or other biocomponents containing NH_2-groups to the activated carriers

Protocol 14 Coupling of biocomponents containing NH_2-groups to the activated carriers

1. Shake the carriers overnight at 4 °C with a solution of protein (5–25 mg/g packed gel) in 0.25 M phosphate buffer pH 7.
2. Filter and wash with buffer followed by 0.1 M aqueous NaCl solution until no unbound protein can be found in the filtrate.
3. Store gels as wet filter cakes at 4 °C.

For labile proteins it is recommended that they be stored under buffered conditions. Proteins containing SH-groups should be kept in presence of stabilizers like dithiothreitol or reduced glutathione. Mercaptoethanol, though widely in use, should be omitted since it evaporates easily and is rather toxic!

Cyanogen bromide-activation method employing carriers with pyridine rings
The conventional cyanogen bromide technique is used widely for activation of polysaccharides. This method, however, suffers from a severe disadvantage: the amount of activated groups cannot be controlled by the amount of reagent or the reaction time, since the reaction proceeds too quickly. If a controlled amount of activated groups for coupling is desired, the following method will provide a very good means to reach this aim. It is an activation of pyridine-containing carriers with cyanogen bromide, which yield aldehyde groups in one step. In this case the amount of aldehyde groups formed can be controlled by the amount of BrCN added [17].

Many pyridine-containing polymers, copolymers, and membranes have been reported. These gels were prepared mostly by direct polymerization of vinyl-pyridine. Because of the hydrophobicity and lack of swelling of these gels in aqueous solutions, they are not suitable to be applied as efficient carriers for enzyme immobilization. Polyvinylpyridine prepared by direct polymerization of vinylpyridine and 1,4-divinylbenzene failed to bind enzymes, and the same may be true for other copolymers. Therefore, we decided to prepare pyridine-containing polymers on carriers that are known to swell in aqueous solutions, such as polyacrylamides and polysaccharides. Controlled-pore glass beads turned out to be suitable as well. The methods of binding pyridine to such carriers are numerous, since it is possible to link almost any functional group of carriers with pyridine derivatives.

Below are some examples to synthesize such carriers:

Enzacryl

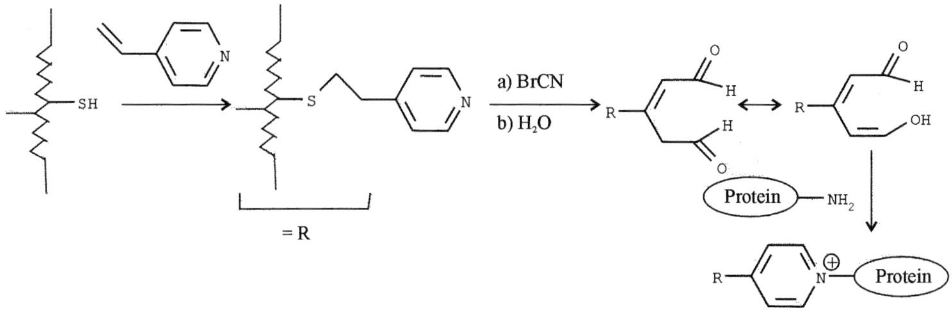

Figure 10 Polythiol based starting material

Protocol 15 Polythiol-based starting material

1. Suspend polythiol carrier (e. g., Enzacryl polythiol) in freshly distilled 4-vinylpyridine (about 20 ml/g gel) and incubate in a tightly stoppered flask at 37 °C on a shaker for at least 3 h.
2. Allow to cool to r.t. and keep for 2 days.
3. Wash the gel with 95% ethanol, followed by absolute ethanol. (An Ellman test may be employed to test if the coupling reaction was quantitative.)
4. Activate the gel (1 g) at r.t. with 1–3 ml cyanogen bromide solution (1 g BrCN / ml of absolute dioxane) while being stirred.
5. After 5 min. of activation, add 20 ml water and stir at r.t. for an additional 20 min.
6. Filter off the gel as quickly as possible and wash with ice-cold water.
7. Couple immediately to about 80–100 mg of the desired biocomponent (e. g., enzyme) dissolved in 20 ml 0.25 M phosphate buffer pH 5, the mixture being stirred for 30 min. at r.t. and afterwards overnight at 4 °C.
8. Wash the gel with 0.1 M NaCl and store as wet gel at 4 °C.

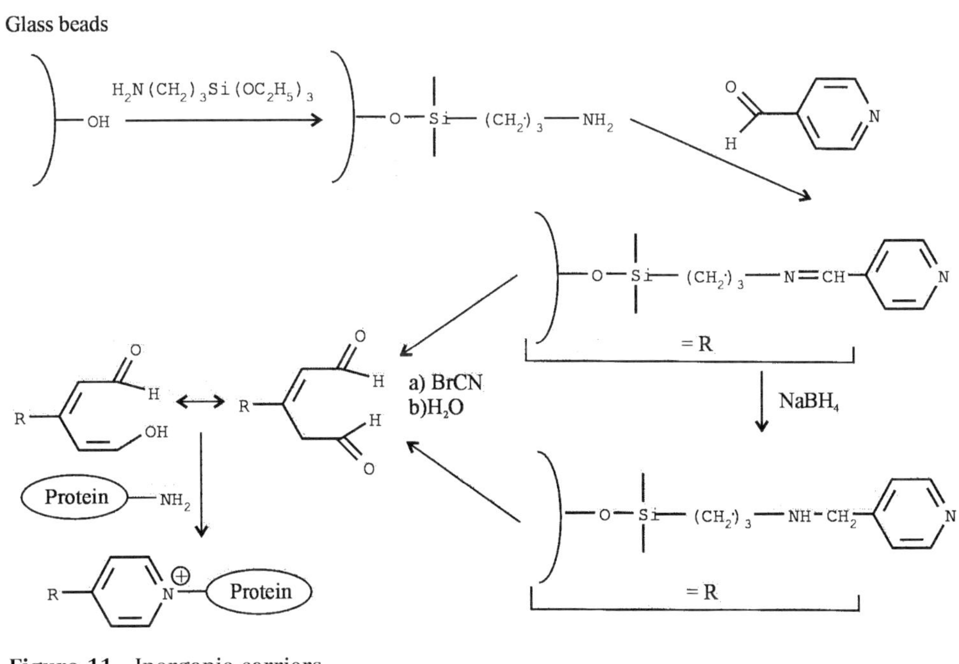

Figure 11 Inorganic carriers

First, the surface of the inorganic carrier must be modified in order to be capable of being derivatized with organic material (e. g., glass surfaces have to be converted to their silanol form). Suggestions for reaching this aim are made in the first step of the following procedure.

Protocol 16 Controlled-pore glass beads

1. Treat controlled pore glass beads (Corning) (e. g., 120/200 mesh, mean pore diameter 1038 A) for 1 h with 3% HNO_3 at 90 °C, rinse with water until the filtrate is neutral and keep for several days in distilled water.

Attention: When glass beads are kept in liquid suspension the suspension should always be degassed in the beginning under mild vacuum, to keep the inner surface free of air, which otherwise may decrease the inner surface. Degassing also is recommended when controlled pore glass is filtered dry by accident and resuspended again in liquid media.

2. Mix the glass beads (10 g) with 50 ml 10% aqueous 3-aminopropyltriethoxysilane and adjust the pH to 3.5 with 6 M HCl (avoid glass electrodes!!). Agitate the suspension gently at 75 °C for 2 h on a shaker.
3. Filter off the coated glass beads (avoid getting them dry!), rinse with water, and dry overnight at 115 °C. If desired, the amount of aminogroups may be calculated from the nitrogen content, determined by elementary analysis. (Usual values are 80–180 mmoles of amino groups per 1 g dry gel.)
4. Treat the amino glass beads thus obtained with 250 ml 2.5% pyridine-4-aldehyde in 0.25 M phosphate buffer, pH 7, stir at r.t. for 1 h, wash carefully with water, and dry in a desiccator over NaOH.

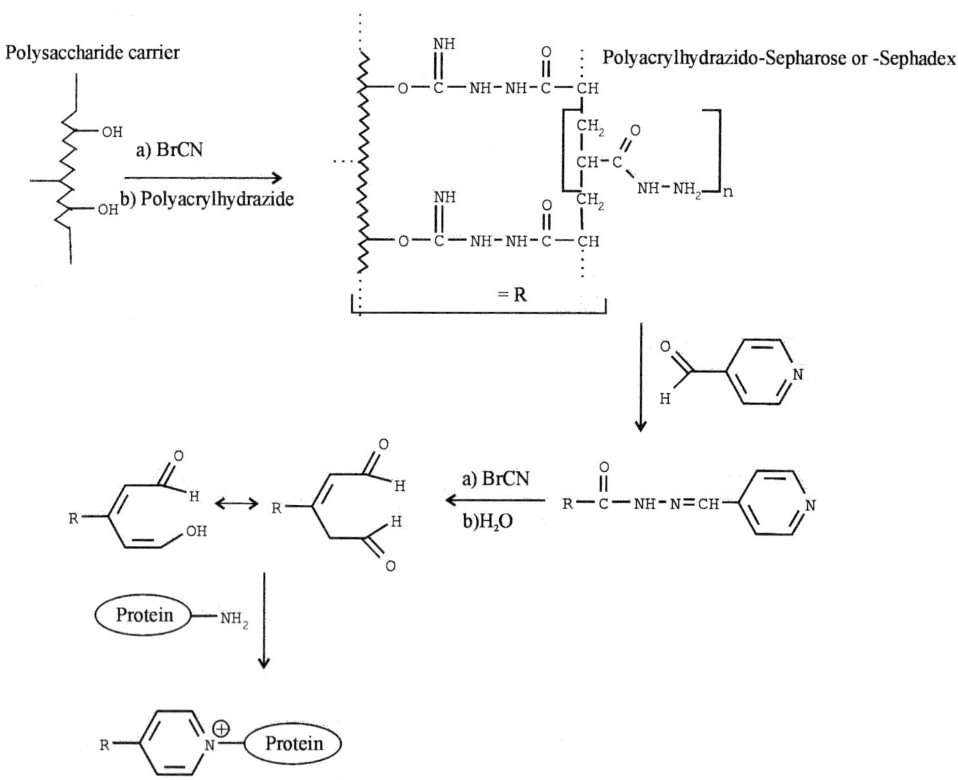

Figure 12 Polysaccharide-based immobilisates

5. The completion of the reaction may be indicated by a negative TNBS test.
6. Moisten aliquots of the beads (300 mg) with dry dioxane (degas quickly under mild vacuum) and treat with cyanogen bromide solution (1 g/ml absolute dioxane) in a closed hood. After 5 min. of activation add 20 ml of 0.2 M borate buffer pH 9 and keep the mixture gently stirred at r.t. for an additional 20 min., keeping the pH constant with 2 M NaOH by manual titration.
7. The coupling procedure to the protein or another amino-group containing biocomponent is the same as mentioned at the beginning of this section.

Protocol 17 Polysaccharide based immobilisates

1. Wash Sepharose Cl-4B (10 g) or other polysaccharide carriers thoroughly with water and suspend in 20–30 ml 2 M aqueous K_2CO_3. Cool the mixture in an ice bath (add also small pieces of ice to the suspension). Add cyanogen bromide solution (1 g/ml DMF) and stir for 90 s.
2. Wash the gel with a small amount of 50% aqueous DMF solution followed by ice water.
3. Stir the activated Sepharose overnight at 4 °C with 30 ml aqueous solution containing 1.7 g polyacrylhydrazide and 1.3 g $NaHCO_3$.
   This procedure introduces reactive hydrazide groups, increases the number of reactive binding sites for further activation and coupling to ligands, and provides good spacers.
4. Wash the gel with 0.1 M aqueous NaCl followed by water until no hydrazide can be found in the filtrate (as indicated by the TNBS test).
5. The amount of free hydrazide groups also may be assayed by TNBS.
6. Dissolve 2 g of this polyacrylhydrazido-Sepharose in 50 ml 2.5% pyridine-4-aldehyde and stir gently overnight at r.t.
7. Filter off the pyridino-gel and wash with water until no aldehyde can be found in the filtrate when tested with the 2,4-dinitrophenylhydrazide test.
8. The activation and biocomponent coupling procedure is the same as mentioned above.

Since the spacers in this gel contain Schiff-base double bonds, which may hydrolyze gradually after prolonged periods, it is possible to treat these gels with $NaBH_4$ to reduce the double bonds.

Protocol 18 Reduction of imine bonds

Suspend 1 g gel in 10 ml 0.25 M phosphate buffer pH 7 and treat at 0 °C with a solution of 50 mg $NaBH_4$ in 1 ml water for 1 h.
Filter and wash carefully with 1 M NaCl solution.

Sodium periodate method of activation for polyhydroxy carriers
Nearly all methods described for the coupling of bioligands to polyhydroxycarriers depend on the initial modification of the gel with CNBr. However, an alternative method exists, that is rapid, simple, and safe and that also results in chemically stable ligand/carrier bonds [18].

The method depends on the oxidation of cis-vicinal hydroxyl groups (of, e. g., agarose) by sodium metaperiodate ($NaIO_4$) to generate aldehyde functions.

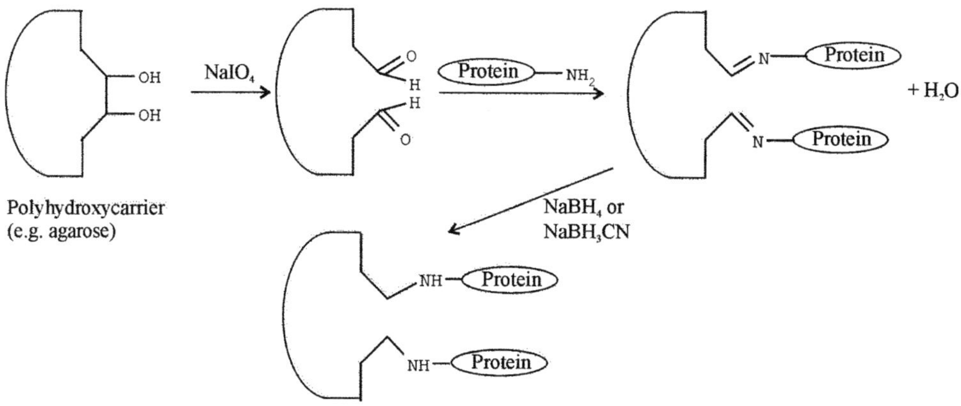

Figure 13

These aldehyde functions react at pH 4 to 6 with primary amines to form Schiff bases, which may be reduced with sodium borohydride ($NaBH_4$) to form stable secondary amines. The reductive stabilization of the Schiff-base intermediate may best be achieved with sodium cyanoborohydride ($NaBH_3CN$). This reagent preferentially reduces Schiff bases (at pH 6 to 6.5), whereas $NaBH_4$ may reduce aldehyde functions as well. However, sometimes it also may be desired to get rid of excess unreacted aldehyde groups in one step.

Nota bene: Since $NaBH_3CN$ selectively reduces Schiff bases, it shifts the equilibrium of the reaction to the right (Fig. 13), thus driving the overall reaction to completion.

Protocol 19 Sodium periodate oxidation

1. Suspend 100 ml of polyhydroxycarriers (e. g., agarose) in 80 ml water and add 20 ml 1 M $NaIO_4$.
2. Gently agitate the suspension on a shaker for about 2 h at r.t.
3. Filter and wash on a sintered glass funnel with 2 l of water.
4. React with a solution of the amino group containing bio-ligand (at pH 5 if the stability of the bioligand allows) for 6–10 h at r.t.
5. Raise the pH to 9 with concentrated Na_2CO_3.

6. Add 10 ml of freshly prepared cold 5 M $NaBH_4$ (in small portions) to the magnetically stirred suspension at 4 °C and keep in the cold for about 12 h. Caution: Avoid excessive foaming.
7. Wash the product in a sintered glass funnel with 2 l 1 M NaCl (without suction) over a 4-h period.
8. Incubate the gel in an equal volume of 1 M NaCl for 15 h.
9. Wash with an additional 2 l of 1 M NaCl over a 2-h period.
10. The filtrates of the wash may be occasionally checked for uncoupled amino material by the TNBS test or with ninhydrin.

2.2 Covalent coupling to inorganic supports

Many inorganic carriers may be used for covalent immobilization, e. g., SiO_2, TiO_2, ZrO_2, Al_2O_3, MgO, sand, Gulsenit, etc. Porous materials have the best characteristics for immobilization of biomolecules, provided that the pores are big enough to permit the entry of the biocomponent and to allow biorecognition without sterical hindrance. Caution: Even inorganic material is not always stable to pH changes. At alkaline pH, controlled-pore glass may become soluble, which is not the case for, e. g., TiO_2. This means that working conditions must be taken into consideration for the choice of the best inorganic carrier.

Silan coupling techniques

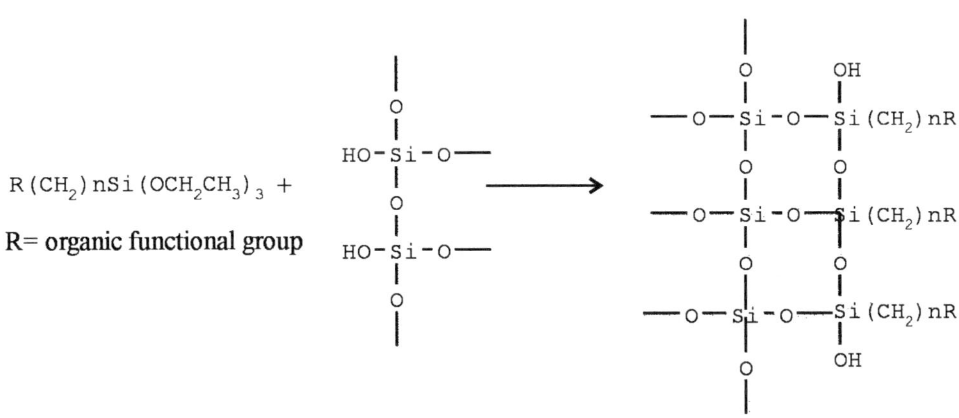

Figure 14 Alkylamine coupling [19]

Protocol 20 Aqueous silanization

1. Add 1 g of clean inorganic support material to 18 ml of distilled water und degas under mild vacuum to remove air from the pores.

2. Add 2 ml of γ-aminopropyltriethoxysilane (10% v/v) and adjust the pH between 3 and 4 with 6 M HCl. Caution: Do not use a glass electrode; it would be silanized too! Keep at 75 °C for 2 h.
3. Filter on a Büchner funnel and wash with about 20 ml distilled water.
4. Dry in an oven at 115 °C overnight.
5. Store the silanized products dry for further use.

If mercapto- or other functional groups are desired, mercaptosilanes may be used as well, following the same procedure.

Nota bene: Whenever dry porous carriers are suspended in liquid media for further use, they have to be degassed in the liquid suspension to remove air and to increase the accessibility for the inner surface.

Protocol 21 Organic silanization

This technique results in higher amine loadings. However, aqueous silanization results in more stable products in most cases.

1. Suspend 1 g of clean carrier in 50 ml 10% solution of the respective silane (e. g. γ-aminopropyltriethoxysilane) in toluene (v/v) and reflux overnight.
2. Filter on a Büchner funnel. Wash with toluene followed by acetone.
3. Air dry first, then keep in an oven overnight at 115 °C.
4. Store the product dry in a sealed vessel.

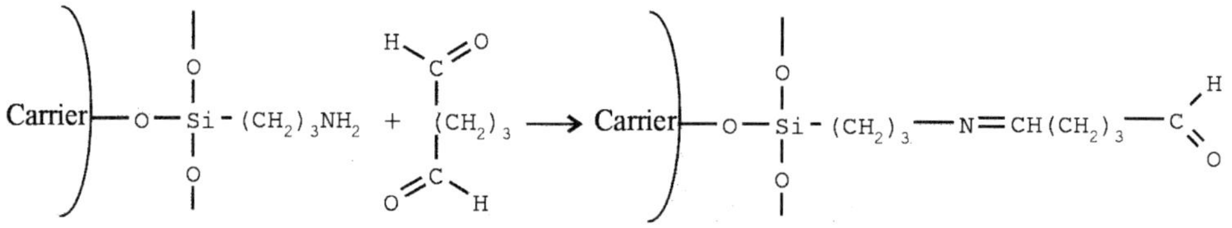

Figure 15 Activation of inorganic aminocarriers via glutaraldehyde to couple biocomponents containing reactive aminogroups

Protocol 22 Activation of inorganic aminocarriers via glutaraldehyde

1. Add 1 g of alkylamine carrier to enough 0.05 M phosphate buffer pH 7.0 to cover them, degas in slight vacuum to remove air from the pores, and decant excess liquid.
2. Suspend the wet, degassed carrier in 25 ml 2.5% solution of glutaraldehyde in 0.05 M phosphate buffer, adjusted to pH 7. Incubate for at least 1 h under shaking.
   Caution: Avoid stirring of highly porous carriers since they may be mechanically destroyed. Use a shaker instead.
3. Filter off and wash exhaustively with ice water. (Avoid drying out the gel and degas again if it happens.)

4. To the carrier thus activated, add the biocomponent (e.g. enzyme) in as small a volume as possible at a minimum of 1% concentration. (Preferably the same buffer may be used as in the activation step.) The quantity of protein added should be 50–100 mg/g wet carrier.
5. Allow 2–4 h for the reaction.
6. Wash with buffer and store wet in a refrigerator until use.

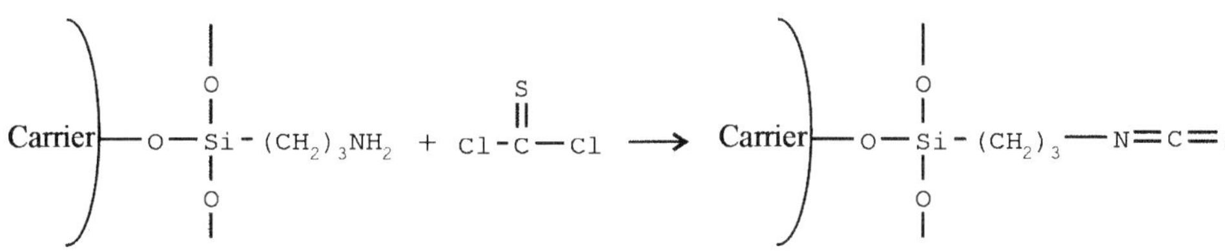

Figure 16 Isothiocyanate coupling

Protocol 23 Isothiocyanate coupling

1. Add 1 g of alkylamine carrier to a 10% solution of thiophosgene in chloroform under a hood (because of the nauseous and toxic nature of thiophosgene) and reflux for 12–15 h.
2. Wash with dry chloroform and vacuum dry in a desiccator.
3. Use this derivative as soon as possible for coupling to a bioligand.
4. Suspend the activated and degassed carrier in small portions to a 1% solution of the respective biocomponent (containing e.g. 50–100 mg of enzyme or other protein per g of dry carrier) in buffer pH 8.5–9.0. If necessary, adjust the pH and maintain until it stabilizes.
5. Incubate under gentle agitation for 2 h.
6. Wash and store as described above.

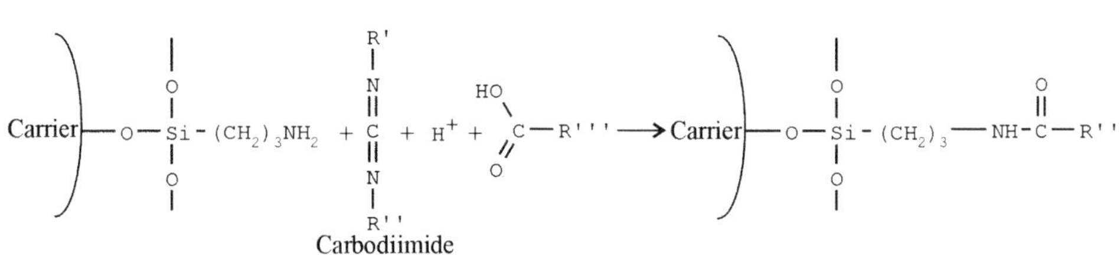

Figure 17 Carbodiimide coupling

This technique has to be used if the protein ligand to be immobilized possesses sterically available carboxyl groups.

Protocol 24 Carbodiimide coupling

1. To 1 g alkylamine carrier, add 50 ml 0.03 M H_3PO_4 adjusted to pH 4.0 (degas if necessary).
2. To this mixture add 100–200 mg of a water-soluble carbodiimide e. g., EDC: 1-ethyl-3-(3-dimethylaminopropyl) carbodiimide hydrochloride
3. Add the biocomponent (50–100 mg/g of carrier) to the mixture if it remains stable under these pH conditions.
4. Allow the mixture to react overnight at 4 °C.
5. Wash and store as described above.

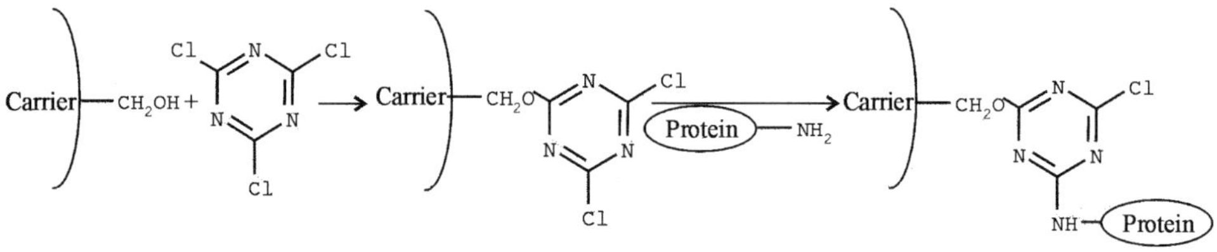

Figure 18 Triacine coupling

Protocol 25 Triacine coupling

1. Suspend 1 g of alkylamine carrier in 10 ml benzene or toluene containing 0.2 ml triethylamine and 0.3 g 1,3-dichloro-5-methoxy-triazine. Incubate for 2–4 h at 45–55 °C on a shaker.
2. Decant and wash with solvent.
3. Dry in an evaporator oven at 100 °C.
4. Couple the biocomponent (e. g., 50–100 mg protein/g activated carrier) dissolved in 0.05 M phosphate buffer pH 8.0, incubating the mixture overnight under gentle agitation at 4 °C.
   If porous carriers are used, do not forget to gently degas the suspension before incubation.
5. Wash with buffer and store wet in the refrigerator.

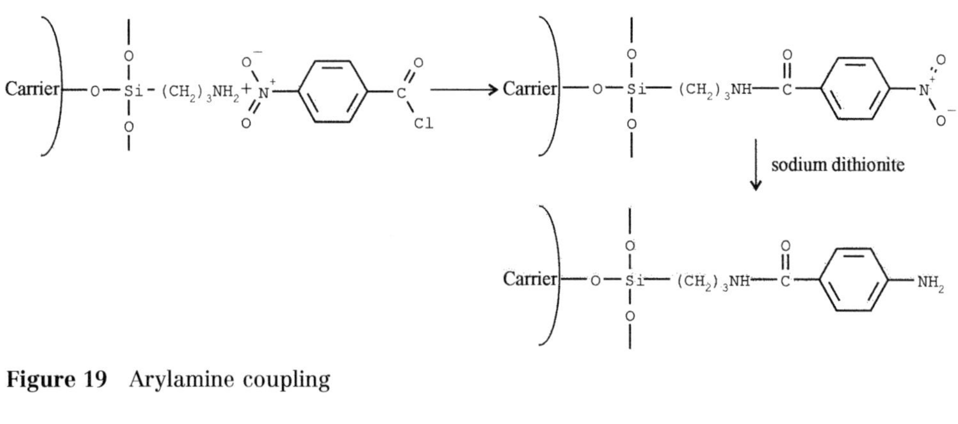

Figure 19 Arylamine coupling

Protocol 26 Preparation of arylamine carriers

The most common use for arylamines is the coupling via azolinkages:

1. Suspend 1 g of alkylamine carrier in 25–50 ml chloroform containing 5% triethylamine (v/v) and 1 g p-nitrobenzylchloride.
2. Reflux for at least 4 h.
3. Wash with chloroform and allow to dry.
4. Boil the gel in a 5% solution of sodium dithionite in water.
5. Dry and store in a dark place for further activation.

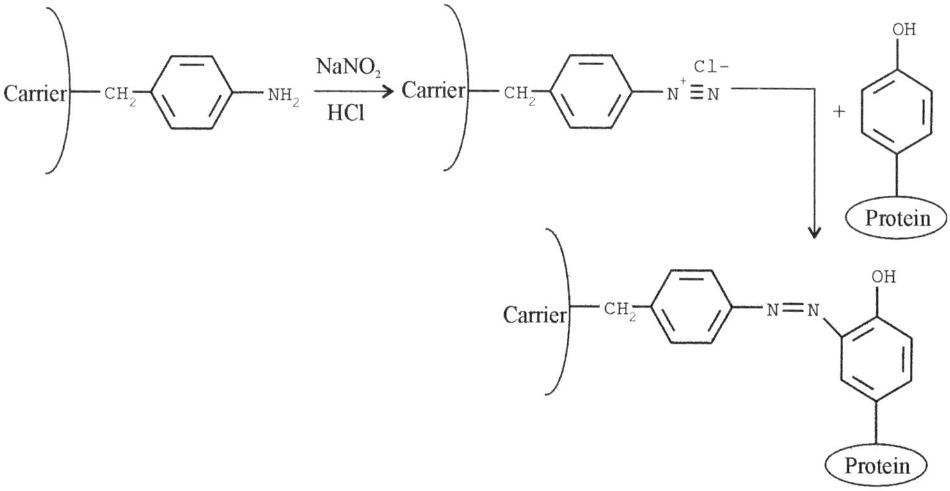

Figure 20 Azo coupling

Protocol 27 Azo coupling

1. Suspend 1 g of arylamine carrier in about 20 ml 2 M HCl, cool in an ice bath, and degas if necessary.
2. To the cold suspension, add 100 mg of solid $NaNO_2$ under shaking.
3. Allow diazotation to proceed for at least 30 min. (The carrier should become yellowish.)
4. Wash on a Büchner funnel with ice water.
5. Add the biocomponent (about 100 mg/g ocarrier) dissolved in an appropriate buffer pH 8.5. The solution of the biocomponent should be 0.5–2% (w/v).
6. For coupling incubate 10–18 h under shaking.

Caution: Maximum coupling and recoverable activity may not be the same. Therefore the optimum coupling time has to be determined.

Preparation of carboxy derivatives and coupling methods: These derivatives can be prepared by the reaction of the alkylamine carrier with succinic anhydride.

Protocol 28 Reaction of the alkylamine carrier with succinic anhydride

1. Suspend 1 g of alkylamine carrier in 25 ml of 0.05 M phosphate buffer pH 6.0 and degas if necessary.
2. Add 1 g of succinic anhydride to this mixture and shake at r.t. for 20 h. Caution: pH adjustment may be necessary over the first few h.
3. Wash with buffer followed by water and dry.

This procedure has the advantage that no cross-linking should occur if handled properly, whereas carbodiimide coupling to alkylamine carriers can cross-link.

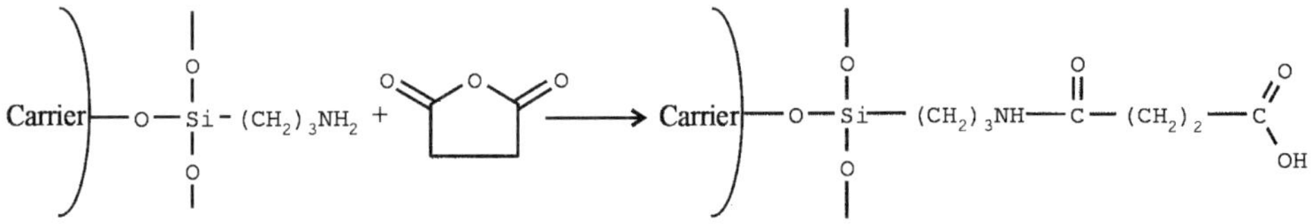

Figure 21 Reaction of the alkylamine carrier with succinic anhydride

Protocol 29 Carbodiimide coupling to the carboxyl derivative

1. Dissolve about 200 mg water-soluble carbodiimide in 25 ml distilled water and add 1 g carboxylated carrier.
2. Adjust to pH 10.0 (degas if necessary) and incubate at r.t. for 2 h. Nota bene: It is also possible to achieve successful activation at pH 4.0
3. Wash the product with distilled water.
4. Add the biocomponent solution for coupling as described above.
5. Coupling under both conditions, pH 4.0–5.0 or pH 9.0–10.0 may give satisfying results.
6. Wash the final product and store wet for use.

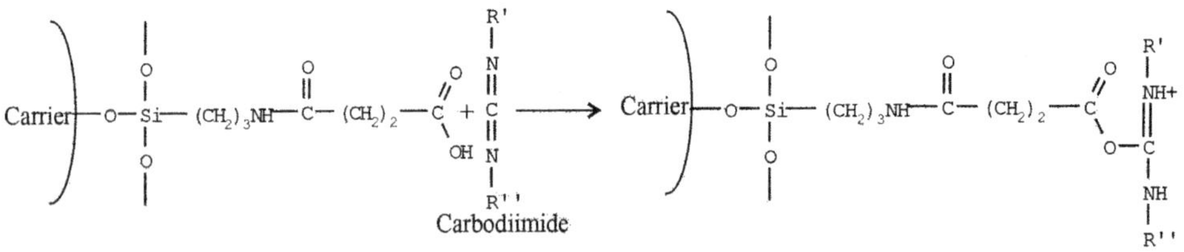

Figure 22 Carbodiimide coupling to the carboxyl derivative

With inorganic carriers, it is possible to prepare the acid chloride from the carboxyl derivative. Use soon after preparation, since it is not very stable.

Protocol 30 Acid chloride coupling

Cover 1 g of carboxylated carrier with a small amount of chloroform and degas if necessary.
Add 50 ml 10% solution of thionyl chloride in chloroform and reflux for 4–6 h.
Filter and dry in a vacuum oven at 60–80 °C.
For coupling of the biocomponent, slightly alkaline conditions turned out to give the best results: Add an appropriate solution of the biocomponent (as already described) buffered to pH 8.0–8.5 and maintain the pH by manual titration with aqueous NaOH solution.
Allow to react for 1–2 h at r.t.
Wash and store at 4 °C until use.

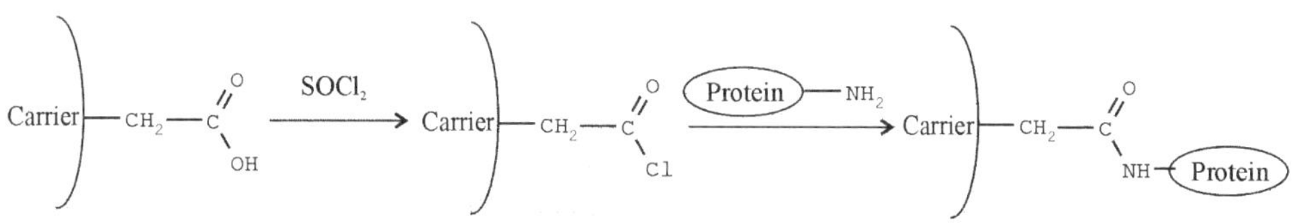

Figure 23 Acid chloride coupling

Protocol 31 Preparation of hydroxysuccinimide esters

This is a very convenient method, since active esters are rather easy to prepare, can be stored for later use, and coupled to enzymes under mild conditions.
Caution: N,N′-dicyclohexyl carbodiimide has to be handled with care; body contact, especially inhalation, must be avoided, since this may lead to anaphylactic shocks!

1. Suspend 1 g of carboxylated carrier in 10 ml dioxane (degas if necessary).
2. Add 200 mg of N-hydroxysuccinimide and 400 mg of N,N′-dicyclohexyl carbodiimide.
3. Allow to react for 4 h at r.t. under gentle shaking.
4. Wash the product with dioxane followed by methanol.
5. Dry in an evacuated oven at 70–80 °C for 1 h.
6. Store refrigerated and desiccated in an amber bottle.
7. Coupling of the biocomponent may be carried out under any conditions keeping the respective biocomponent the most stable.

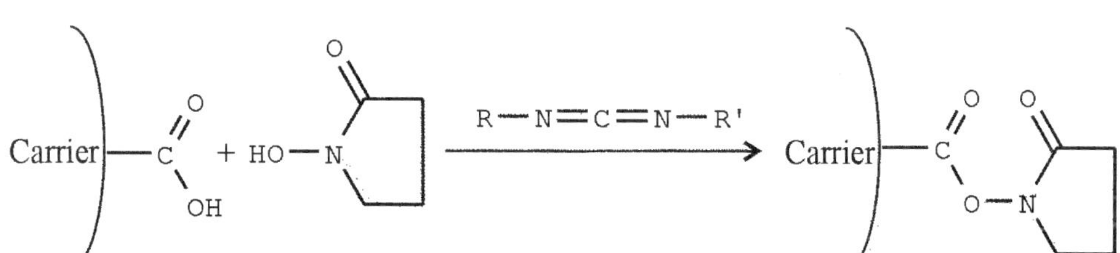

Figure 24 Preparation of hydroxysuccinimide esters

Activation of inorganic supports with γ-mercaptopropyl-trimethoxysilane: Preparation of supports containing surface sulfhydryl groups may be useful, if coupling via sulfhydryl linkages is desired. Biocomponents immobilized that way may be removed easily from the matrix with the help of reducing and HS-containing reagents, retaining their biological activity.

Protocol 32 Activation with γ-mercaptopropyl-trimethoxysilane

1. Suspend 1 g of cleaned porous carrier in sufficient water to enable degassing under mild vacuum.
2. Add 5 ml 10% aqueous solution (v/v) of γ-mercaptopropyl-trimethoxysilane, previously adjusted to pH 5.0 with 6 M HCl.
   Caution: Do not use a glass electrode!
3. Reflux for 4 h.
4. Wash with distilled water.
5. Heat the product in an oven at 120 °C for 4 h.

2.3 Immobilization of biomolecules onto metal surfaces

A wide variety of protein- or nucleic-acid-based biosensors has been developed, since the demand for microbiosensors in medicine and biotechnology has increased dramatically during the last decade. Therefore, it became very important to attach biomolecules covalently to metal surfaces, e. g., gold, silver, or platinum. Thin film technology especially is able to provide high purity and reproducibility for the electrode surface. Therefore, surface chemistry is of eminent importance. Below covalent coupling techniques for metal carriers are introduced.

Nota bene: The activation techniques of the silanized metal surfaces may be used for other carrier-material as well [20, 21]. Before coupling, it is necessary to activate the metal surface.

Protocol 33 Activation of a platinum surface

1. For purification, reduce platinum surfaces by using either 1% sodium dithionite or 15% $TiCl_3$/HCl or 5% $FeSO_4$/6 M HCl or 8.5% $NaBH_4$ for incubation. (30 min., 56 °C)
2. Now oxidize in a solution of 2.5% potassium dichromate in 15% nitric acid for 30 min. at 56 °C.
3. Rinse the oxidized electrodes with water, followed by acetone.
4. Dry and derivatize immediately with silane.

Protocol 34 Coupling with 3-amino- or mercapto-propyl-triethoxysilane

1. Incubate the oxidized metal surfaces in a 5% aqueous solution of the respective silane (pH 3.5) at 37 °C for 30 min.

2. Clean the sililated surface with water, followed by ethanol.
3. Alternative procedure: Incubate the metal in a 7% solution of the respective silane in 50% aqueous acetone (pH 4) at 37 °C for 30 min.
4. To enhance the silane layer stability, further cross-link the layer on the metal surfaces by drying at 110 °C for 15 min.

Protocol 35 Activation with p-Quinones

1. Incubate the silanized metal surface with a 1% solution of p-chloranil in toluene. In addition to p-chloranil, other p-halogenanils and related quinones also may be used to activate the silanized metal surfaces.
2. Rinse with toluene, followed by acetone and water, and use immediately for coupling of the aminogroup containing biocomponent.

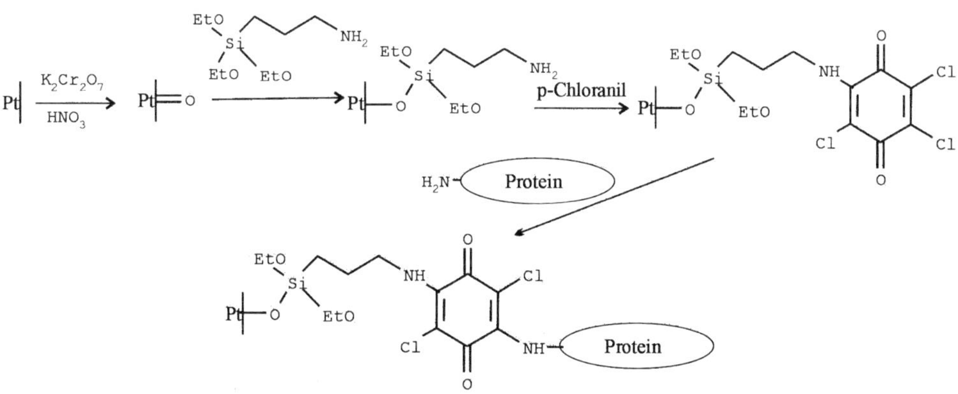

Figure 25 Activation with p-Quinones

Protocol 36 Activation with BTCAD

1. Incubate aminosilan coated metal surfaces with a 1% solution of BTCAD in anhydrous tetrahydrofurane (THF) for 30 min. at r.t.
2. Rinse with THF and water and couple immediately with the biocomponent in buffer solution (see below).

Caution: If the solution of the biocomponent contains anhydride-hydrolyzing impurities, a further activation of the carboxyclic acid groups may be achieved by using water-soluble carbodiimides, e. g., incubating the metal carrier in an aqueous solution of N-cyclohexyl-N′-[2-(N-methylmorpholino)-ethyl-]-carbodiimide-4-toluene sulfonate (CDI) (60 mg/ml) for 30 min. at r.t. After rinsing with bidistilled water, couple with the biocomponent as listed below.

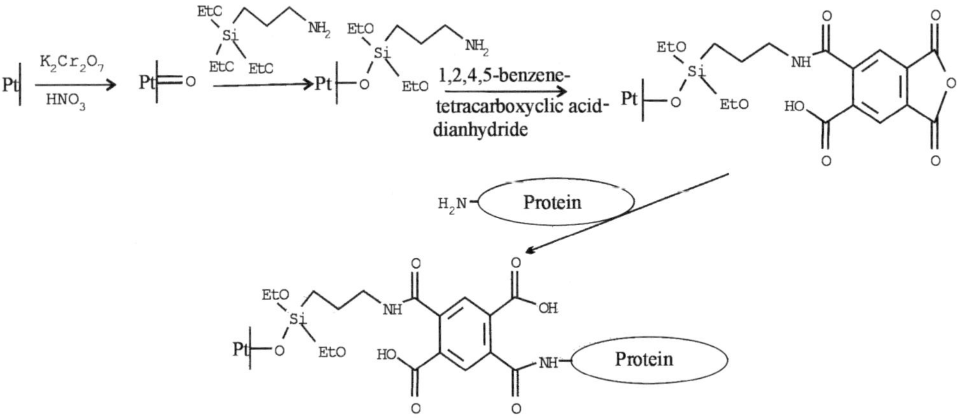

Figure 26 Activation with 1,2,4,5-benzene tetracarboxyclic acid dianhydride (BTCAD)

Protocol 37 Activation by succinic anhydride

1. Incubate aminosilane-coated carriers with succinic anhydride (1% in anhydrous THF) for 30 min. at r.t.
2. Rinse with THF and water and activate with CDI as listed above.
3. Couple immediately with the biocomponent.

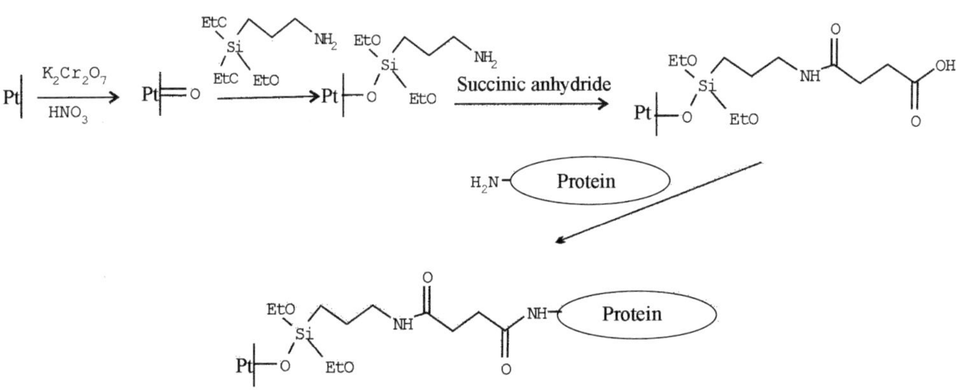

Figure 27 Activation by succinic anhydride

Protocol 38 Activation by diazotization

1. Immerse the BTCAD-activated carrier immediately in a 2% solution of p-phenylenediamine in anhydrous THF for 30 min. at r.t.
2. Rinse the carrier after amide-bond formation with THF, 1% acetic acid, and water.
3. Activate by diazotization with the help of 0.25 M $NaNO_2$ /1 M HCl at 0 °C for 40 min.
4. Rinse with water and couple immediately to the biocomponent.

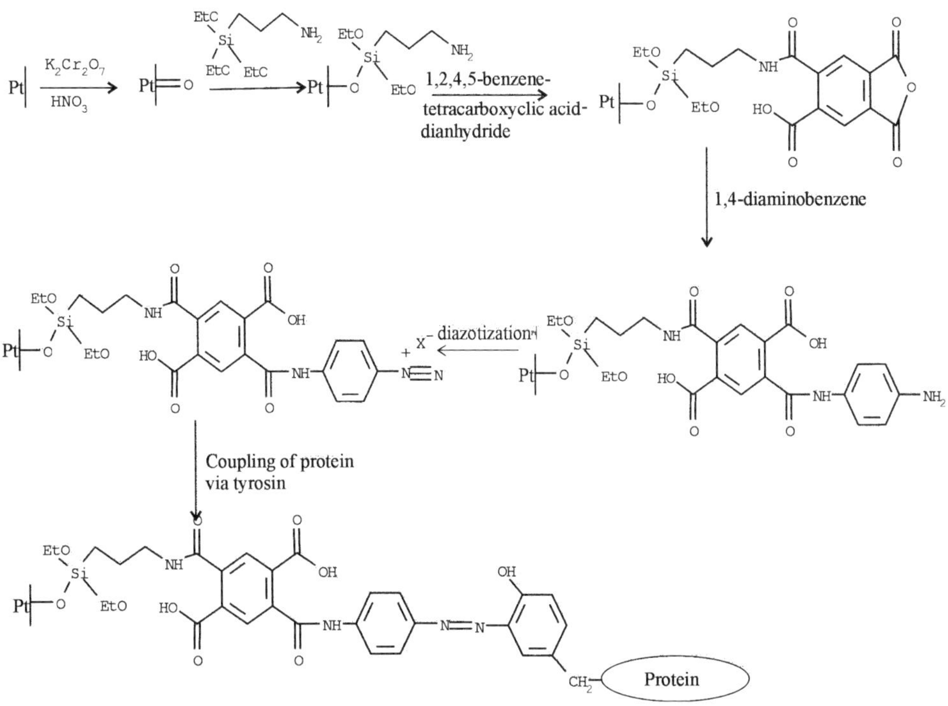

Figure 28 Activation by diazotization

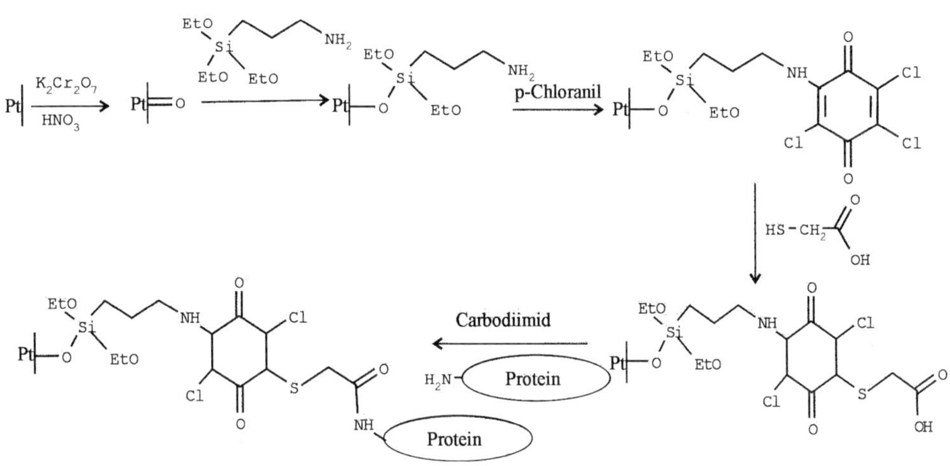

Figure 29 Activation with chloranil, thioacetic acid, and carbodiimide

Protocol 39 Activation with chloranil, thioacetic acid, and carbodiimide

1. To activate the silanized carrier surface, incubate with a solution of chloranil (1% in toluene) for 30 min. at 25 °C.
2. Wash with toluene and acetone and react immediately with thioacetic acid (10% solution buffered to pH 6) for 30 min. at r.t.

3. Couple with carbodiimide as listed above.
4. Immerse the activated carrier in a solution of amino-groups containing biocomponent (e. g. 5 mg/ml protein in 0.1 M phosphate buffer pH 7) for 2 h.
5. Rinse several times with 4 M saline to eliminate adsorbed protein and store wet at 0 °–5 °C.

2.4 Immobilization of biocomponents onto nylon

Nylon is readily available in a wide variety of physical forms, such as films, membranes, powders, hollow fibers, and tubes. The material is mechanically strong and nonbiodegradable. Therefore, it is possible to expose it to biological media without impairment of its structural integrity for prolonged periods. Beyond that, some of the nylons (such as nylon 6 or nylon 66) are relatively hydrophilic, providing an optimal microenvironment for many immobilized biocomponents such as proteins [22].

Covalent binding of ligands to nylon carriers involves three separate operations:

- Nylon is partially depolymerized by cleaving amide linkages
- Either the amino groups or the carboxyclic groups thus released are activated
- The biocomponent is allowed to react with the activated nylon derived from the second step

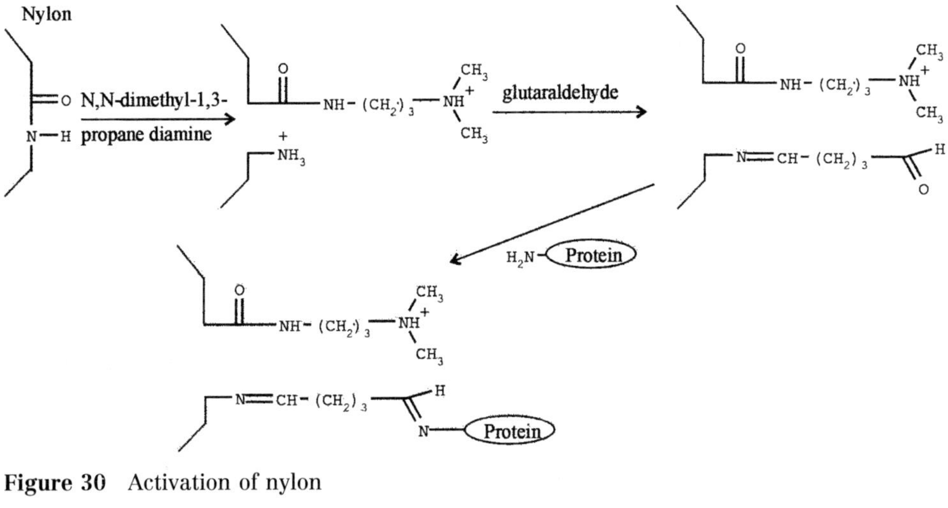

Figure 30 Activation of nylon

Protocol 40 Etching to increase the inner surface

1. Fill a 3 m length of nylon 6 tube, 1 mm bore, with a mixture of 18.6% (w/w) $CaCl_2$, 18.6% (w/w) water in methanol and incubate at 50 °C for 20 min.

This process etches the inside surface of the nylon tube by dissolving the regions of amorphous nylon. Thus, the available surface area and its wettability will be increased.

2. Purge the amorphous nylon from the tube by perfusion with water for 30 min. Flow rate: 5 ml/min.

Protocol 41 Nonhydrolytic cleavage of the nylon

1. Dry the tube (pretreated according to Procedure 1) by perfusing it with methanol for 30 min. at a flow rate of 2 ml/min.
2. Fill with N,N-dimethyl-1,3-propane diamine and incubate for 12 h at 70 °C.
3. Remove excess amine by washing through the tube with water for 12 h at a flow rate of about 2 ml/min.

Protocol 42 Hydrolytic cleavage of the nylon

1. Perfuse the tube (pretreated according to Procedure 1) for 40 min. with 3.65 M HCl at 45 °C and a flow rate of 2 ml/min.
2. Wash through with water for 12 h at a flow rate of about 2 ml/min.

Protocol 43 Activation of nylon, pretreated according to Procedures 2 or 3

1. Activate the free amino groups, which are released in the above treatments by perfusion of the tube with 5% (w/v) glutaraldehyde in 0.2 M borate buffer, pH 8.5, for 15 min. at 20 °C and a flow rate of 2 ml/min.
2. Wash the tube free of excess bifunctional agent by perfusion for 10 min. with 50 ml of 0.1 M phosphate buffer, pH 7.8.
3. For coupling, fill the tube immediately with a solution of the amino-group-containing biocomponent (e. g., 0.5–1.0 mg protein/ml) in the same buffer and incubate for 3 h at r.t.
4. Remove the excess unbound biocomponent by washing through with 1 l of 0.5 M NaCl in 0.1 M phosphate buffer, pH 7, at a flow rate of 5 ml/min.
5. Store, filled with buffer, at 4 °C.

O-Alkylation

This activation method is very useful, if it is desired to retain the mechanical strength of the support, which is partially hydrolized by the methods already described in this chapter. A very good means to reach this aim is, to carry out O-alkylation of nylon using triethyloxonium salts. Triethyloxonium tetrafluoroborate must be prepared first.

Protocol 44 Preparation of triethyloxonium tetrafluoroborate

1. Slowly add 50 ml of 10% (v/v) 1-chloro-2,3-epoxy-propane in dry ether to 200 ml of 5% (v/v) boron trifluoride diethyl etherate in dry ether.
2. Stir under reflux for 1 h and then at r.t. for another 3 h.

3. Wash the precipitated triethyloxonium tetrafluoroborate three times by decantation with 100 ml aliquots of dry ether.
4. Dissolve the precipitate in dry dichloromethane to a final concentration of 10% (w/v).
5. This solution can be used for up to 48 h for the O-alkylation of nylon 6 tubes.

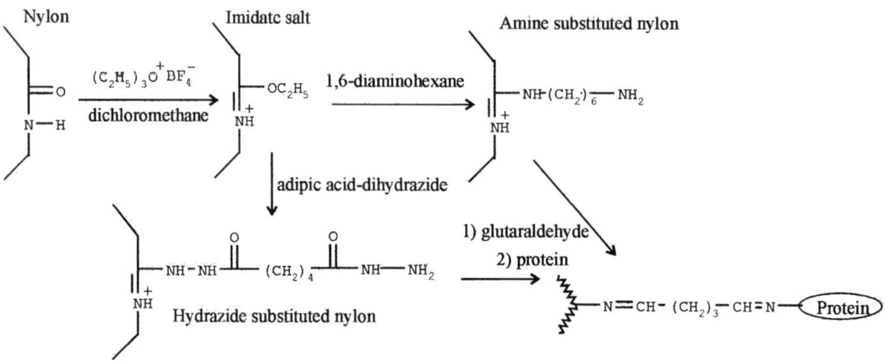

Figure 31 O-Alkylation of nylon for covalent binding of ligands:

Protocol 45 O-Alkylation of nylon

1. Fill a 3 m length of nylon tube with the triethyloxonium tetrafluoroborate solution and incubate at r.t. for 15 min.
2. Remove excess alkylating agent by suction.
3. Wash through the tube for 2 min. with dioxane.
4. Fill the O-alkylated nylon tube immediately with a solution of either the amine (10%, w/v 1,6-diaminohexane in methanol) or the acid hydrazide (3%, w/v, adipic dihydrazide in formamide) and incubate for 3 h at r.t.
5. Thereafter, excess amine or hydrazide is removed by perfusing the tube with water for 12 h.
6. Both derivatives can be activated with glutaraldehyde and coupled to biocomponents as described in the preceeding procedures.
   Alternatively these derivatives can be reactivated with a bisimidate:
7. Dry the tube by perfusion with methanol for 10 min.
8. Fill with a 2% (w/v) solution of diethyladipimidate dihydrochloride in 20% (v/v) N-ethylmorpholine in methanol and incubate for 45 min. at r.t.
9. Wash the tube through with 50 ml of methanol for 5 min.
10. Immediately fill with the solution of the biocomponent (e.g. 0.5–1 mg protein/ml) in 0.1 M phosphate buffer pH 8, and incubate for 3 h at 4 °C.
11. Remove excess of noncovalently bound biocomponent by perfusion with 0.5 M NaCl in 0.1 M phosphate buffer, pH 7.

This method does not significantly impair the mechanical strength of the nylon. Beyond that, the modified polyamide backbone carries no charged groups resulting from the modification procedures.

N-Alkylation

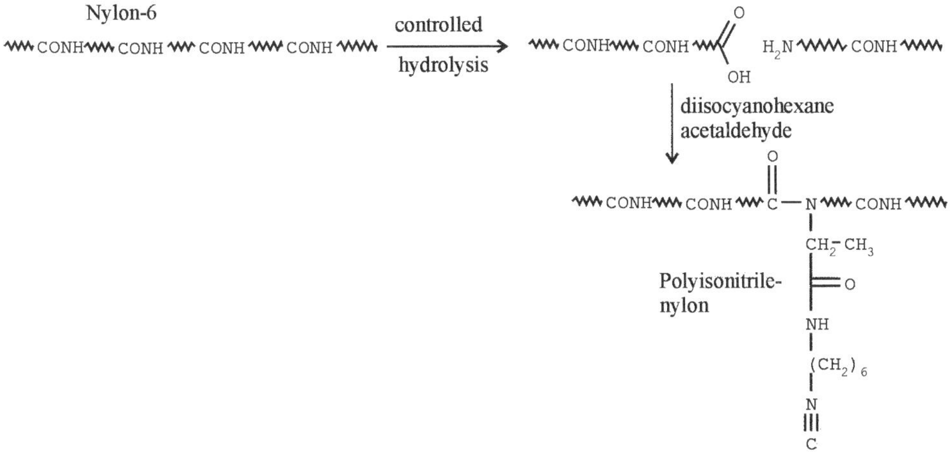

Figure 32 N-Alkylation of nylon

For the preparation of polyisonitrile-nylon as given below, some reagents must be synthesized first:

Protocol 46 1,6-Diisocyanohexane

Synthesis of N,N′-Diformyl-1,6-diaminohexane

1. Dissolve 1,6-diaminohexane (60 g, 0.5 moles) in ethyl formate and reflux the mixture for 2 h. (170 ml, 3.3 moles) under stirring over ice.
2. Upon completion of dissolution a white precipitate is formed.
3. Add additional 50 ml ethyl formate and reflux the mixture for 2 h.
4. Gradually solid N,N′-diformyl-1,6-diaminohexane will separate at the bottom of the flask.
5. Cool the reaction mixture, decant the liquid, and transfer the solid to a rotary evaporator to remove residual solvent.
6. The product (about 60 g, 70% yield, m.p. 105 °C) is used in the next procedure without further purification.

Synthesis of 1,6-Diisocyanohexane

1. Dissolve p-toluene sulfonyl chloride (150 g, 0.8 moles) in 300 ml of pyridine (KOH dried) under stirring.
2. Add N,N′-diformyl-1,6-diaminohexane (30 g, 0.2 moles) in small portions with stirring. The proceeding of the reaction is monitored by an increase in temperature and a color change from yellow to brown and finally to black.
3. After dissolution is complete, stir for 1.5 h at r.t.
4. Add water (250 ml) and crushed ice.
5. When temperature is lowered, extract with three 150 ml portions of ether.
6. Wash the combined ether extract with water (three 250 ml portions).

7. Dry over anhydrous Na_2SO_4.
8. Remove ether by evaporation.
9. Distill the crude oil in vacuum (pressure < 0.1 mm Hg)
10. The fraction distills at 85–92 °C (9 g, 30% yield)
11. Store in closed vials at –5 °C.

Protocol 47 Preparation of polyisonitrile-nylon (powder)

1. Suspend commercial nylon 6 pellets stepwise in a 20% methanolic solution of $CaCl_2$ (30 g/l) and stir at r.t. until a homogenous and extremely viscous solution is obtained.
2. Add the nylon solution dropwise under vigorous stirring into a large excess of water.
3. Filter off the powder thus obtained, wash with water, resuspend in water and homogenize.
4. Separate the powder again, wash with water followed by ethanol/ether, and dry in the air.
5. Remove moisture and solvent traces in a vacuum desiccator over phosphorous pentoxide.
6. Grind the dry powder in a mortar. The powdered particles are spherical and range in diameter from 0.2 to 0.7 mm. The mean carboxyl content will be about 25 mmoles/g dry nylon powder.
7. For controlled hydrolysis of the nylon powder suspend 10 g in 3 M HCl (300 ml) and stir for 4 h at r.t.
8. Separate the powder on a suction filter, wash with plenty of water, ethanol, and ether, and dry in the air.
9. Remove solvent traces and moisture in a desiccator over phosphorous pentoxide.
10. Store the partially hydrolized powder in a closed vessel.
11. The carbonyl content of nylon 6 powder thus obtained will be about 60–70 mmoles/g.
12. To obtain polyisonitrile-nylon, suspend partially hydrolized nylon powder (2 g) in 80 ml isopropanol.
13. Add acetaldehyde (20 ml, 0.3 moles) or isobutyral (32 ml, 0.3 moles) followed by 1,6-diisocyanohexane (8 ml, 0.06 moles) and stir in a closed vessel for 24 h at r.t.
14. Separate the resulting polyisonitrile-nylon powder on a suction filter and wash with isopropanol (50 ml) followed by ether (200 ml).
15. Air dry and remove solvent traces in a desiccator over phosphorous pentoxide.
    Store the polyisonitrile nylon powder at –5 °C in a dark, stoppered vial over silica gel.
16. The mean carboxyl content of polyisonitrile nylon is about 20 mmoles/g. The mean isocyanide content is 40–50 mmoles/g.

Protocol 48 Preparation of polyaminoaryl-nylon

1. Dissolve p,p′-diaminodiphenylmethane (2 g, 0.01 moles) in 200 ml of methanol and add 0.5 ml (0.005 moles) of isobutyral.
2. Suspend polyisonitrile-nylon powder (2 g) in this solution and add 1 ml of glacial acetic acid (0.017 moles). Stir in a closed vessel for 24 h at r.t.
3. Separate the polyaminoaryl-nylon on a suction filter and wash with dimethylformamide, methanol, and ether.
4. Dry in the air.

The diazotization capacity of this polyaminoaryl-nylon will be about 20 mmoles/g.

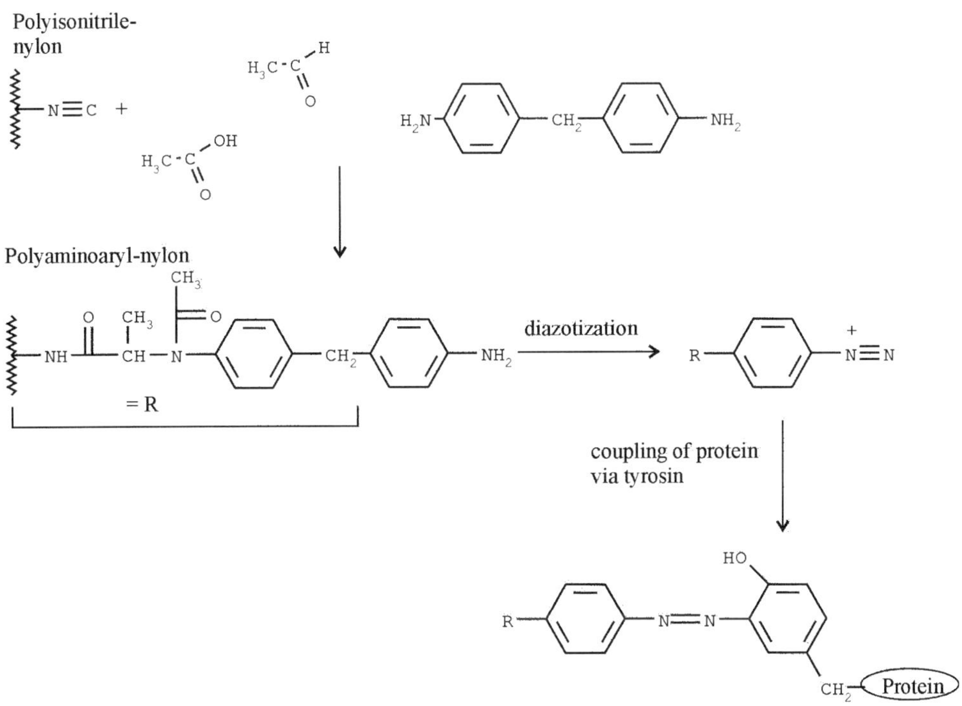

Figure 33 Coupling of protein to N-alkylated-nylon

Protocol 49 Coupling of proteins to polyisonitrile-nylon through the amino-groups of the ligand

1. Suspend polyisonitrile-nylon (50 mg) in 2 ml of cold 0.1 M sodium phosphate buffer, 0.5 M in sodium acetate, pH 7.
2. Add a cold aqueous solution of protein (5–10 mg in 1 ml) followed by 0.1 ml acetaldehyde.
   (Caution: The acetaldehyde – b.p. 21 °C – should be pipetted with a precooled pipette to prevent bubble formation)
3. Stir overnight in a closed vessel at 4 °C.

4. Separate the immobilisate on a filter. Wash with water, 1M KCl, 0.1 M in $NaHCO_3$ and again with water.
5. Keep wet at 4 °C

Protocol 50 Coupling of proteins to polyisonitrile-nylon through carboxyl groups on the ligand

1. Suspend polyisonitrile-nylon (50 mg) in 1 ml cold Tris HCl buffer pH 7.0.
2. Add a cold solution of protein in the same buffer (2–10 mg in 1 ml) followed by 0.1 ml acetaldehyde (pipetted from a precooled pipette) and stir overnight at 4 °C.
3. Wash and resuspend in water (4 ml) as described above.

Protocol 51 Coupling of proteins to polyaminoaryl-nylon

1. Suspend polyaminoaryl-nylon (100 mg) in cold 0.2 M HCl (7 ml) and add aqueous sodium nitrite (25 mg in 1 ml) dropwise.
2. Stir for 30 min. over ice.
3. Separate the diazotized polyaminoaryl-nylon (red brown!) on a suction filter and wash with cold water, cold 0.1 M phosphate buffer pH 8, and resuspend in 6 ml of the same buffer.
4. Add a cold aqueous solution of protein (5–15 mg in 2–3 ml) under stirring.
5. Keep overnight under stirring at 4 °C.
6. Separate the immobilisate by filtration and wash with water; 1 M KCl, 0.1 M in $NaHCO_3$ and water.
7. Resuspend in water (5 ml) and store at 4 °C.

3 Remarks concerning the choice of coupling techniques

The characteristic of the respective biocomponent determines the choice of the coupling technique. Enzymes unstable at pH values > 8 must be immobilized with the help of techniques most effective at low pH values (e. g., glutaraldehyde or in some cases carbodiimide). Similarly, proteins easily denatured at lower pH values must be immobilized with techniques most effective at alkaline conditions.

There are also other factors to be considered: To carry out site directed immobilization, it might be useful to vary the functional groups of the biocomponent involved in coupling. Beyond that, groups in the active or biorecognitive sites that may be capable of binding to the activated carrier have to be protected before coupling. Coupling temperature, ionic strength, composition, and all other parameters that could denature the biocomponent or interfere with the coupling reactions must be taken into consideration. Self-digesting enzymes

(e. g., proteases) have to be immobilized at low temperature (0–4 °C) for the shortest period of time possible to minimize autolysis.

4 Useful tests for assaying functional groups

4.1 TNBS test for assaying amino- or hydrazido groups

TNBS forms highly chromogenic derivatives with primary amines or hydrazide groups [23].

If sulfhydryl groups are also present in higher amount, interferences might be observed.

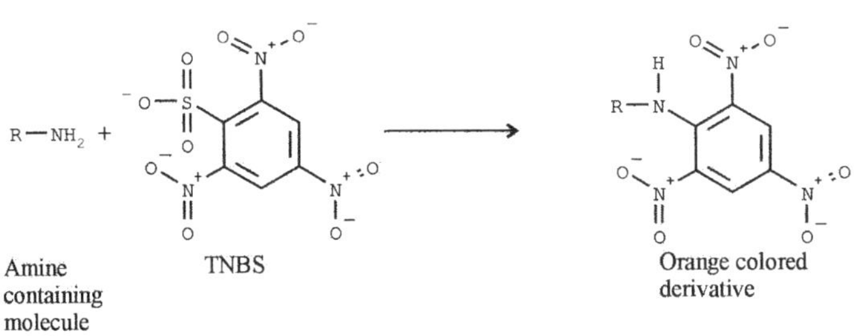

Figure 34

Protocol 52 TNBS test for assaying amino- or hydrazido groups

Reagent A: (always prepare freshly)

Dissolve 10 mg TNBS/ml H_2O.
Mix TNBS solution with the same volume of a saturated sodium borate solution.

Reagent B:

Prepare a 0.1 M aqueous adipic dihydrazide solution (17.4 g/l)

1. Suspend 200 ml gel (containing amino groups) in 2 ml of reagent A and stir for 30 min. at r.t.
2. Filter on a Büchner funnel and wash with 23 ml 0.2 M NaCl solution (or water). The combined liquid contains the excess unreacted TNBS.
3. Incubate aliquots of the wash (25, 50, 100 ml) with 1 ml reagent B and 1 ml saturated borate solution for 20 min. at 37 °C.
4. Measure on a spectrophotometer at 500 nm.
5. Estimate the content of reactive amino groups from the difference of initial TNBS amount (= Reagent A) and the excess unreacted TNBS ($\varepsilon = 1.65.10^4$).

Protocol 53 Procedure for assaying soluble primary amines

1. Dissolve the compound to be assayed in 0.1 M sodium hydrogencarbonate buffer pH 8.5. For large molecules like proteins, take 20–200 mg/ml, for small molecules take 2–20 mg/ml.
2. Dissolve TNBS in 0.1 M sodium hydrogencarbonate buffer pH 8.5 at a concentration of 0.01 (w/v). The solution has to be prepared freshly.
   Nota bene: TNBS stock solutions may be prepared in ethanol at a concentration of 1.5%. This solution is stable to long term storage. Dilute as needed in bicarbonate buffer to the required concentration.
3. Add 0.5 ml TNBS solution to 1 ml of each sample solution and mix well.
4. Incubate at 37 °C for 2 h.
5. Add 0.5 ml of 10% SDS and 0.25 ml of 1 M HCl to each sample.
6. Measure the absorbance of the samples at 335 nm.
7. The determination of the number of amino groups in a particular sample may be done with the help of a standard calibration curve employing an amine containing compound assayed under identical conditions.

4.2 o-Phthaldialdehyde reagent (OPA) to detect amino groups via a fluorescent product

OPA reacts with amines to form a fluorescent product in presence of 2-mercaptoethanol [24].

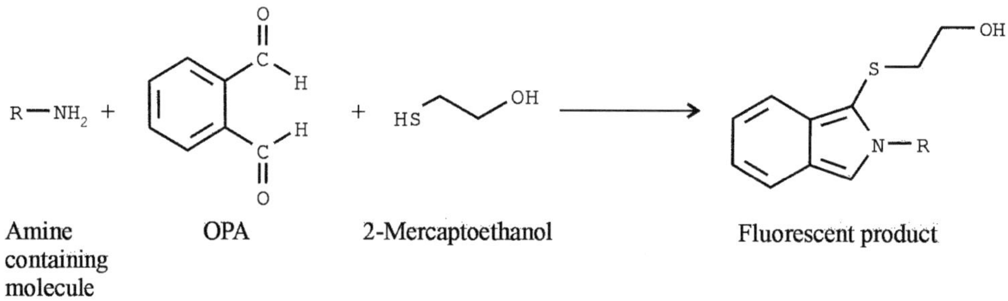

Figure 35

Protocol 54 OPA to detect amino groups via a fluorescent product

1. Prepare the samples dissolved in a no-amine-containing buffer or water at an expected concentration level within the standard curve range (see below)
   Nota bene: This method can tolerate the presence of most buffer components, denaturants and detergents without quench effects.
2. Add 2 ml of OPA reagent (Pierce) to 200 ml of the respected standard or sample and mix well.

3. Measure fluorescence of each sample and standard using an excitation wavelength of 360 nm and an emission wavelength of 436 nm.
4. Determine the respective sample concentration by comparison with the standard curve. (Concentration range between about 500 ng/ml and 1 mg/ml)

4.3 Ellmann's test for determination of sulfhydryl groups [25, 26]

Ellmann's reagent (5,5′-dithiobis(2-nitrobenzoic acid)) reacts with sulfhydryl groups under mild alkaline conditions, releasing the highly chromogenic 5-thio-2-nitrobenzoic acid (TNB), which can be quantified by its absorbance at 412 nm.

Ellmann's reagent also can be used as a derivatization reagent.

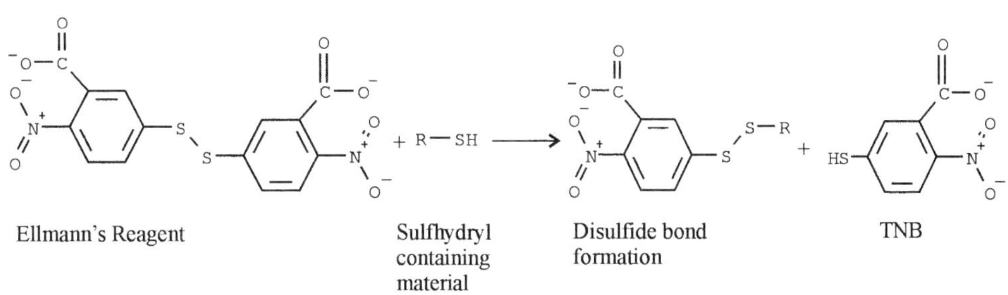

Figure 36

Protocol 55 Ellmann's test for determination of sulfhydryl groups

1. Dissolve Ellmann's reagent (Pierce) in 0.1 M sodium phosphate buffer pH 8 at a concentration of 4 mg/ml.
2. Prepare a set of standards by dissolving a stock solution of 3.5 mg (= 2 mM) cystein/ml in 0.1 M sodium phosphate buffer pH 8 down to at least 0.125 mM.
3. To 2 ml sample (or an appropriate amount of SH-carrier) in phosphate buffer pH 8 add 400 ml of Ellmann's reagent and mix well.
4. Incubate at r.t. for 15 min.
5. Measure the absorbance of each solution at 412 nm.
6. Plot the absorbance *versus* cystein concentration for each of the standards.
7. Calculate the sulfhydryl concentration of the samples by comparison with the standard curve.

5 Troubleshooting

Notes on the activation with glutaraldehyde: The exact manner in which glutaraldehyde reacts was uncertain for a long time. We studied this reaction in our lab thoroughly. Therefore, I want to give some hints, since this may improve coupling techniques.

Whenever colorless amino-carriers are activated with excess glutaraldehyde, color changes to a faint yellow or light brown are observed that cannot be the result of the simple reaction mechanisms given in the literature (and that also are used in this chapter as a simplified mechanism, which is easier to draw). Quiocho [27] showed that aldol condensations and cross-linking reactions occur that are responsible either for the observed colors or for observed aging effects, especially when glutaraldehyde is used as a cross-linking agent.

Examples are given in the following reaction scheme:

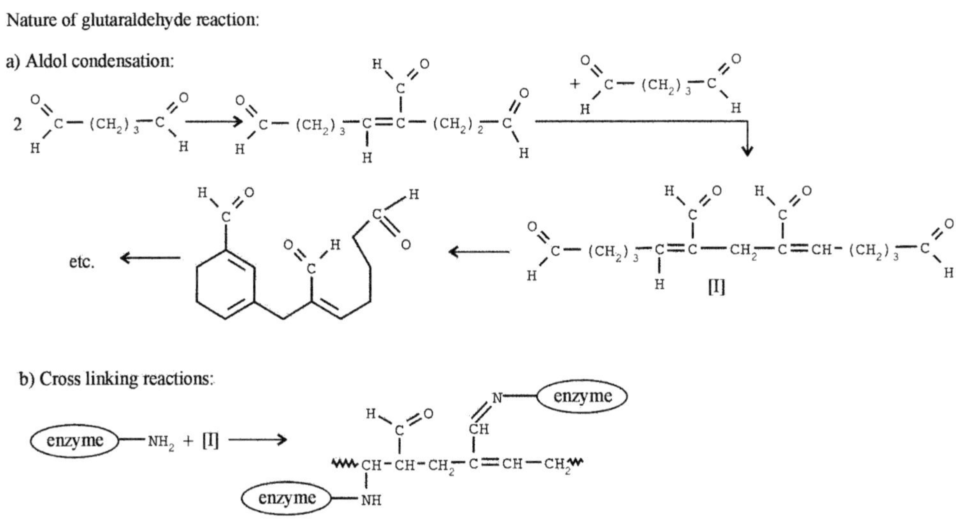

Figure 37

Together with the editor of this book, we found that the pH of the reaction mixture decides what really happens. An outline of these findings is given in Fig. 38.

However, it should to be pointed out that glutaraldehyde activation and cross-linking are still the best and easiest ways to immobilize biocomponents.

Caution: Glutaraldehyde causes pain when coming into contact with skin or other body surfaces and may act as a hapten when incorporated!

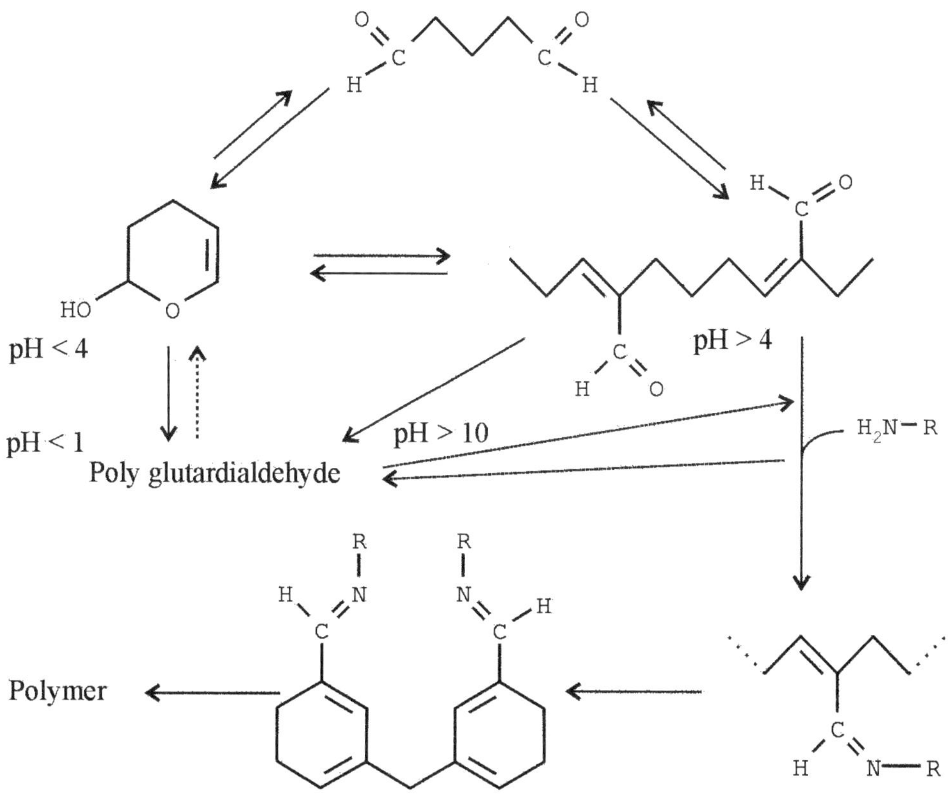

Figure 38

References

1 Chu BCF, Kramer FR, Orgel LE. (1986) Synthesis of an amplifiable reporter RNA for bioassays. *Nucleic Acids Res* 14: 5591–5603

2 Ghosh SS, Kao PM, McCue AW, Chappelle HL (1990) Use of maleimide-thiol coupling chemistry for efficient synthesis of digonucleotide-enzyme conjugate hybridisation probes. *Bioconjugate Chem* 1: 71–76

3 Sheehan JC, Preston J, Cruickshank PA (1965) A rapid synthesis of oligonucleotide derivatives without isolation of intermediates. *J Am Chem Soc* 87: 2492–2493

4 Sheehan JC, Cruickshank PA, Boshart GL (1961) A convenient synthesis of watersoluble carbodiimides. *J Org Chem* 26: 2525–2528

5 Yamada H, Imoto T, Fujita K, et al. (1981) Selective modification of aspartic acid-101 in lysozyme by carbodiimide reaction. *Biochemistry* 20: 4836–4842

6 Chase JW, Merrill BM, Williams KP (1983) F sex factor encodes a single stranded DNA binding protein with extensive sequence homology to E. coli SSB. *Proc Natl Acad Sci, USA* 80: 5480–5484

7 Chu FS, Fred Chi C, Hinsdill RD (1976) Production of antibody against ochratoxin A. *Appl Environ Microbiol* 31: 831–835

8 Chu FS, Lan HP, Fan TS, Zhang GS (1982) Ethylenediamine modified bovine

serum albumine as protein carrier in the production of antibody against mycotoxins. *J Immunol Methods* 55: 73–78
9 Chu FS, Ueno I (1977) Production of antibody against aflatoxin B1. *Appl Environ Microbiol* 33: 1125–1128
10 Williams A, Ibrahim IA (1981) A mechanism involving cyclic tautomers for the reaction with nucleophiles of the water-soluble peptide coupling reagent 1-ethyl-3-(3-dimethylaminopropyl)carbodiimide. *J Am Chem Soc* 103: 7090–7095
11 Gilles MA, Hudson AQ, Borders CL (1990) Stability of water soluble carbodiimides in aqueous solution. *Anal Biochem* 184: 244–248
12 Staros JV, Wright RW, Swingle DM (1986) Enhancement by N-hydroxysulfosuccinimide of water-soluble carbodiimide-mediated coupling reactions. *Anal Biochem* 156: 220–222
13 Garbarek Z, Gergely J (1990) Zero-length cross-linking procedure with the use of active esters. *Anal Biochem* 185:131–135
14 Domen PL, Nevens JR, Mallia AK, et al. (1990) Site directed immobilization of proteins. *J Chromatogr* 510:293–302
15 Pittner F, Miron T, Pittner G, Wilchek M (1980) Immobilization of protein on aldehyde containing gels. I. Activation of hydrazide gel with glutaraldehyde. *J Solid Phase Biochem* 5(3):147–166
16 Porath J (1974) General methods and coupling procedures. In: WB Jakoby, M Wilchek (eds): *Methods in Enzymology.* Academic Press, London, 13–30
17 Pittner F, Miron T, Pittner G, Wilchek M (1980) Immobilization of protein on aldehyde containing gels. II. Activation of pyridine rings with cyanogen bromide. *J Solid Phase Biochem* 5(3):167–180
18 Parikh I, March S, Cuatrecasas P (1974) Topics in the methodology of substitution reactions with agarose. In: Jacoby WB, Wilchek M (eds): *Methods in Enzymology* 34(B) Academic Press, NY, 77–102
19 Weetall H (1976) Covalent coupling methods for inorganic support materials. *Methods Enzymol.* 44: 134–138
20 Moser I, Schalkhammer Th, Mann-Buxbaum et al. (1992) Advanced immobilization and protein techniques on thin-film biosensors. *Sensors and Actuators* B7: 356–362
21 Pittner F. et al. (1992) Construction of electrochemical biosensors: Coupling techniques and surface interactions of proteins and nucleic acids on electrode surfaces. In: G Costa, S Miertus (eds.) *Proceedings of the Conference on Trends in Electrochemical Biosensors.* World Scientific Publishing Co. Pte. Ltd., Singapore, 69–84.
22 Hornby WE, Goldstein L (1974) Immobilization of enzymes on nylon. In: Jacoby WB, Wilchek M (eds): *Methods in Enzymology* 34(B), Academic Press, NY, 118–134
23 Cuatrecasas P (1970) Protein purification by affinity chromatography. Derivatizations of agarose and polyacrylamide beads. *J Biol Chem* 245:3059
24 Fried VA, Ando ME, Bell AJ (1985) Protein quantitation at the picomole level: An O-phthaldialdehyde pre-TSK column-derivatization assay. *Anal Biochem* 146:271–276
25 Ellmann GL (1959) Tissue sulfhydryl groups. *Arch Biochem Biophys* 82:70–77
26 Riddles PW, Blakely RL, Zerner B (1979) Ellmann's reagent: 5,5′-dithiobis(2-nitrobenzoic acid) – a reexamination. *Anal Biochem* 94:75–81
27 Quiocho FA (1976) Immobilized proteins in single crystals. In: Mosbach K (ed.): *Methods in Enzymology,* Academic Press, NY, 44:546–558

2 Fluorescence Techniques

Christian Mayer and Thomas G. M. Schalkhammer

Contents

Methods and Tools in Biosciences and Medicine
Analytical Biotechnology, ed. by Thomas G.M. Schalkhammer
© 2002 Birkhäuser Verlag Basel/Switzerland

1 Introduction

Fluorescence is a spectrochemical analysis method in which the molecules of the analyte are excited by irradiation at a certain wavelength and the emitted radiation at a longer wavelength is measured. This analysis method is widely accepted and is a powerful technique that is used for a variety of medical, pharmaceutical, environmental and biotechnological applications. The fields are diverse, e. g., detection of oil in water, oil sample fingerprinting, bacterial viability tests, chlorophyll measurement, monitoring of algae, membrane structure analysis, proliferation assays, DNA/RNA quantification, fluorescent tracers, brightening agents, protein quantification, enzyme assays, protein conformation studies, vitamins, toxin analysis, fluorescent proteins (e. g., GFP, YFP, and RFP), antibiotic testing, and many more.

2 Basis

The measurement of fluorescence called fluorometry is an analytical tool for quantitative and qualitative analysis. The process of fluorescence is defined as the very fast molecular absorption of light energy at one wavelength and a more or less instantaneous re-emission at a longer wavelength. Only a few molecules fluoresce naturally; therefore, a fluorescent molecule often is added for formation of a fluorescent compound. The characteristics of fluorescent compounds are an excitation spectrum (the wavelength and efficiency of light absorption) and an emission spectrum (the wavelength and number of photons emitted). These spectra are very specific for one fluorescent compound and do not match to others. Therefore, fluorometry is a highly specific analytical technique.

Fluorescence is measured by a device, the fluorometer, that selects the wavelength of light required to excite the compound of interest from the internal light source. It also selects the wavelength to measure the emitted light. The number of photons emitted (the signal) is proportional to the concentration of the analyte. The capability to select light for excitation and emission is given by the use of either monochromators or optical filters (filter fluorometers).

2.1 Advantages of fluorescence

The reasons to choose fluorometry as an analysis method are extraordinary sensitivity high specificity, selectivity, and simplicity at moderate costs.

Specificity/Selectivity: Because there are only a few compounds showing fluorescence it is easy to choose the measurement parameters (excitation and emission wavelengths) in a way effective to exclude interferences. Non-targeted components with intrinsic fluorescence very unlikely emit at the same wavelengths. This is a big advantage compared to spectrophotometric techniques, where many of the sample materials absorb light in a broad spectral range. Contrary to photometric techniques, fluorescence is a positive signal!

Sensitivity: The detectability is usually about 0.1 part per billion, but it can be as low as several parts per trillion (the limits depend strongly on the sample background). This is 3 to 6 orders of magnitude better than the detection limit of a spectrophotometer. Even single fluorescent molecules can be resolved given high quality equipment.

Wide concentration range: The calibration curve of fluorescence is linear to sample concentration over a very large range. In some applications fluorescence can be measured over 3 to 6 orders of magnitudes of concentration without sample dilution.

Simplicity/Speed: Because of simplicity and sensitivity of the technique, minimal sample preparation is needed resulting in a very quick analysis. Additionally, the small sample size and speedy process allow *in vivo* measurements.

Low Cost: Reagent and instrumentation costs are low compared to many other analytical techniques. Small sample size and minimal sample preparation contribute strongly to the cost efficiency of this technique.

3 Methods

3.1 Reagents

Alexa Fluor® 647 goat anti-mouse IgG (H+L) *2 mg/mL* [Molecular Probes, A-21235]
Alexa Fluor® 680–R-phycoerythrin goat anti-mouse IgG (H+L) conjugate *1 mg/mL* [Molecular Probes, A-20983]
Anti-tubulin (bovine), mouse monoclonal 236–10501 [Molecular Probes, A-11126]
Avidin-FITC labeled [Sigma, A 2901]
6-((6-((biotinoyl) amino) hexanoyl) amino) hexanoic acid, succinimidyl ester (biotin-XX, SE) [Molecular Probes, B-1606]
BODIPY FL-phallacidin [Molecular Probes, #B-607]
Boric acid [Sigma, B9645]
BSA [Sigma, A 3059]
Calcium Green™-1, AM *cell permeant* [Molecular Probes, C-3011]
Calcium Green-2 [Molecular Probes, C-3730]
5-(and-6)-carboxy SNAFL®-1, diacetate [Molecular Probes, C-1271]
7-dimethylaminocoumarin-4-acetic acid, succinimidyl ester (DMACA, SE) [Molecular Probes, D-374]
Dabcyl (4-(4′-dimethylaminophenylazo) benzoic acid) succinimidyl ester [Molecular Probes, D-2245]
DMF [j.t.baker, 7400]
DMSO [j.t.baker, 7033]
DTT dithiothreitol [Sigma, D9163]
EDTA Disodium salt: Dihydrate [Sigma E, 5134]
EnzChek® Elastase Assay Kit [Molecular Probes, E-12056]

Fluorescein Diacetate [Sigma, F 7378]
Fluorescein-5-isothiocyanate (FITC Fluorescein isothiocyanate Isomer I) [Sigma, F-4274]
Fura Red™, AM *cell permeant* [Molecular Probes, F-3020]
5-iodoactamidofluorescein [Molecular Probes, I-3]
NAD^+: β-Nicotinamide adenine dinucleotide monohydrate from yeast [Sigma, 43407]
HCl [j.t.baker, 6081]
NaCl [j.t.baker, 0278]
$NaHCO_3$ [j.t.baker, 0263]
NaH_2PO_4 * H_2O [j.t.baker, 0303]
Na_2HPO_4 [j.t.baker, 0306]
NaOH [j.t.baker, 0402]
Nigericin Sodium salt [Sigma, N 7143]
$MgCl_2$ [j.t.baker, 4003–01]
Paraformaldehyde [j.t.baker, 1157]
PBS: Dissolve 0.36 g NaH_2PO_4 * H_2O, 1.02 g Na_2HPO_4, and 8.77 g NaCl in 750 mL deionized water, adjust pH with 1 M NaOH or 1 M HCl if necessary, and bring the volume to 1000 mL with deionized water.
Propidium Iodide [Sigma, P 4170]
6-propionyl-2-dimethylaminonaphthalene (prodan) [Molecular Probes, P-248]
Rhodamine 6 G [Sigma, 83691]
Silver nitrate [Sigma, S1179]
1 M sodium bicarbonate solution: Dissolve 8.4 g $NaHCO_3$ in 100 mL deionized water; the pH should be about 8.3–8.5.
SYBR Green [Molecular Probes, S-7580]
SYBR Green I [Molecular Probes, S-7563]
TE buffer: 1 mM EDTA, 10 mM Tris-HCl, pH 8.0
TBE (89 mM Tris base, 89 mM boric acid, 1 mM EDTA, pH 8)
Tris: Tris (hydroxymethyl) aminomethane [Sigma, 93304]
Triethylammonium acetate (solution) [Sigma, 17898]

3.2 Principles and examples

The fluorescence process

As already cited, only a few molecules (generally polyaromatic hydrocarbons or heterocycles) fluoresce. Therefore, fluorescent probes or labels often are used, which are fluorophores designed to localize a target of interest within a specific region of a biological specimen. In all cases, fluorescence is the result of a three-stage process. The process responsible for the fluorescence of molecules is illustrated by the simple electronic-state diagram (Jablonski diagram) [1–4].

Stage 1: Excitation

Light sources such as xenon lamps or lasers connected to filter/monochromator systems supply photons of an appropriate wavelength (energy $h\upsilon_{EX}$). A photon is absorbed by the molecule, changing it from the ground state to one of many vibrational levels in one of the excited electronic singlet states, as indicated in Figure 1 (usually the first excited singlet state S_1'). This process distinguishes fluorescence from chemiluminescence, in which the excited state is populated as a result of a chemical reaction.

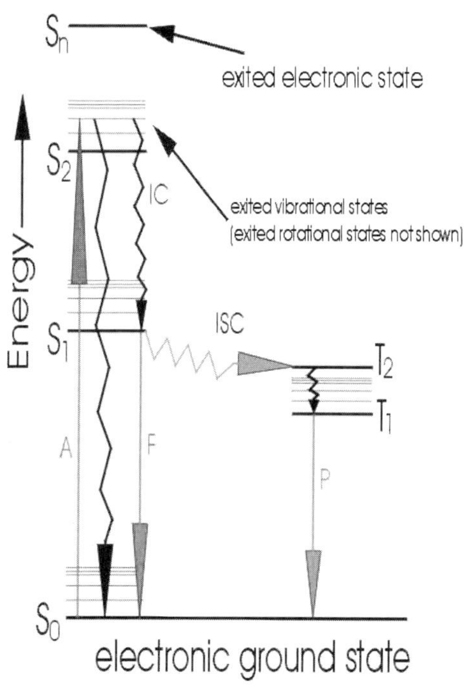

Figure 1

Stage 2: Excited-state lifetime

This excited state exists only for a finite time (typically 1–10*10^{-9} seconds). The molecule undergoes conformational changes and also can interact with its environment. Within this exited lifetime; a part of the additional energy is distributed mostly to vibrations ("internal conversion"), resulting in a relaxed excited state (S_1). In this state, the lowest vibrational level, the molecules choose their future. Most of the molecules (non-fluorophores, but also a part of excited fluorophores) continue with internal conversion until they reach the ground state (S_0). Other processes such as collisional quenching, fluorescence energy transfer and intersystem crossing (see Stage 3) also may depopulate S_1. In the case of fluorescence, the exited state is emptied by the emission of light.

The ratio of emitted photons to the number of excited fluorophores (absorbed photons) is called fluorescence quantum yield, separating the fluorophores in "weak" (near 0%), and "strong" (near 100%) ones.

Stage 3: Fluorescence emission
A photon of energy $h\upsilon_{EM}$ is emitted, returning the fluorophore to its ground state S_0. Because of the described energy loss, this emitted photon has less energy than the exciting one ($h\upsilon_{EX}$) and therefore a larger wavelength. The energy (and wavelength) difference of the 2 photons is the reason for the sensitivity of fluorescence. Contrary to absorption, it allows the measurement of photons at a very low background (excitation photons are blocked by filters and setup geometry). This energy difference ($h\upsilon_{EX} - h\upsilon_{EM}$) is called the Stokes shift.

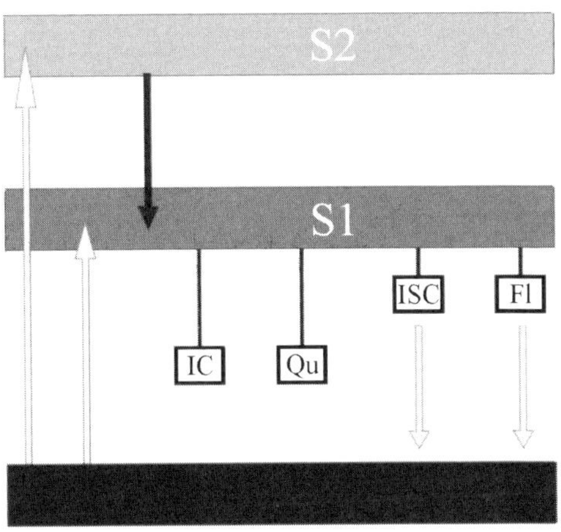

Figure 2

Quantum yield

$$\Phi = \frac{k_F}{k_{IC} + k_{Qu} + k_{ISC} + k_F}$$

Life time

$$\tau = \frac{1}{k_{IC} + k_{Qu} + k_{ISC} + k_F}$$

Processes competing with fluorescence
Other processes can relax the molecule from state S_1 to the ground state S_0. The most important pathways are:

IC = Internal Conversion (all zigzag arrows in Figure 1) is defined as direct vibrational coupling between the ground and excited electronic states (vibronic level overlap) or quantum mechanical tunneling (no overlap). This rapid process (10^{-12} s) results in two consequences: 1) fluorescence and other phenomena start in the relaxed exited state and 2) it competes effectively with fluorescence in most molecules, defining the quantum yield.

Qu = Collisional deactivation (external conversion) leads to non-radiative relaxation.

ISC = Intersystem crossing: If the energy states of the singlet state overlap those of the triplet state, vibrational coupling between the two states leads to **phosphorescence**. (Phosphorescence is the relaxation from the triplet excited state to the singlet ground state with emission of light, due to the principal forbidden transition a very slow process) [5].

Other processes, which may compete with fluorescence, are excited-state isomerization, photoionization, photodissociation and acid-base equilibria [6, 7].

Fluorescence spectra

The fluorescence process is repetitive; the same fluorophore can be repeatedly excited and detected, but the fluorophore can be destroyed irreversibly by photobleaching. In solution broad energy spectra are observed as a result of the interactions of the fluorophore with its environment. With few exceptions, the fluorescence excitation spectrum of a single fluorophore species in dilute solution is identical to its absorption spectrum. The fluorescence emission spectrum is independent of the excitation wavelength because of the emission from the "relaxed" excited state (S_1).

With the solvent influences, the energy balance for the fluorescence process should be written as:

$E_{fluor} = E_{abs} - E_{vib} - E_{solv\ relax}$,

where E_{fluor} is the energy of the emitted light, E_{abs} is the energy of the light absorbed, and E_{vib} is the energy lost by the molecule from vibrational relaxation. $E_{solv.relax}$ is defined as the reorientation of the solvent cage of the molecule in the excited state as well as when the molecule relaxes to the ground state.

Protocol 1 Principle determination of fluorescence spectrum with Rhodamine 6 G or NAD^+ as example

1. Using a micro spatula, transfer a very small amount of Rhodamine 6 G or NAD^+ into a fluorescence cell, fill the cell with water, and shake properly. Empty and refill the cell with water until the solution appears colorless.
2. Measure the absorbance spectrum using a spectrophotometer.
3. Use the peak maximum obtained from the absorbance measurement and define it as excitation wavelength. Scan the emission spectrum and determine the maximum emission wavelength.
4. In a few cases, the observed absorbance peak is not the most efficient one for excitation. Therefore, use the obtained emission maximum wavelength and scan for the proper excitation wavelength (with a few exceptions this will be identical to the absorbance wavelength).

Remarks: Many modern fluorometers offer the possibility of a synchronous search for all excitation and emission wavelengths. In this case Steps 1 and 2 can be skipped. Nevertheless, it is wise, especially for beginners, to do it the "old-fashioned way", because the absorbance spectrum will give hints if the used concentration is practical.
Troubleshooting: In the case of absorbance peaks higher than 1 or emission peaks out of scale, dilute. In the case of no absorbance peak, add additional fluorophore.
In the case of multiple absorbance peaks (e. g., for NAD^+), investigate all peaks as excitation wavelength starting with the highest one. In this case (and if the absorbance does not match the excitation wavelength), it may be necessary to execute step 3 again with the new excitation wavelength. In the case of absorbance but no fluorescence peaks, dilute, fluorescence is the more sensitive method, and your dilution may be in a concentration range with problems discussed in this section.

Fluorescence intensity
Fluorescence intensity is dependent on two parameters: how much light can be absorbed and how much can be emitted. The extent of excitation is dependent on the Lambert-Beer law (as is absorbance), defined by molar extinction coefficient, optical pathlength, and solute concentration. The extent of emission is defined by the quantum yield. This results in behavior similar to absorbance; meaning linearity at diluted concentrations and non-linear behavior at concentrated ones. Additionally, fluorescence is subject to other phenomena at higher concentrations, so that the working curve does not end in a plateau but runs through a maximum.

The influences are as follows:

1) Re-absorption of emitted radiation, which decreases the quantum yield and shifts the emission maximum to longer wavelength (additional decrease's due to the fact that one normally does not change the measuring wavelength.

2) Formation of dimers: (two fluorophores in the ground-state couple) $F + F \rightarrow F_2$ This shifts the absorption maximum to a longer wavelength, thereby decreasing the fluorescence intensity (at the given wavelength) [8]

3) Formation of excimers: (an activated fluorophore couples with one in the ground state) $F^* + F \rightarrow FF^*$ this excimer can fall apart into $F^* + F$, but also into $F + F$ (+hv) (two non-excited fluorophores resultig from the emission of a photon or non-radiation relaxation) [9].

All these aspects lead to the following fluorescent calibration graph:

The working range (linear part) of a fluorophore can be predicted by the formula: $c_{max} = 0.05 / (2.303 * \varepsilon_{(\lambda)})$. If the concentration is bigger than c_{max} the fluorescence is no longer proportional to the concentration but to quantum yield and light intensity.

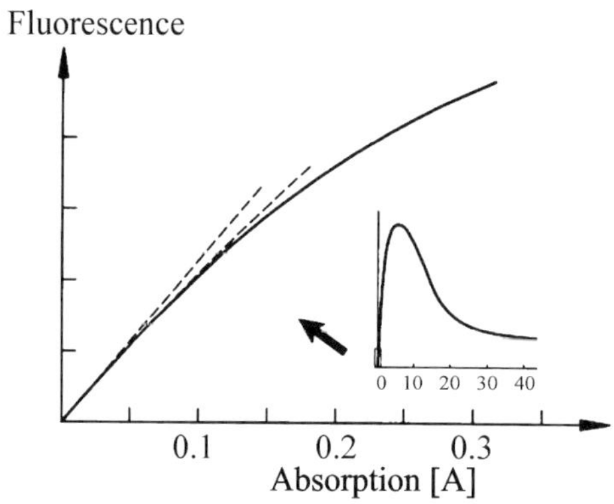

Figure 3

Protocol 2 Calibration graph (with Rhodamine 6 G)

1. Dissolve 10 mg Rhodamine 6 G in 10 ml water (1 mg/ml). Using that as a stock solution, setup a series of 1 + 9 dilutions of 10 orders of magnitude (10 ml each vial).
2. Using the parameters obtained in protocol 1, measure the intensity of all dilutions and plot a diagram.
3. Find the linear part of the curve and make additional dilutions in that range, adding their values to the diagram to produce a working line.
4. Measure the intensity of an unknown sample, using the working line it determines the samples concentration of fluorophore.

Troubleshooting: Intensity values do not form a line / especially afterwards added measurement values do not fit to the curve / "Maxima" not only at high but also at low concentrations: Always measure the most diluted fluorophore first; if you have to measure a less diluted fluorophore afterwards, be sure that your cell is washed properly (several times). If your fluorophore is light sensitive be sure that it is stored properly (in the dark). Fluorophore does stick to the cell wall and is not removed after several washing steps: change solvent to remove it (e. g. acetone).

Fluorescence instruments

There are several types of devices using fluorescence to be used in various platforms:

Spectrofluorometers measure the *average* properties of bulk samples in ml volumes. In micro-cuvettes and flow-through-cells, a volume in the μl range is accessible (flow-spectrofluorometers).

Microplate readers are specific for the microtiter plate format and measure the *average* properties of bulk samples at ~100–200 μl volumes.

Flow cytometers are related to flow-spectrofluorometers but focus on the fluorescence of (living) cells or micro-particles. Some allow these cells within a sample to be identified and quantified and even separated.

Fluorescence scanners are used for macroscopic objects to show fluorescence as a function of space (in 2–3 dimensions). This technique is used to quantify protein and nucleic acids on DNA/protein chips, electrophoresis gels, and chromatograms.

Fluorescence microscopes are used for micron-sized objects to analyze fluorescence as a function of space. This technique is often used to identify fluorophore on or inside a living cell (e. g., coupled to a protein of interest or expressed by the cell, e. g., GFP).

Because all types of fluorescence measuring devices share some essential parts, the setup of these parts is given here.

Light source: A lamp (e. g., 150 W xenon lamp) or a laser provides the energy to excite the compounds of interest. Lamps emit a broad range of light, more wavelengths than those required exciting the compound. Often, a sophisticated setup of mirrors and slits is used to focus a beam of that light toward the target. Figure 4 shows some setups of mirrors used in modern fluorimetric devices.

Excitation filters (in modern devices, a monochromator working with slits

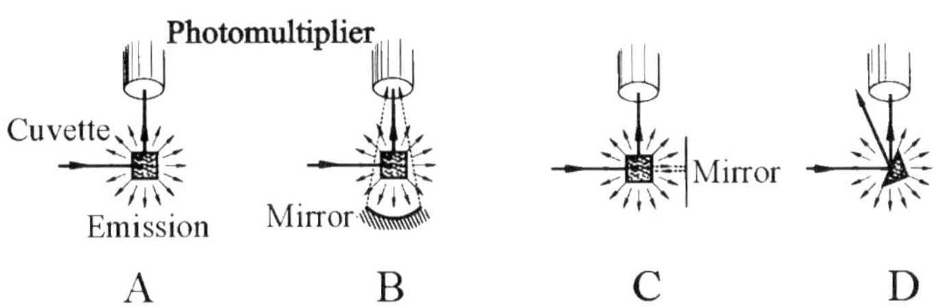

Figure 4

and a concave diffraction grating) are used to block out the wavelengths of light not needed and thus disturb the measurement. This setup results in a small band of light energy leaving the excitation slit and exciting the sample. The better the excitation wavelength is defined, the more energy is lost from the source; thus, a compromise of wavelength-precision *versus* energy input for the detector often is necessary.

Sample cell hold the sample of interest. Except in the case of measuring surface fluorescence, the cell material must not absorb at the excitation or emission wavelength of the component of interest. The design of the “cells” considering the above mentioned techniques might vary widely.

Emission filters (in modern devices, a monochromator) block scattered light, remove stray light with the excitation wavelength, and block fluorescence from other molecules. The setup includes slits and mirrors to guide the emitted light toward the detector.

Light detectors are either photomultiplier tubes or photodiodes. Both detectors transform the light intensity into a digital readout. Because the light sources are not capable of emitting the same energy over all wavelengths, a second light path is needed, which measures the excitation light, and enables the operator to define a baseline (which is the defined as the zero line). Figure 5 compares diodes and photomultipliers.

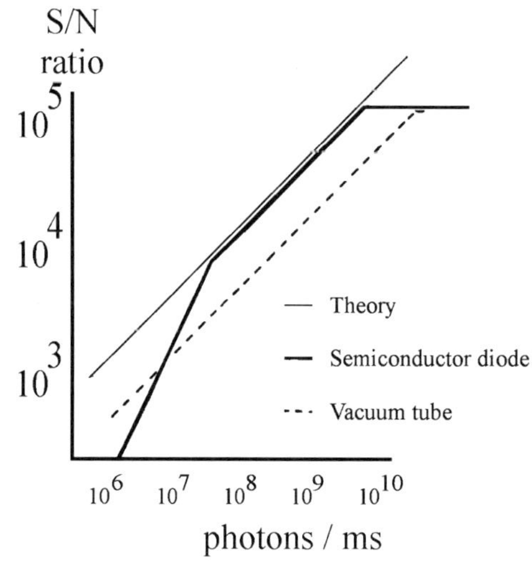

Figure 5

Detector	dark current	quantum yield
Vacuum photocell	10^5 photons/s	0.15
Si – photodiodes	10^7 photons/s	0.9

Fluorophores

Figure 6 shows one of the most typical fluorophores (a polyaromatic one), the fluorescein. For a few devices, their capability of producing light of some few wavelengths determines more or less the fluorophore to be used. Other samples determine by their composition the fluorophore, which has to be measured, but very often it is the choice of the operator to pick the most promising one. Besides considerations for sample composition and background signals, in all cases the fluorescence output (height of the signal) will be the most interesting value. This

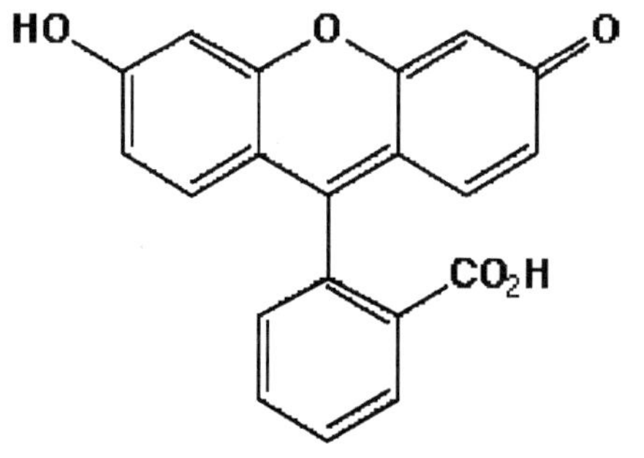

Fluorescein

Figure 6

output is dependent on the absorption capabilities of the fluorophore of interest, described by the absorbance coefficient (ε). Equally important for the output is the ratio with which the fluorophore emits photons and its ability to undergo repeated excitation/emission cycles. The emission efficiency is quantified by the quantum yield (QY) for fluorescence. Therefore the fluorescence intensity per molecule is proportional to the product of (ε) and QY. The range of these parameters among commercially used fluorophores is 5000 to 200,000 $cm^{-1}M^{-1}$ for ε and 0.05 to 1.0 for QY. Some proteins (e. g. some Phycobiliproteins [10, 11]) have multiple fluorophores on each protein and consequently have much larger extinction coefficients (on the order of 2.10^6 $cm^{-1}M^{-1}$) compared to low molecular weight fluorophores. Table 1 gives some commercially used fluorophores with their excitation and emission wavelengths. One should always keep in mind that these wavelengths as well as the absorption constant and the quantum yield are determined under specific environmental conditions and may change if these conditions are changed (e. g., use of ethanol instead of water).

Many of these fluorophores are synthesized as chemically reactive molecules. This gives the operator the possibility to label non-fluorescing components like many proteins, DNA or drugs to measure them with this sensitive technique.

The need for new fluorescent probes is given by the growing number of sophisticated applications, often asking not only for sensitivity but also for photo-stability, as well as the possibility to measure with background fluorescence (due to the sample composition). The synthesis of new fluorescent probes, which are excited in the near infrared, results in the possibility of lifetime-based sensing of a variety of analytes. Such probes display lifetimes in the range of nanoseconds to microseconds. Some of these promising, highly photostable fluorophores are metal-ligand complexes. This long lifetime allows for fading out the prompt auto-fluorescence and therefore increaser the sensitivity in all applications with background fluorescence. The long lifetime enables one to measure rotational motions of high-molecular-weight or membrane-bound proteins or to perform polarization immunoassay of high-molecular-weight analytes. Additionally with such long wavelength probes, one can avoid the use of complex lasers and use simple and robust laser diodes as the light source for time-domain, frequency-domain, or steady-state fluorescence.

As mentioned above fluorescence detection sensitivity is severely compromised by background signals, which cannot always be avoided by introducing new fluorophores. These background signals arise from unbound or non-specifically bound probes ("reagent background") or from endogenous sample components ("auto-fluorescence"). Detection of auto-fluorescence can be minimized either by selecting filters that reduce the transmission of the other components relative to the component of interest or by selecting probes that absorb and emit at longer wavelengths. Narrowing the fluorescence detection bandwidth does not help, it increases the resolution of all fluorophores, but it also compromises the overall fluorescence intensity detected. Using probes that can be excited at above 500 nm minimizes the auto-fluorescence of cells, tissues, and biological fluids. An additional aspect of using such fluorophores

Table 1 Table of Fluorophores

Fluorophore	Ex (nm)	Em (nm)	Notes
Fluorophores as labels			
Hydroxycoumarin	325	386	
Cascade Blue	375–400	423	
Lucifer yellow	425	528	
NBD	466	539	
R-Phycoerythrin (PE)	480–565	578	
Red 613	480–565	613	
Fluorescein	495	519	(pH sensitive)
BODIPY-FL	503	512	
Cy3	512–552	565–615	Standard fluorophore
TRITC	547	572	
X-Rhodamine	570	576	
Lissamine Rhodamine B	570	590	
Texas Red	589	615	
Cy5	625–650	670	Standard fluorophore
Cy7	743	767	
Allophycocyanin (APC)	650	660	
Fluorophores binding to nucleic acids			
Hoechst 33342	343	483	AT-selective
DAPI	345	455	AT-selective
Hoechst 33258	345	478	AT-selective
SYTOX Blue	431	480	DNA
Chromomycin A3	445	575	CG-selective
Mithramycin	445	575	
SYTOX Green	504	523	DNA
SYTOX Orange	547	570	DNA
Ethidium Bromide	493	620	
Acridine Orange	503	530–640	DNA and RNA
Thiazole Orange	510	530	
Propidium Iodide (PI)	536	617	
Fluorescent Proteins (quantum yield around 0.2–0.6)			
Y66F	360	508	
Y66H	360	442	
GFPuv	385	508	
ECFP	434	477	
Y66W	436	485	
S65A	471	504	
S65C	479	507	
S65L	484	510	
S65T	488	511	
EGFP	489	508	
EYFP	514	527	
DsRed	558	583	

in such dense media is the advantage of decreasing the light-scattering when increasing the excitation wavelength.

Photobleaching

Photobleaching is defined as irreversible destruction of the excited fluorophore. Photobleaching becomes the limiting factor for fluorescence applications if high-intensity illumination conditions are used. The destruction of the fluorescence capabilities of a fluorophore means a chemical reaction (with other fluorophores or other molecules), which gives multiple pathways of destruction in labeled biological solutions. In all cases, photobleaching originates from the triplet excited state, which is created from the singlet state (S_1) via intersystem crossing.

The only effective way to avoid photobleaching is to reduce the intensity of the excitation light. This means a decrease in detection sensitivity; therefore, the sensitivity is (if necessary) enhanced by low-light detection devices such as CCD cameras or by high-numerical aperture objectives combined with the widest emission bandpass filters compatible with satisfactory signal isolation. Another possibility is to use a more photo-stable fluorophore. An example for this is the substitution of fluorescein by Molecular Probes' Alexa Fluor 488, a dye that provides greater photostability and is compatible with standard fluorescein optical filters [12]. Many other reagents are also more photo-stable; a limiting aspect is always the compatibility with the experiment, e.g., the survival of cells is a limiting aspect when studying pathways inside these cells. In general, it is difficult to predict the necessity for and effectiveness of such countermeasures because photobleaching strongly depends on the fluorophore's environment [11–15].

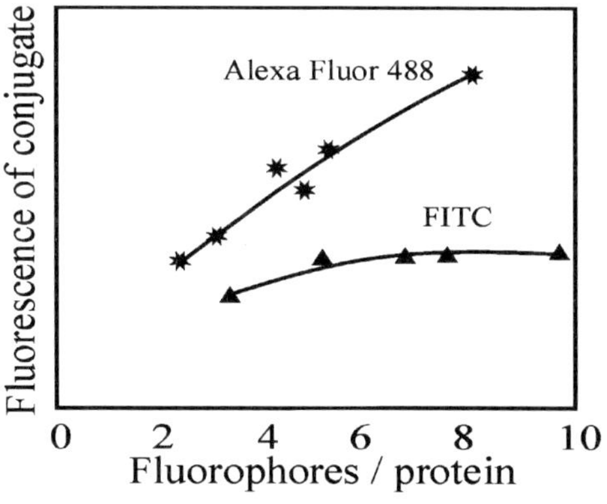

Figure 7

Relative fluorescence of anti–IgG conjugates prepared from the Alexa Fluor 488 dye and fluorescein isothiocyanate (FITC)

Multicolor labeling
Sometimes there is the need to monitor different biochemical functions, e. g., in fluorescence microscopy, flow cytometry, or DNA sequencing. A solution to this problem is the introduction of two or more fluorophores called a multicolor labeling experiment. In order to separately measure the two fluorophores their emission spectra must be distinguishable, meaning their emission peaks must not overlap. Consequently, fluorophores with narrow spectral bandwidths are needed. The ideal combination of fluorophores for this multicolor labeling means a strong absorption at the same wavelength but separated emission spectra. This need limits the application of this technique because there are only a few combinations of fluorophores in compliance with the requirements. However modern devices can generate several wavelengths at the same time (even if laser light is needed) and therefore are capable of detecting several fluorophores at the same time as long as their emission spectra do not overlap [16, 17].

Protocol 3 describes the detection of cell viability, which often is used as a quantitative parameter to determine the cytotoxicity of exogenous substrates or drugs and cytotoxic cellular interactions. With the help of flow cytometry and fluorophores, dead cells can be removed from a process. Propidium iodide (PI), a fluorescent dye when intercalated in DNA penetrating only dead or dying cells, is used in combination with non-fluorescent fluorescein diacetate (FDA), which is processed into a fluorescent fluorescein by esterases in cells that possess intact membranes. This results in different colors of healthy and dead cells.

Protocol 3 Cell viability detection using flow cytometry

1. Prepare the single-cell suspension in buffer (PBS pH 7.2 + 0.1% BSA) and adjust to 10^6 cells/ml.
2. Dissolve FDA in 1 ml of ethanol to a stock concentration of 10 mM (keep in the dark, 0 °C). Dilute 40 µl stock solution in 10 ml buffer to obtain the working solution.
3. Dissolve PI in 1 ml of buffer to a concentration of 10 mM (keep in the dark, 0 °C).
4. Tune laser to 488 nm. Establish PMT voltage settings for each fluorescence channel and set compensations for Fluorescein.
5. Assess auto-fluorescence signals for unstained cells.
6. Add 10 ml of FDA working solution and 30 ml of PI working solution.
7. Incubate at room temperature (r.t.) for 3 min. and introduce into the flow cytometer.
8. Flush the tubing with dilute bleach (FDA is absorbed to plastic), then flush through with water to remove the bleach.
9. Measure at 519 (for Fluorescein formed from FDA) and 630 nm for PI.

Remarks: Because of the activation of fluorescence inside the cell, no removal steps of excess fluorophore are needed.

Troubleshooting: Be sure the auto-fluorescence is properly obtained and that signals are compensated accordingly. Prepare single-color controls for compensation (non-viable positive control cells fixed in 1% paraformaldehyde may serve).

Ratiometric measurements
This type of application makes use of a shift in the absorption or emission spectrum. Thus, only one fluorophore is needed. Typically, this type of measurement showed be applied if the fluorophore exists in a free and an ionized form. Both forms must fluorescence, and the free and ion-bound forms must exhibit some differences in their emission or excitation spectra. The molecular basis for this effect is a reconfiguration of the fluorophores' pi-electron system that occurs upon protonation (or deprotonation). If the result of such structural changes is a large shift in one of the fluorophores characteristics, pH sensitivity is given, and the fluorophore can be used as pH indicator, the ratio of the optical signals can be used to monitor the association equilibrium and to calculate ion concentrations. Because both forms are observed, such measurements can eliminate distortions of data caused by photobleaching and variations in probe loading and retention, as well as by instrumental factors such as illumination stability.

Examples for such dyes are the calcium indicators fura-2 and indo-1 and the pH indicators BCECF, SNARF, and SNAFL [18–22].

Protocol 4 Ratiometric measurement of calcium

1. Dissolve Calcium Green-2 to a stock solution of 50 μM. Keep in the dark at 0 °C
2. Dissolve calcium chloride to a stock solution of 1 mM. Prepare a dilution series in the range to be investigated (nM to μM).
3. Prepare working buffer: 50 mM Tris/HCl pH 7.0
4. Set instrument to excitation wavelength 535, emission 550 nm
5. Mix 0.1 ml Calcium Green-2 stock solution, 0.9 ml calcium dilution and 2 ml working buffer in a 3 ml fluorescence cell. Determine emission intensity for all dilutions.
6. Plot signal *versus* concentration (working curve)
7. Measure unknown sample, insert signal value into working curve, and obtain concentration.

Remarks: The "different emission spectrum" in this case is due to the higher fluorescence intensity when calcium has bound to the fluorophore.

Troubleshooting: Calcium sample in mM range: Keep dilution series, but dilute sample. Shifting values at sample measurement: Be sure that the sample buffer is a weak buffer and auto-fluorescence is low. (Check spectrum of sample, adapt protocol from 0.9 to 0.1 ml for sample/calcium dilution [decreases detection limit!])

Protocol 5 Intracellular pH measurement using a confocal microscope

1. Prepare a 1 mM stock solution of 5-(and-6)-carboxy SNAFL®-1, diacetate in DMSO (store in dark and dry at –20 °C)
2. Prepare cells in serum-free, amine-free, and amino-acid-free medium (depending on cells) and adjust to $5*10^6$ cells/ml.
3. Prepare calibration measurements: Cell fractions are made pH permeable (usually with nigericin, a H/K ionophore) and flushed with solutions of different pHs
4. Dilute stock to 20 µM SNAFL-ester in serum-free, amine-free, and amino-acid-free medium.
5. Mix cells containing medium (calibration and to investigate cells) with SNAFL-ester medium 1:1, incubate at r.t. for 30 min. (cell esterases produce SNAFL), and wash cells.
6. Set instrument parameters to simultaneous double-excitation at 488 (acid form) and 568 (basic form). Collect emission at 525 (acid form, green) and 615 (basic form, red)
7. Ratio the green and red emissions by performing an arithmetic division of one image by the other.
8. With data gained from the calibration cell measurements, a standard pH curve is generated and the pH of the investigated cells is read out from that curve. The calibration has to be done for each experiment; it is valid only under the same conditions.

Remarks and troubleshooting: Be aware that useful pH range of SNAFL is 6.2–7.8. Be sure that the gain ranges of both channels cover the expected dynamic range of both emissions. The conditions (SNAFL concentration, incubation time) may vary and have to be optimized for each cell type. High dye concentrations and/or long incubation times will cause accumulation of the dye in some cell compartments, which may lead to measurement errors.

Protocol 6 Intracellular calcium measurement using a confocal microscope

1. Prepare a stock solution containing 1 mM acetoxylmethyl ester of Calcium Green and 0.5 mM acetoxylmethyl ester of Fura-Red in DMSO (store in the dark and dry at –20 °C)
2. Dissolve calcium chloride to a stock solution of 1 mM. Prepare a dilution series in the range to be investigated (nM to µM).
3. Prepare cells in serum-free, amine-free and amino-acid-free medium (depending on cells) and adjust to $5*10^6$ cells/ml.
4. Prepare calibration measurements: Cell fractions are made calcium permeable (usually with ionomycin, 5 µM) and flushed with solutions of different calcium chloride concentrations.
5. Dilute stock to 10 µM Calcium Green-ester in serum-free, amine-free, and amino-acid-free medium.

6. Mix cells containing medium (calibration and to investigate cells) with Calcium Green/Fura Red-ester medium 1:1, incubate at r.t. for 30 min., and wash cells.
7. Set instrument parameters to excitation at 488. Collect simultaneous double emission at 525 and 615 or use appropriate filters.
8. Ratio the green and red emissions by performing an arithmetic division of one image by the other.
9. With data gained from the calibration cell measurements, a standard working curve for calcium concentration is generated and the unknown calcium concentration of the investigated cells is read out from that curve. The calibration has to be done for each experiment, it is valid only under the same conditions.

Remarks: The calibration measurement also can be made with one cell sample, which is several times newly calcium calibrated. Between each measurement, cells have to be washed twice in buffer for the next-point calibration and allowed 3–5 min. equilibration.

Troubleshooting: Be sure that the gain ranges of both channels cover the expected dynamic range of both emissions. The conditions (fluorophore concentration, incubation time) may vary and have to be optimized for each cell type. High dye concentrations and/or long incubation times will cause accumulation of the dye in some cell compartments, which may lead to measurement errors.

Signal amplification

The straightforward way to enhance fluorescence signals is to increase the number of fluorophores available for detection. However, simply increasing the probe concentration is often quite counterproductive. It may cause changes in the probe's chemical and optical characteristics, and can even lead to lower signals that are due to dimer and excimer formation. Therefore, some more sophisticated methods to amplify fluorescent signals are introduced to circumvent these limitations [23–26]

- avidin–biotin secondary detection techniques
- antibody–hapten secondary detection techniques
- enzyme-labeled secondary detection reagents
- probes that contain special fluorophores (e. g. phycobiliproteins)

All these techniques share the capability to introduce a new "strong" fluorophore (with high quantum yield) or even a bigger number of strong fluorophores, but also exhibit some limitations. Increased labeling (resulting in more than one fluorophore per protein) of antibodies, as well as the avidin-biotin system, reduces their specificity and affinity. Additionally, for proteins in general, increased labeling leads to precipitation of the protein. For membranes, for example, increased labeling changes the membrane permeability.

Furthermore, at high degrees of substitution, the limitations due to excimer and dimer formation or the re-absorption of emitted light become dominant, resulting in decrease in efficiency. Figure 8 shows that with an increasing

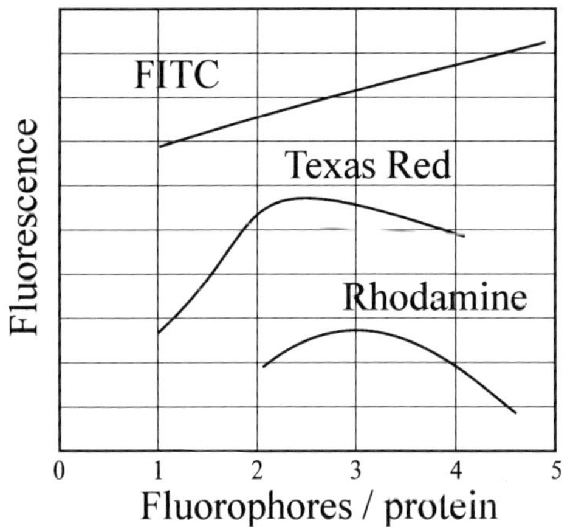

Figure 8

number of fluorophores per protein the fluorescence signal sometimes even decreases. It also indicates that the concentrations where these effects occur are specific for each fluorophore.

Protocol 7 Biotinylation of proteins

1. Dissolve the salt-free and amine-free (dialyze if necessary) protein at 5–15 mg/mL in 0.1 M sodium bicarbonate
2. Freshly prepare immediately before use 20 mg/mL biotin-XX ester solution (e. g., 2 mg biotin-XX succinimidyl ester dissolved in 0.1 mL dry DMSO)
3. Add the following amount of biotin-ester solution to your protein solution: ml (biotin-ester solution) = mol protein * 0.284.
4. Stir gently at r.t. for 1–1.5 h.
5. Separate protein from excess biotin by gelfiltration.

Remarks and troubleshooting: For the separation of proteins by gelfiltration, the proper filtration material has to be used. For very small proteins, it may be better to use dialysis or other methods.

Protocol 8 Avidin–biotin secondary detection technique for cell surface proteins

1. Prepare buffer: phosphate-buffered saline (PBS pH 7.2) containing 0.1% sodium azide and 2% BSA.
2. Prepare the cell suspension in buffer and adjust to 2 × 10^7 cells/ml.
3. Dilute biotin-labeled antibody in buffer in aliquots 50 µl (the amount of antibody may vary strongly for different test; we recommend an antibody concentration of 1 mg/ml).
4. Add 50 µl of the cell suspension to each aliquot and incubate for 30 to 45 min. on ice.
5. Centrifuge at 200 * g for 3 min. at 4 °C. Remove the suspernatant.

6. Washing step: Add 100 µl cold buffer, mix, centrifuge, and remove the supernatant. Repeat this step one or two times.
7. Dilute FITC-labeled avidin in buffer (the amount varies by the amount of antibody used. For an antibody concentration of 1 mg/ml, we recommend 1 mg/ml).
8. Add 50 µl of FITC-labeled avidin and incubate for 30 min. on ice.
9. Centrifuge at 200 * g for 3 min. at 4 °C. Remove the supernatant.
10. Washing step: Add 100 µl cold buffer, mix, centrifuge and remove the supernatant. Repeat this step one or two times.
11. Resuspend cells in 500 µl of buffer containing 1% paraformaldehyde and mix samples thoroughly.
12. Set excitation to 488 nm and assess auto-fluorescence signals for unstained cells.
13. Analyze at 519 nm on a flow cytometer.

Remarks: The conditions (fluorophore concentration, incubation time) may vary and have to be optimized for each cell type. The protocol gives a description of FITC, but a broad range of other avidin-linked fluorophores (e. g. Texas Red, Rhodamine) can be used when adapting steps 11 and 12. In principle, the protocol also can be used to investigate proteins inside a cell. The antibodies/avidin have to be microinjected into the cells.

Protocol 9 Using the secondary antibody detection techniques from cell surface proteins with phycobiliproteins as example

1. Prepare buffer: phosphate-buffered saline (PBS pH 7.2) containing 0.1% sodium azide and 2% BSA.
2. Prepare the cell suspension in buffer and adjust to 2×10^7 cells/ml.
3. Dilute antibody in buffer in aliquots 50 µl (the amount of antibody may vary strongly for different test; we recommend an antibody concentration of 1 mg/ml).
4. Add 50 µl of the cell suspension to each aliquot and incubate for 30 to 45 min. on ice.
5. Centrifuge at 200 * g for 3 min. at 4 °C. Remove the supernatant.
6. Washing step: Add 100 µl cold buffer, mix, centrifuge, and remove the supernatant. Repeat this step one or two times.
7. Dilute phycoerythrin-labeled antibody (Alexa Fluor® 680-R-phycoerythrin goat anti-mouse IgG (H+L) conjugate) in buffer (the amount varies by the amount of antibody used. For an antibody concentration of 1 mg/ml, we recommend 1 mg/ml).
8. Add 50 µl of phycoerythrin-labeled antibody and incubate for 30 min. on ice.
9. Centrifuge at 200 * g for 3 min. at 4 °C. Remove the supernatant.
10. Washing step: Add 100 µl cold buffer, mix, centrifuge, and remove the supernatant. Repeat this step one or two times.

11. Resuspend cells in 500 µl of buffer containing 1% paraformaldehyde and mix samples thoroughly.
12. Set excitation to 488 nm and assess auto-fluorescence signals for unstained cells.
13. Analyze at 700 nm on a flow cytometer.

Remarks: The conditions (fluorophore concentration, incubation time) may vary and have to be optimized for each cell type. In principle, the protocol also can be used to investigate proteins inside a cell. The antibodies have to be microinjected into the cells.

Multi-photon excitation

A practical problem of fluorophores absorbing in or near the ultraviolet is the lack of laser-based devices to produce this wavelength at reasonable costs. If two or more photons are used to excite the fluorophore of interest, a long wavelength laser, which is available in various designs from various suppliers, can do this. This approach is, moreover, significantly more cost-effective.

Additionally, multi-photon excitation has become of interest because of the possibility of localized excitation suppressing background fluorescence by several orders of magnitude. Such a two-photon excitation is, e. g., observed from alkanes. Among various applications, alkane fluorescence may become a new technique in biophysics to study lipid chemistry [27–29].

External factors

Many environmental factors influence the fluorescence of a particular fluorophore. The three most common are

- pH
- Polarity (of the solvent)
- Proximity and concentrations of quenching species

As mentioned above, fluorescence spectra are strongly influenced by their environment. A pH dependence of the absorption and/or emission spectra (shifting maxima) may not always be disturbing, as mentioned above; it offers the use of some fluorophores as pH indicators.

A dependence of fluorescence characteristic on solvent polarity is most often observed with fluorophores that have large excited-state dipole moments. For these fluorophores, the characteristics shift to longer wavelengths in polar solvents. This can be disturbing, but it should be kept in mind that in physiological and biological applications one is confronted with all kinds of environmental changes, including polarity changes. This especially includes such differences in cell compartments or polarity changes due to the vicinity of proteins, membranes, and other biomolecular structures. As for pH-sensitive fluorophores, this also can be used to investigate such differences in the environmental polarity, for example, a protein's interior. Representative fluorophores for these investigations are, e. g., the aminonaphthalenes such as prodan, badan, and dansyl [30–33].

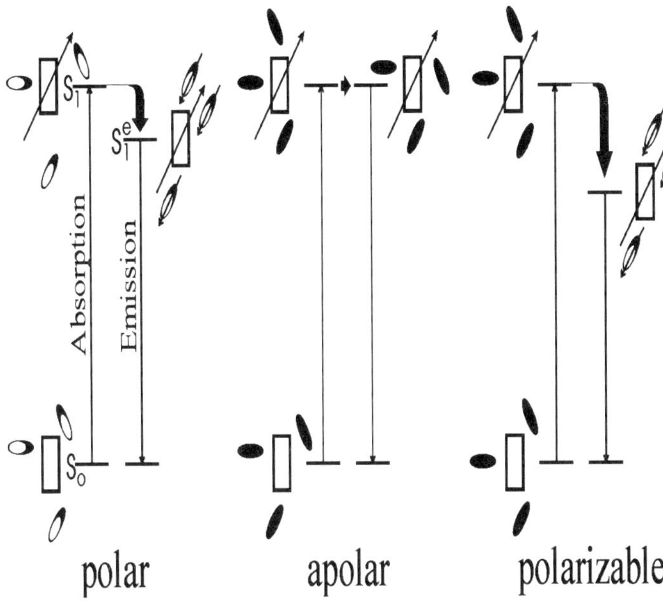

Figure 9

Protocol 10 Polarity measurement using prodan

1. Dissolve prodan in ethanol to a concentration of 1 μM.
2. Set excitation to 361 nm and measure emission spectrum (maximum should be at 498 nm).
3. Add increasing amounts of water and measure emission spectrum (maximum shifts toward higher wavelengths).
4. Add increasing amounts of dioxan and measure emission spectrum (maximum shifts toward lower wavelengths).

Remarks and troubleshooting: Be aware of the fact that prodan dissolves in water only in small amounts if you want to investigate very polar solutions. When investigating very non-polar solvents like octane, these solvents do not mix with water or ethanol; it is therefore necessary to start with DMF or dioxan as solvent.

Quenching can be defined as a bimolecular process that reduces the fluorescence quantum yield without changing the fluorescence emission spectrum. The formation of complexes with extrinsic quenchers can lead to non-fluorescent ground-state species. Collisional quenching may also result in the formation of transient excited-state interactions, resulting in a non-radiation relaxation of the fluorophore. The most ubiquitous quenchers are paramagnetic, such as O_2 and heavy atoms such as iodide. These quenchers reduce fluorescence quantum yields in a concentration-dependent manner. Some proteins can also efficiently quench some fluorophores; the reason seems to be due to charge-transfer interactions with aromatic amino-acid residues [34–36].

Consequently, it is possible to raise antibodies against such fluorophores, which can highly, specifically quench a fluorophore and thereby suppress background fluorescence of a disturbing fluorophore. Another useful application of the quenching is due to the fact that quenching normally is caused by collisional interactions. Therefore, it is possible to obtain information on the diffusion rates of quencher and fluorophore. Another use of the quenching process, which could be described as internal quenching, is a new powerful tool for PCR, called molecular beacons. This technique will be discussed in detail under "molecular beacons", p. 82.

Another important effect on the fluorescence of a label is the binding of the probe to its target. In some cases this binding does not include a covalent bond, but rather some deformation of the fluorophore due to electrostatic forces changing its fluorescence. While this still can be seen as "environmental" influence (similar to polarity influences), the covalent bond of a fluorophore induces a change of the molecule. So it is not surprising that binding of a probe, e. g., to a protein, does influence the quantum yield. In some ideal cases, an effectively non-fluorescent dye can become a fluorophore with a high fluorescence quantum yield when bound to a particular target. Examples for that are Molecular Probes' ultrasensitive SYBR, SYTO, PicoGreen, and many others.

An even more dramatic approach is to use an enzyme to convert a non-fluorescent dye by cleavage into a fluorescent product. This can be used as a very sensitive detection method of enzymatic activity. The use of such enzymatic activity, including the conversion from a non-fluorescent due to auto-quenching to a fluorescent dye, is discussed in the next section.

Protocol 11 Staining dsDNA in gels using fluorescent dyes with SYBR Green I

1. Run gel (Acrylamid or agarose gel).
2. Dilute SYBR Green I to a final dilution of 1:10,000 in TBE.
3. Shake gel in dye solution in a plastic box for 20 min. for thin gels, and for 60 min. for higher percentage gels 60 min.
4. Set excitation to 497 nm and measure emission spectrum at 520 nm (alternatively, an excitation of 254 nm can be used).

Remarks and troubleshooting: The dye is provided as solution in DMSO. Dilution in TE or TAE is also possible; the pH in all cases should be 8. Because of light sensitivity, cover the dye during shaking in all cases. Do not use glass boxes for staining, because the dye absorbs to glass. If it is necessary for further investigation, the dye can be removed by ethanol precipitation.

Fluorophore–fluorophore interactions

Fluorescence signals tend to decrease at high concentrations even without external quenchers. The reason for that is at least partly self-quenching, defined as the quenching of one fluorophore by another. Self-quenching occurs not only at high fluorophore concentrations but also at high labeling densities. Studies of self-quenching indicate that the mechanism involves energy transfer to non-fluorescent dimers. The quenching can be an advantage in some experiments; for example, if heavily labeled and therefore highly quenched biopolymers are introduced into cells, they show almost no background signal. When these polymers are cut to pieces by enzymes, it results in dramatic fluorescence signals of the now well-separated fluorophores, showing the position and concentration of that enzyme in a cell [37, 38].

Protocol 12 Protein activity studies with Elastase (using the Molecular Probes Elastase kit)

1. Prepare a 1 mg/mL stock solution of the DQ elastin substrate in deionized water and working solutions of 100 μg/ml.
2. Prepare 1*Reaction Buffer. Dilute 10* Reaction 1 to 10.
3. Prepare 100 U/mL porcine pancreatic elastase stock solution and 0.4 U/ml working solution in 1*Reaction Buffer (for positive controls).
4. Mix 1* Reaction Buffer and DQ elastin working solution 1:1.
5. Dilute the enzyme under investigation in 1*Reaction Buffer. Because varying enzyme activities, prepare a number of different enzyme dilutions.
6. Mix enzyme solution 1:1 with elastin-buffer obtained from step 4.
7. Mix porcine pancreatic elastase work solution 1:1 with elastin-buffer obtained from step 4 as positive control, and use the mentioned buffer mixed 1:1 with 1*Reaction Buffer as negative control.
8. Incubate for 30 min. at r.t. in the dark.
9. Set excitation to 505 nm and measure emission spectrum at 515 nm.
10. Measure the fluorescence intensity of samples and positive and negative controls.
11. If necessary, correct the measured values by subtracting the values of the negative controls.
12. The samples with results in the range of the positive control should be used. The activity of their elastase has to be recalculated with the dilution factor.

Remarks and troubleshooting: Because various investigation interests, there may be the need for shorter or longer incubation times. For shorter incubation times, increase the amounts of enzymes and for longer incubation times, decrease them.

A fluorophore-fluorophore interaction discussed above, which includes only one type of fluorophore, is the formation of dimers. If these dimers are capable of fluorescence, meaning they exist in an excited-state, they are named excimers, which exhibit altered emission spectra compared to the single fluorophores.

Another fluorophore-fluorophore interaction well distinct from multicolor labeling is the fluorescence resonance energy transfer (FRET or Förster-Transfer) [39, 40]. FRET is not simply one fluorophore emitting radiation and another fluorophore picking up the emitted light. Rather, FRET means an excited-state interaction in which emission of one fluorophore is coupled to the excitation of another. Because of the need of interaction, the effect is strongly distance-dependent. The scheme points out this Förster-Transfer:

$$F_1 + h\nu_1 \rightarrow F_1^*$$
$$F_1^* + F_2 \rightarrow F_2^* + F_1$$
$$F_2^* \rightarrow F_2 + h\nu_2$$

The transfer rate $k_t = (1/\tau) \times [R/R_0]^{-6}$ gives the distance dependence, whereby R is the distance donor to acceptor, R_0 is the critical donor – acceptor distance (about 1–5 nm), and τ is the fluorescence lifetime.

Because of the proportion of the interaction energy of two dipoles to R^{-3} and the proportion of the probability of energy transfer to energy^2, the transfer efficiency (E_t) is given as

$$E_t = R_0^6 / (R^6 + R_0^6)$$

Protocol 13 Using FRET for the investigation of protein assemblies

1. Dissolve the salt-free and amine-free (dialyze if necessary) antibodies directed against the proteins of interest in separate samples at 10 mg/ml in 0.1 M sodium bicarbonate pH 9.
2. Freshly prepare immediately before use 10 mg/ml 7-dimethylaminocoumarin-4-acetic acid, succinimidyl ester (DMACA, SE) in DMSO
3. Mix one of the antibody solutions 10:1 with the DMACA solution. Stir at r.t. in the dark for 1 h.
4. Freshly prepare immediately before use 10 mg/ml fluorescein-5-isothiocyanate (FITC) in DMSO.
5. Mix the other antibody solution 10:1 with the FITC solution. Stir at r.t. in the dark for 1 h.
6. Prepare 10 mg/ml BSA in 0.1 M sodium bicarbonate pH 9 as negative control.
7. Mix BSA solution 10:1 with the FITC solution. Stir at r.t. in the dark for 1 h.
8. Separate protein from excess dye by gelfiltration or dialysis.
9. Incubate proteins under investigation with fluorescence labeled antibodies (the antibodies should be in excess if possible).
10. Set instrument parameters to: excitation 373 nm, emission 525 nm.
11. Measure fluorescence intensity and compare with negative controls. The occurrence of FRET means an assembly of the proteins.

Remarks and troubleshooting: BSA should not assemble to one of the proteins, otherwise, the negative control will give positive results. A negative answer (no FRET) does not necessarily mean that there is no assembly. It can also prove that the antibodies destroy the proteins' ability to form a complex.

Self-quenching, FRET, and excimer formation all depend strongly on the interaction of adjacent fluorophores. Therefore, these techniques can be used to monitor molecular assembly or fragmentation processes. Examples for that are detection of membrane fusion, ligand–receptor binding, and nucleic acid hybridization, which is shown in Figure 10, whereby the stages are:

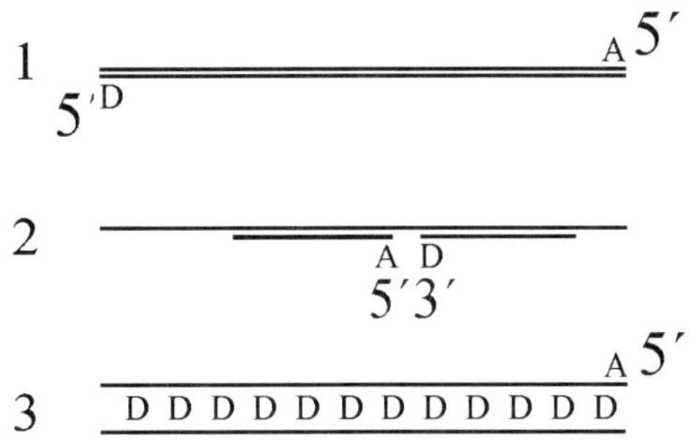

Figure 10

1. Hybridization of short oligonucleotide fragments
2. Hybridization of oligonucleotides to long DNA-sequences
3. Detection of hybridization using intercalating dyes

This technique enables detection directly in solution and therefore a high speed of hybridization. The detection limit of this method is 10^{-18} Mol

Protocol 14 Using FRET for the investigation of protein assemblies of cell surface proteins

1–7. Steps from protocol 13.
8. Prepare the cell suspension in PBS pH 7.2 buffer and adjust to 4 × 10^6 cells/ml.
9. Mix labeled antibody solutions with cells 1:1:1 and incubate for 30 min.
10. Centrifuge at 200 * g for 3 min. at 4 °C. Remove the supernatant.
11. Washing step: Add 100 µl cold buffer, mix, centrifuge and remove the supernatant. Repeat this step once or twice.
12. Resuspend cells in 500 µl appropriate buffer (allowing the cells to live).
13. Set instrument parameters to: excitation 373 nm, emission 525 nm.
14. Measure fluorescence intensity and compare with negative controls. The occurrence of FRET means an assembly of the proteins.
15. To investigate phase or environmental dependent behavior of cells, take measurements over a longer time period or change solution parameters.

Remarks and troubleshooting: BSA should not assemble to one of the proteins, otherwise, the negative control will give positive results. A negative answer (no FRET) does not necessarily mean that there is no assembly. It can also prove that the antibodies destroy the proteins ability to form a complex. The conditions (fluorophore concentration, incubation time) may vary and have to be optimized for each cell type. In principle, the protocol also can be used to investigate proteins inside a cell. The antibodies have to be microinjected into the cells.

3.3 Selected applications

Oil in water detection and monitoring

Fluorometry is used to detect crude oil, gasoline, diesel, and kerosene. Aromatic solvents and refined petroleum products (containing aromatics) also are detected with this technique. Because of the fast measurements, applications include emergency response, pollution prevention, leak detection, treatment verification, and process control. Detection limits range from µg/l to mg/l.

Oil sample fingerprinting

Petroleum is a complex mixture of thousands of different organic compounds formed from a variety of organic materials. Because of the infinitely variable natural factors during formation, there exists distinct chemical differences between oils. Therefore, oil from one crude oil field is readily distinguishable from another; differences due to differences in the refinery processes enable one to distinguish oils from even the same crude oil field. Thus, all petroleum oils, to some extent, have chemical compositions that differ from each other. In fingerprinting the oil samples are initially analyzed. In case of the need to know whether an unknown sample is what it promises or denies to be (quality or, e. g., environmental disaster), an analysis of the unknown sample is made. This provides the answer if it matches the initial analysis, even when there has been a serious amount of altering due to environmental factors.

Bacterial viability

Very often one is confronted with the question of whether bacteria are still alive or what the percentage of living bacteria in a mixed population is. A two-color fluorescence assay quantitatively distinguishes between live and dead bacteria in minutes, even in samples containing a range of bacterial types.

Chlorophyll measurement and algae monitoring

Fluorometry is used to measure chlorophyll, the photosynthetic pigment present in all forms of plants. Because the algae photo-systems also work with chlorophyll or similar structures, chlorophyll determinations also monitor their concentration. With this monitoring, the composition and ecological status of lakes, rivers, reservoirs and ocean waters are controlled. The online detection of algae in drinking water and wastewater is another field of application of fluorometric measurements.

DNA/RNA quantification

Quantification of DNA and RNA is a prelude to many practices in molecular biology. The high sensitivity and selectivity of fluorometric assays allows a reliable and accurate quantification of DNA or RNA.

Fluorescent tracer studies
Fluorometry using one of many non-toxic and environmentally harmless dyes is applied to measure water flows. By using this technique, one is able to model surface- and groundwater systems, to trace contaminants, and to detect leaks.

Protein quantification, histamine analysis, vitamin assays, aflatoxin analysis, antibiotic sensitivity testing, enzyme activity and cell proliferation
Fluorometry is used with a wide variety of commercial assays for accurate quantification of proteins, drugs, vitamins, and toxins in solution. Furthermore, it is used for enzyme activity studies or for sensitive cell counting and proliferation.

This method enables one to measure radioactivity without the use of radioactivity measuring devices. It makes use of an imaging plate, which is able to transfer the high-energy radiation into a fluorimetric measurable signal. Such imaging plates are coated with highly dispersed crystals, e. g., barium fluorohalide phosphor. The energy emitted by a radioactively labeled sample is transferred to these crystals, and stored as trapped electrons, creating an absorption band at about 600 nm. Exposure to Ne-He (633 nm) laser light releases the trapped electrons to the conduction band. The trapped electrons recombine with the holes trapped by Eu^{2+}; the result is an emission of light of about 400 nm. The emitted blue light is measured fluorimetrically. The described process is known as photo-stimulated luminescence. This Imaging Plate technique provides a fast and accurate tool for quantitative analysis of alpha, beta, and gamma-emitting isotopes.

Fluorescence lifetime imaging
Fluorescence lifetimes are mostly independent from photobleaching or probe concentration. But it is well known that the lifetimes of many fluorophores are altered by the presence of analytes such as Ca^{2+}, Mg^{2+}, Cl^{-}, pH or K^{+}. In the Fluorescence Lifetime Imaging Method (FLIM) a fluorescence microscope shows as image contrast what is derived from the fluorescence lifetime at each point in the image. FLIM is therefore capable of providing chemical imaging of intracellular ions. Because FLIM does not require wavelength-ratiometric probes, it allows quantitative ion imaging using visible wavelength illumination. For the fluorescence lifetime measurement of a fluorophore it can be excited by one- or coherent two-photon excitation. By measuring lifetime and intensity it is possible to study conformational dynamics of DNA.

Multidimensional confocal fluorescence spectroscopy of single molecules in solution
As discussed, a fluorophore has specific characteristics other than the absorption and emission spectra. If one is able to measure information such as fluorescence lifetime and fluorescence anisotropy in addition to these characteristics and signal intensity in one experiment, the measurement becomes

sensitive down to single-molecule detection and identification. Such measurements open up a wide range of new opportunities for ultra-sensitive analytical applications in chemistry, biology, and medicine, e. g., to monitor the conformational dynamics of biomolecules such as enzyme function.

The setup of such measurements includes a confocal epi-illuminated fluorescence microscope, a polarizing beam-splitter in conjunction with dual channel detection, open detection volumes, and burst integrated fluorescence (BIFL). BIFL is a real-time spectroscopic technique, that records the arrival time of a signal photon relative to the exciting laser, as well as the time lag to the preceding photon. Using this setup it is possible to identify and quantify different single-dye molecules in aqueous solution. The combination of multidimensional confocal fluorescence spectroscopy with fluorescence correlation spectroscopy (FCS), for example, can be used for studies of conformational dynamics of biological processes, as described below.

As in all single-molecule experiments, accuracy must be very near 100%, in other words, practically all fluorescence photons must be detected during such a single-molecule process. Because of fluorescence saturation, signal-to-background ratio, and photobleaching, a "good" fluorophore has to be chosen, meaning that it has a high-absorption coefficient, high fluorescence quantum yield, and low photobleaching; otherwise, it is not possible to simply increase the amount of exciting light. (Radiation power in such experiments may be well above 1 kW/cm^2, which is already a photobleaching problem, e. g., for Rhodamines or Coumarins.

Fluorescence polarization

The time spread of molecular rotation and of the relaxation process form the basis of fluorescence polarization. As described above, the very fast absorption process of a fluorophore is followed by vibrational relaxation, which is very fast, followed by the emission of radiation to relax the molecule completely. This process is still fast (about 10^{-8}s) but is much slower than the two precursory processes as indicated in Figure 11. It leaves a time gap big enough that rotation of the fluorescent molecule occurs. Illuminating with and measuring the emission of vertically polarized light shows dependence to the rotation speed of the fluorophore. Because of the dependence of the rotation to the mass of a molecule, a small molecule labeled with a fluorophore can be monitored as it binds to other molecules of similar or greater size [4, 41–43].

Fluorescence polarization **(P)** is described by the following equation:

$$P = (V - H) / (V + H)$$

V is the vertical component of the emitted light, and h equals the horizontal component of the emitted light of a fluorophore excited by vertical plane polarized light. The polarization unit P (normally given in mP) is dimensionless and not dependent on the concentration of the fluorophore or the intensity of the emitted light. This is the fundamental advantage of using this technique. As

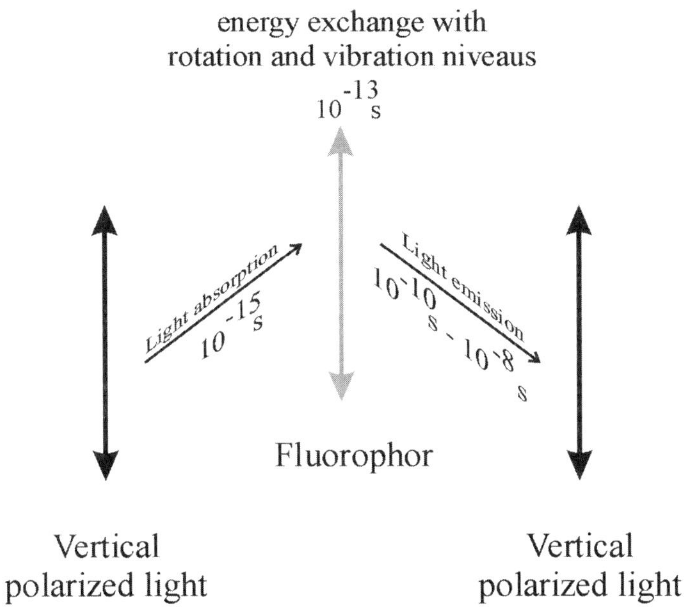

Figure 11

can be seen from the formula, the maximum of P is 1, meaning 1000 mP. This theoretical value could be achieved by fixing all molecules in the direction of the vertical polarized light. If the molecules cannot rotate (because of fixation), the output will be 100% vertical polarized light. However, in solution because of the random orientation of the molecules, the maximum ("limiting polarization", P_0) is 500 mP. P_0 means that there is no rotation at all before the relaxation process occurs.

Molecular bases are the directions in which a fluorophore absorbs or emits light, called the excitation and emission dipole. These dipoles may, but do not have to, be parallel as indicated in Figure 12. For measurements, these dipole directions have to be taken into account; for the theoretical explanations, we assume them to be parallel. Illuminating fluorophores in solution with vertical

Figure 12

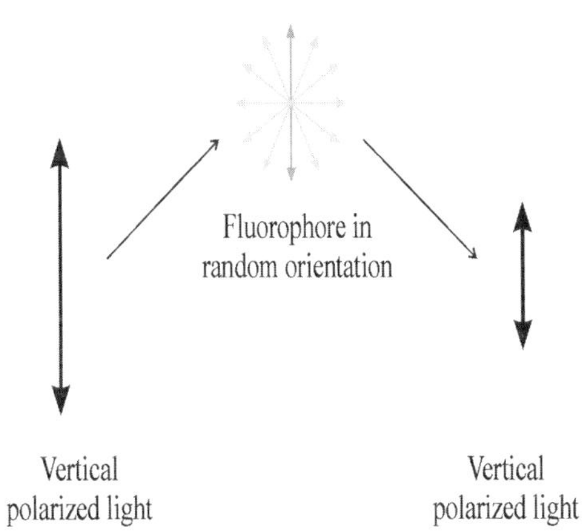

polarized light results in the absorption in varying amounts, depending on how these fluorophores are aligned relative to the vertical plane. (Fluorophores with their excitation dipoles arranged perpendicular to the plane of polarized light absorb no light at all; fluorophores with their excitation dipoles parallel to the plane of polarized light absorb maximally). Because of the statistical distribution of the fluorophores excitation dipoles aligned relative to the vertical plane, their emitted light has a horizontal component, without the fluorophores rotating. This theoretically resultant polarization defines P_0 as 500 mP. In practice, the molecules are not only randomly distributed, they also rotate. This means that the output polarization is between 0 and 500 mP. The value is dependent on how far the molecule has rotated during the fluorescence lifetime of the excited state. Because the size of a molecule defines its rotational speed, the fluorescence polarization becomes dependent on the size of the molecule, enabling this technique to monitor size changes such as binding events. (Because of the definition of P_0 polarization values greater than 500 mP indicate systematical errors, like scattered light or incorrect calibration.)

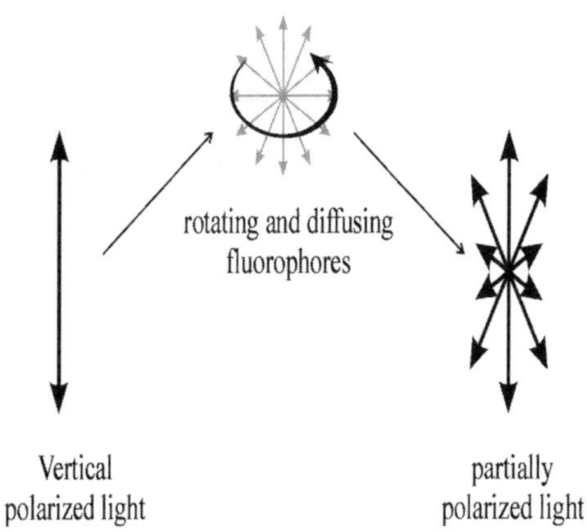

Figure 13

The relationship among the observed polarization, the limiting polarization, the fluorescence lifetime of the fluorophore, its rotational relaxation time and the temperature dependence is given by the Perrin equation:

$1/P - 1/3 = (1/P_0 - 1/3)*(1 + k\tau T/V\eta)$

Where k is the Boltzmann's constant; T is the absolute temperature; τ is the fluorescence lifetime; η is the viscosity, and V is a geometrical factor (for a spherical molecule, it's the volume). The equation shows that with decreasing fluorescence lifetime and temperature the polarization is near P_0, meaning it increases towards a maximum.

Advantages of fluorescence polarization

Fluorescence polarization is a fast technique, with highly reproducible results. Assays are easy to construct because the tracers do not have to respond to binding by intensity changes. The technique can work in colored solutions, even suspensions, and is relatively insensitive to instrument changes such as drift or gain settings.

Applications

As discussed above, the polarization of a fluorophore depends on the environment of the fluorophore and the size of the molecule the fluorophore is attached to. These properties make fluorescence polarization a powerful technique in biochemical systems [44–46].

Interaction of a fluorophore with macromolecules A small fluorophore has a characteristic polarization free in solution, which will depend primarily on the lifetime of fluorescence emission. When the fluorophore binds to a macromolecule, its rate of rotation in solution will now be characterized by the rate of rotation of the macromolecule. Because of the slower rotation, and increase in polarization is observed. Thus, the binding of the small molecule to the macromolecule may be followed by polarization changes. This approach is especially useful when a biochemical "normal" molecule like NADH is a fluorophore itself. Then the interaction with a protein can be observed, which proves to be very useful in this case because of NADH binds to many enzymes, and their reaction schema can be monitored.

Interaction of macromolecules In many cases, the interaction between two proteins has to be monitored, while none of the proteins is fluorescence active. In this case a fluorophore is covalently linked to one protein, which determines the fluorophore's polarization. If complexation with other proteins occurs, the measured polarization will increase because of the slower rotation of the complex relative to that of the single protein. Sensitivity considerations suggest that, if possible, the smaller of the two proteins should be fluorescently labeled. Using, fluorescence polarization, it is possible to follow dissociation of multimeric proteins, e. g., the calcium-induced dissociation of the tetrameric human plasma factor XIIIa.

Protocol 15 Using fluorescence polarization for the investigation of protein assemblies

1. Dissolve the salt-free and amine-free (dialyze if necessary) protein of interest at 10 mg/mL in 0.1 M sodium bicarbonate pH 9.
2. Freshly prepare immediately before use 10 mg/ fluorescein-5-isothiocyanate (FITC) in DMSO.
3. Mix the protein solution 10:1 with the FITC solution. Stir at r.t. in the dark for 1 h.
4. Prepare 10 mg/mL BSA in 0.1 M sodium bicarbonate pH 9 as negative control.
5. Mix BSA solution 10:1 with the FITC solution. Stir at r.t. in the dark for 1 h.
6. Separate protein from excess dye by gelfiltration or dialysis.
7. Incubate the second protein under investigation with fluorescence labeled protein in ratio 2:1.

8. Set instrument parameters to excitation 488 nm, emission 535 nm.
9. Measure fluorescence polarization of fluorescence labeled protein alone and then of protein mixture and compare with negative controls. An increase in polarization means protein assembly.

Remarks and troubleshooting: The mixing ratio of non-labeled – labeled protein may be varied, keeping in mind that "free" (because of excess) labeled protein disturbs the measurement.

Protocol 16 Using fluorescence polarization for the detection of protein activity with FITC as example

1. Dissolve the salt-free and amine-free substrate (the substrate itself must have an amine group) in appropriate buffer (e. g., 0.1 M sodium bicarbonate pH 9 if possible).
2. Freshly prepare immediately before use 10 mg/ fluorescein-5-isothiocyanate (FITC) in DMSO.
3. Mix the substrate solution 10:1 with the FITC solution. Stir at r.t. in the dark for 1 h.
4. Prepare 10 mg/mL BSA in 0.1 M sodium bicarbonate pH 9 as negative control.
5. Mix BSA solution 10:1 with the FITC solution. Stir at r.t. in the dark for 1 h.
6. Separate substrate from excess dye by gelfiltration or dialysis.
7. Incubate the protein under investigation with fluorescence-labeled substrate in molar ratio 2:1.
8. Set instrument parameters to excitation 488 nm, emission 535 nm.
9. Measure fluorescence polarization of fluorescence-labeled substrate alone and then of protein mixture and compare with negative controls. A decrease in polarization means that the protein cleaves the substrate.
10. Using the negative control and known concentrations of the protein under investigation, a baseline can be produced, which enables one to determine an unknown protein activity.

Remarks and troubleshooting: The smaller the substrate, the more the need for other purification methods will appear. If the substrate's mass matches the fluorophore's mass the mixing ratio (step 3) should be 1:1. The procedure will have no sense if the loss of mass due to the enzymatic reaction is very small compared to the substrates mass. (e. g., the loss of an ester group of a 10'000 kD protein will affect the polarization so poorly that it can not be measured).

Fluorescence correlation spectroscopy (FCS)

Fluorescence correlation spectroscopy is a correlation analysis of statistical fluctuation of the fluorescence intensity, which can be applied using fluorescence polarization. With the help of this unique and sensitive technique, one can obtain the physical parameters, that control the fluctuations through this analysis. It is used to analyze the concentration fluctuations of fluorophores in solution and measure, e. g., the size of micelles and liposomes. In this case, the fluorophore is covalently linked to the fatty chain. Time-resolved fluorescence anisotropy experiments proved, e. g., that the main depolarization mechanism is due to internal restricted motion and translational diffusion within the micelle. These two motions are determined because the rotational correlation time of the micelle as a whole can be calculated from FCS measurements [47–49].

Confocal fluorescence microscopy

A major limitation of classical fluorescence microscopy is the difficulty of examining thicker objects, such as entire cells or tissue sections, because of light emitted and scattered by the out-of-focus tissue. The thicker the objects, the more disturbances are caused by the image-forming elements, which are above or below the desired focal plane. Simply blocking out these elements is no solution, because their spatial arrangement is needed for a biologist focusing on morphological details. Confocal fluorescence microscopy solves some of these problems; it enables observation of microscopic structures within thick (tens or hundreds of microns) specimens.

In conventional microscopy, the complete object field is evenly and steadily illuminated. In confocal microscopy, the incident light, a laser beam, is focused by an objective lens into a cone-shaped beam so that the maximal brightness of the beam strikes one spot at a chosen depth in a specimen. Because of this setup, and by placing small apertures in the light path, almost all of the out-of-focus fluorescence is blocked, allowing detection of just that particular point of interest. Next, the fluorescent light emitted from this spot is focused to an image point where a pinhole aperture is placed. To achieve all these necessities of incoming and outgoing light, a dichroic beam splitter (a selective mirror that reflects shorter wavelengths, e. g., below 530 nm, but transmits wavelengths above 530 nm) deflects the beam into an objective lens, which focuses the beam onto the object. The setup is shown in Figure 14, and also indicates that light, which is emitted from regions located above or below the in-focus plane, comes to focus elsewhere and is therefore almost totally prevented from reaching the light detector. Light, which is emitted from the focus plane, passes the pinhole in the image plane and reaches the detector, located behind the beam splitter and the aperture. This light detector, normally a photomultiplier, generates a corresponding analogue electrical signal, depending on the intensity of light. The numerical apertures of the objective lenses and the aperture diameter characterize the system and determine the theoretical lateral and axial resolution in confocal microscopy.

Scanning the light source across the tissue in the X and Y directions results in a high-resolution image of a thin slice of the tissue. The laser beam is moved two-dimensionally into the focus plane by deflection of two galvanometric mirrors. Being governed by computer-controlled stepper motors enables scanning a series of such optical 'slices' through the thickness (Z-direction) of the specimen. Practical thicknesses of such section slices are approximately 0.4 μm.

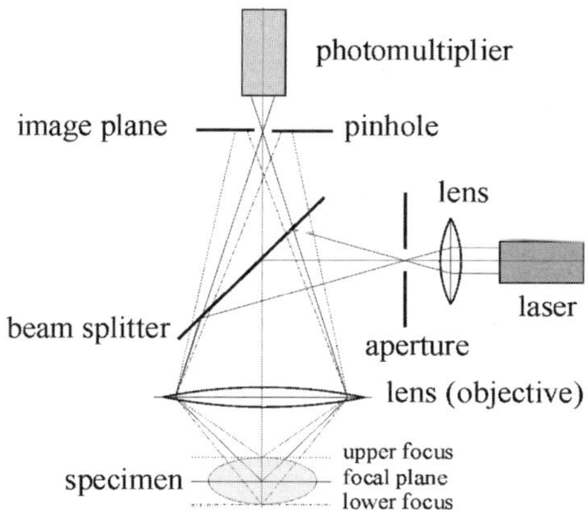

Figure 14

Thus, the three-dimensional arrangement of complex structures can be studied in detail, and constructed digitally within minutes. [50, 51]

Besides the need for a photostable fluorophore, there will always be the need for a strong (meaning bright) one because the background is never zero and signal enhancement is therefore limited. Working with bright fluorophores results in the use of a smaller aperture with a favorable signal-to-noise ratio, lower laser power, and shorter exposure times.

Fluorophores preferred by the authors are indocarbocyanine (Cy3) and fluorescein isothiocyanate (FITC)

An additional aspect must be considered, when more than one fluorophore is used in the investigation. The emissions of the fluorophores have to be well separated; overlapping of peaks is a problem because of the normally poor filter equipment of most devices. To check these problems, examine a single stained section with each fluorophore. If you observe a "breakthrough" from one fluorophore into the other's filter set, you have a problem. An example for that is the use of FITC and TRITC (tetramethyl rhodamine isothiocyanate) in the same experiment, which should be avoided.

Detailed protocols are given in the chapter on immunological application of confocal microscopy (see chapter 4).

Green fluorescing protein (GFP)

As mentioned above, many fluorophores and various techniques have been used to investigate the structure and molecular pathways of cells. But they all share one disadvantage: the introduction of these fluorophores into cells means cell death or at least extreme cell stress and is therefore mainly used to investigate cells that were killed before the experiment. Some methods have been described to introduce fluorophores and monitor their fluorescence in living cells, but either the introduction methods are complicated and work-intensive or just a few targets are possible (mainly enzymes, which modify the fluorophore). Green fluorescent protein (GFP) is a completely different approach to investigate cell pathways, proteins and structure, using the introduction of this fluorophore on the DNA level. Because many fluorophores are not synthesized in cells, or at least are not attached to a special protein, this unique approach made the GFP a valuable and favored tool for many investigations into living cells. Examples for these are the use of GFP as a marker for gene expression or the rapid detection of gene transfer and as a tag to study protein localization. The presence/absence/localization and concentration of GFP can be observed without interruption or termination of an experiment as is required with the detection of most other commonly used markers.

GFP is a 238 amino-acid protein that acts as an energy-transfer acceptor under physiological conditions in the jellyfish *Aequorea victoria*, deriving excitation energy from the emission of blue light via a calcium-binding-activated photoprotein, aequorin.

Its wild-type absorbance/excitation peak is at 395 nm with a minor peak at 475 nm. GFP is capable of producing a strong green fluorescence, emission peak at 508 nm, which can be monitored using standard ultraviolet microscope technology. Interestingly, excitation at 395 nm leads to a decrease over time in the 395 nm excitation peak and a reciprocal increase in the 475 nm excitation band [52].

The gene for GFP has been isolated and has become a useful tool for making expressed proteins fluorescent by creating chimeric genes composed of those of GFP linked to genes of proteins of interest. Thus, one may have an *in vivo* fluorescent protein, which may be followed in a living system, as GFP does not appear to interfere with cell growth and function. Its expression allows the isolation of stable cell lines expressing an uncharacterized protein for which there is no specific antibody or in which an epitope tag would produce an altered function. There have been several recent developments for the use of GFP and several different color variants.

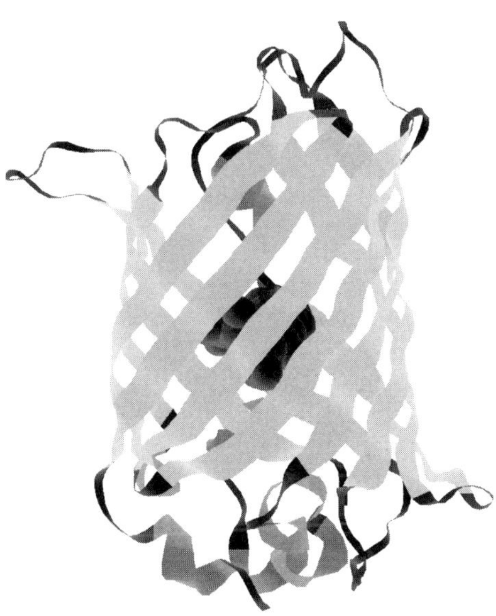

Figure 15

GFP exists as a dimer. As indicated in Figure 15 the monomers have the shape of a cylinder, comprising 11 strands of β-sheet on the outside with an α-helix inside and short helical segments on the ends of the cylinder, which is a unique motif. The fluorescent groups are located inside the cylinders on the central helix, providing protection of the fluorophore from collisional quenching or photochemical damage e. g. by oxygen. The motif also causes overall stability and resistance to unfolding by heat and denaturants. Denaturation, for example, requires treatment with 6 M guanidine hydrochloride at 90 °C or pH of < 4.0 or > 12.0.

Responsible for the fluorescence of the GFP is an internal Ser-Tyr-Gly sequence, which is post-translationally modified to a 4-(*p*-hydroxybenzylidene)-imidazolidin-5-one structure [53]. Molecular oxygen is proposed to be needed for the modification process; on the other hand oxygen quenching rate suggests that oxygen is quite well excluded from interactions with the fluorophore.

GFP exhibits two ground and two excited states; a set of polar interactions around the fluorophore accommodates proton rearrangements, enabling interconversions of the two states due to proton transfer. Over a non-denaturing range of pH, increasing pH leads to an increased excitation at 475 nm, concurrent with a reduction in fluorescence by 395 nm excitation [54]. The setup of amino-acid side chains in the vicinity of the fluorophore is responsible for different fluorescence characteristica of certain mutants and enabled to produce a number of other stable proteins with fluorescence emissions of other wavelengths.

Control of the dimerization of GFP enabled introduction of FRET into the experiments, allowing the study of protein assemblies inside cells under mild conditions. The basis for these FRET studies is to avoid dimerization, as also GFPs that are different in their emission and absorption characteristics can form a dimer. Such dimers would simulate an assembly because of the occurring FRET.

Fluorescent GFP has been expressed in many types of cells, including bacteria, yeast, mammalian cells, plants, and drosophila [55–59]. GFP tolerates N- and C-terminal fusion, and therefore functions as a tag to a broad variety of proteins, many of which have been shown to retain native function. With the protein-targeting sequences intact, specific intracellular localization, e. g., to the nucleus or mitochondria, can be achieved. Therefore, GFP is used as a reporter of gene expression, protein-protein interactions, and numerous other applications.

The initial problem using GFP was the excitation and emission spectra of the wild-type protein for fluorescence microscopy, as the spectra do not match the common filter sets or lasers in confocal microscopy. As mentioned above, extensions at the C-or N-terminus do not disrupt the motif structure of the GFP. Modifications of side chains can change the spectral properties of GFP, because of the changes of the environment inside the cylinder. Because of the availability of *E. coli,* clones expressing several mutants of the GFP gene were constructed shifting the spectra to 490/509, which is ideal for FITC filter sets.

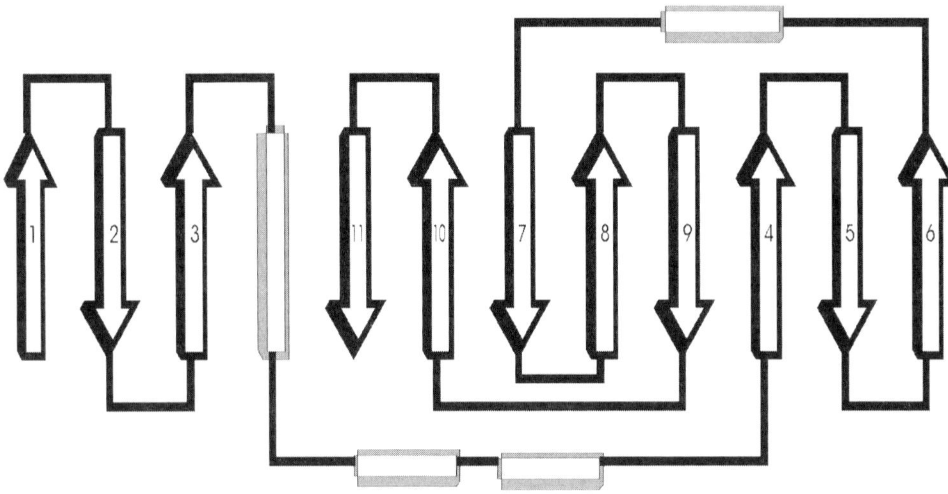

Figure 16

Furthermore, mutant GFPs, which exhibited increased fluorescence, as well as several different color forms of GFP were constructed, which dramatically expanded their biological applications, e. g., to use mutant GFP in fluorescence activated cell sorting (FACS). These new mutants of GFP are named, like the Green Fluorescent Protein according to their emission wavelengths, BFP (blue-), CFP (cyan-), RFP (red-), and YFP (yellow-fluorescent protein). One can now actually make double labeled specimens expressing fluorescently labeled proteins [60–62].

In order to gain significant wavelength shifts or enhancements, most mutations have been successfully explored in regions of the sequence adjacent to the fluorophore, i. e., in the range of positions 65–67, however, amino-acid substitutions significantly outside this region also affect the protein's spectral character.

Protocol 17 Protein investigation in living cells using GFP

1. Clone the protein of interest into a commercially available vector containing GFP and multiple cloning sites in frame to the GFP tag. Use cloning enzymes according to the instructions available by the supplying companies.
2. Transfect and select adherent cells according to commercial standard protocol (e. g., Effectene by Quiagen). GFP enables one to check its expression with a fluorescence microscope using excitation at 488 and emission at 511 nm (FITC or GFP filters). It is wise to check that the day after transfection and during the selection procedure.
3. Grow positive selected adherent cells on coverslips.
4. Mount the coverslips to special mounting rings and cover the cells with medium to keep them alive.

5. Set instrument parameters to excitation 395 or 475 nm, emission 509 nm (use 488 nm laser and FITC or GFP filters).
6. Observe the living cells under fluorescent or confocal microscope to check location and dynamics of the protein of interest.

Remarks: If using mutants of GFP, the wavelength/filter-sets have to be adapted.

Molecular beacons
Molecular beacons (MB) are single-stranded DNA (ssDNA) probes that can report the presence of specific nucleic acids. The short and synthetic MBs are composed of a hairpin-shaped oligonucleotide that contains both a fluorophore and a quencher group. This internally quenched fluorophore acts like a switch, which is normally "off". The fluorophore's fluorescence is restored when the MB binds to a target nucleic acid. This event opens the hairpin structure, separates the fluorophore and quencher, and turns the "switch" on [62–73]. In order to serve as a signal transduction mechanism for molecular recognition, the MBs must be designed in such a way that the loop portion of the molecule is a probe sequence complementary to a target, nucleic acid molecule. Then a specific molecular recognition event, the hybridization of a nucleic acid strand to its complement target can take place. To act in the described way the crucial formation of the hairpin structure is achieved by arm sequences flanking either side of the probe. These arm sequences are complementary to each other, therefore anneal to form the MB's stem, but are unrelated to the target sequence. These arms are normally composed of five to seven base pairs (see Fig. 17).

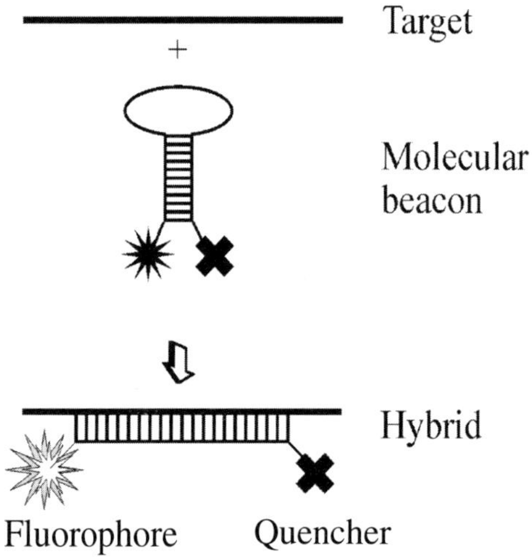

Figure 17 Molecular beacon: These molecules are non-fluorescent, because the fold-back-hybrid keeps the fluorophore close to the quencher. When the beacon-sequence in the loop hybridises to its target, forming a rigid double helix, the quencher separates from the fluorophore, restoring fluorescence.

Attached to one of these arms is a fluorophore; the quencher is attached to the end of the other arm. Without a target, the quencher and fluorophore are in close proximity to each other and no fluorescence occurs. The physical basis is the above-discussed fluorescence energy transfer, normally direct energy transfer or in a few cases fluorescence resonance energy transfer (FRET). In

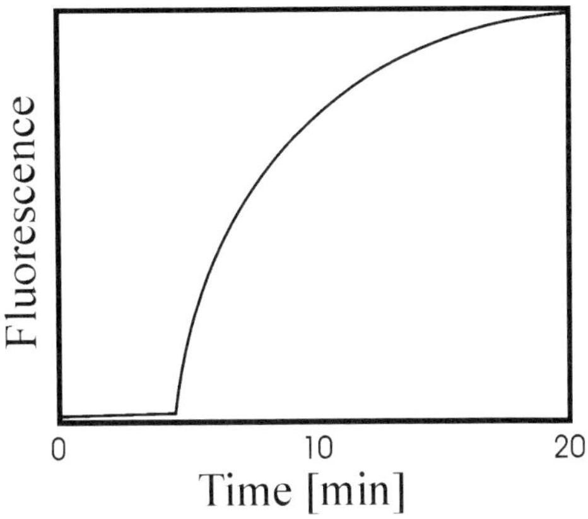

Figure 18

both cases, the fluorescent dye serves as an energy donor and the quencher serves as acceptor, emitting the energy that it receives as heat (or emitting on a larger wavelength in case of FRET).

Theoretically, the hybridization of a nucleic acid strand to its complement strand is one of the most specific molecular recognition events known. However, in practice, the length of the probe is limited because hybridization also needs time. Additionally, the formation of other hairpin structures (due to complementary sequences in the probe) would cause more complicated hybridization protocols. On the other hand, the probe has to be long enough to overpower the annealing of the arm sequences. Such a properly designed molecular beacon will therefore open the stem structure during the formation of a hybrid with the target DNA. Then the fluorophore and the quencher are moved away from each other, and fluorescence occurs. Figure 18 shows the spontaneous fluorescence response of molecular beacons to the addition of target. The signal /background ratio is roughly 200. Because dimethylaminophenylazobenzoic acid (DABCYL), a non-fluorescent chromophore, has been found to efficiently quench a large variety of fluorophores, it serves as the universal quencher for any fluorophore in molecular beacons.

Advantages of MB probes

Maybe the biggest advantage of these probes is that it is not necessary to remove unbound excess probes in order to observe hybridized probes. That is very useful in situations where it is either possible or not desirable to isolate the probe-target hybrids from an excess of the hybridization probes, such as in real-time PCR. Such dyes, which are activated as a result of enzymatical cleavage, or polarity changes have been discussed above; however, their applicabilities are limited. Because of the unlimited number of sequences, practically all nucleotide targets can be investigated. Molecular beacons offer the possibility to label nucleic acids within living cells. By using different colors, multiple targets can be detected in the same solution.

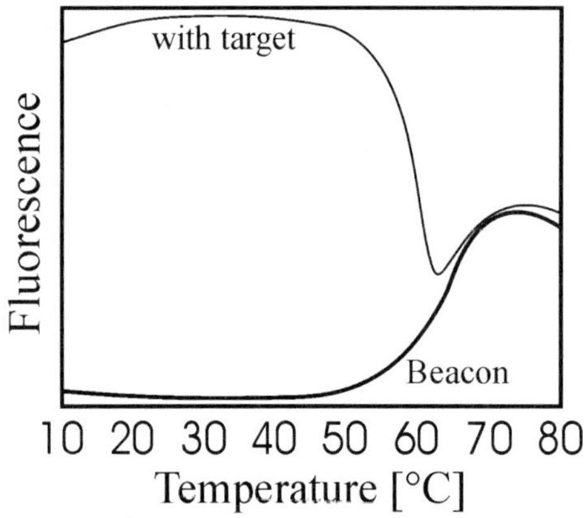

Figure 19

Another major advantage of MBs is their molecular recognition specificity. Owing to their structure, the recognition of targets by molecular beacons is so specific that single-nucleotide differences can be readily detected. The basis for this is the counteracting of the stem hybrid, which does not open if the target does not fit perfectly, thereby discriminating DNA, that differs from one another by a single nucleotide. Figure 19 gives an impression of the beacons behavior under different thermal conditions. Fluorescence occurs because of the denaturation of the hybrid above 60 °C. Below 50 °C, the free probe shows the small background fluorescence, whereas the target-bound form is strongly fluorescent.

Applications of MB probes

Molecular beacons are used for a variety of investigations, even including some protein investigations. MBs simply added to a sealed PCR-tube enable the real-time monitoring of DNA/RNA amplification without disturbing the PCR. This is due to the fact that the MB hybridizes at the annealing temperature (i. e. the step where the fluorescence is measured) and dissociates at elevated temperatures, which keeps the MB from interfering with polymerization [65].

With the help of microinjection into living cells, molecular beacons can be used to detect DNA/RNA in living cells [66, 67]. By using two different MB probes, single-nucleotide variations in DNA can be investigated. One probe with fluorophore 1 and the loop sequences specific for the wild-type allele, and the second probe with fluorophore 2 and the loop sequences specific for the mutant allele indicate wild-type/mutant or heterozygote.

Although MB probes were originally designed for nucleic acid studies, they can be used to investigate proteins. The basis for such measurements is the fact that some proteins destroy the hairpin structures as a result of binding to the beacon, which gives the same result, meaning fluorescence, as hybridization to a target. An example for that is the *E. coli* single-stranded DNA binding protein (SSB) [68].

To construct highly sensitive biosensors, MBs can be immobilized using the avidin biotin system. Critical is the introduction of the avidin–biotin bridge, which must not disturb the hybridization; therefore, biotin is normally linked via a spacer to the quencher side. This enables the setup of MB probe arrays for simultaneous multiple analyte detection [69–71].

Protocol 18 Synthesis of molecular beacons: quencher

1. Dissolve 50–250 nanomoles of DNA containing a sulfhydryl group at its 5′ end and a primary amino group at its 3′ end in 500 µl of 0.1 M sodium bicarbonate, pH 8.5.
2. Dissolve 20 mg of dabcyl-succinimidyl ester in 100 µl DMF.
3. Add dabcyl-ester solution to the oligonucleotide solution in 10-µl aliquots at 20-minute intervals. Stir for at least 12 h.
4. Spin in a microcentrifuge for 1 min. at 10,000 rpm. Pass supernatant through a Sephadex G-25 column to remove excess dabcyl-ester using 0.1 M triethylammonium acetate, pH 6.5, as column buffer. Filter the eluate through a 0.2 µm filter.
5. Purify oligonucleotides by HPLC, using a C-18 reverse-phase column. Elute with a linear gradient of 15% to 60% acetonitrile in 0.1 M triethylammonium acetate, pH 6.5. Collect the peak that absorbs at 260 nm and 491 nm.
6. Precipitate fraction adding salt to 0.5 M (in the water fraction) and ethanol to 70% volume, incubate 30 min. at -20 °C, spin for 10 min. at 13,000 rpm, discard supernatant, wash two times with 70% ethanol at 4 °C and dry pellet for 10 min. in air.
7. Dissolve pellet in 250 µl 0.1 M triethylammonium acetate, pH 6.5.

Remarks and troubleshooting: It is always wise to work with sterile solutions and materials when working with DNA to prevent degradation by DNAses. If you introduce another quencher, the absorption parameters have to be adapted. Low yields in coupling reactions indicate a wrong buffer pH.

Protocol 19 Synthesis of molecular beacons: fluorophore

8. Add 10 µl of 0.15 M silver nitrate to the DNA-Quencher solution, incubate for 30 min., add 15 µl of 0.15 M DTT and shake for 5 min. Spin for 2 min. at 10,000 rpm and transfer the supernatant to a new tube.
9. Freshly prepare immediately before use 40 mg 5-iodoactamidofluorescein in 250 µl 0.2 M sodium bicarbonate, pH 9.0.
10. Add fluorophore solution to supernatant and incubate the mixture for 2 h.
11. Run Sephadex G-25 column and HPLC described in steps 4 and 5. The collected absorbing fractions should be checked for 509 nm emission or the fractions should be collected according to that emission, (if the HPLC is equipped with a fluorescence detector).
12. Precipitate the collected fractions as described in step 6 and dissolve the pellet in 100 µl TE buffer.

Remarks and troubleshooting: If you introduce another fluorophore, the absorption parameters have to be adapted. Low yields in coupling reactions indicate a wrong buffer pH or degraded dyes. Always use fresh fluorophores.

Recently, controlled-pore glass columns became available, which are capable of introducing the dabcyl-quencher group at the 3′ end of an oligonucleotide, enabling the synthesis of molecular beacons completely on a DNA synthesizer. This enabled a number of companies to produce and sell custom Molecular Beacon Probes. Table 2 gives a list of companies licensed to sell these beacons.

Company	Location	World Wide Web Address
Biolegio BV	The Netherlands	http://www.biolegio.com
Biosearch Technologies	California	http://www.biosearchtech.com
BioSource International	California	http://www.biosource.com
Eurogentec	Belgium	http://www.eurogentec.be
Gene Link	New York	http://www.genelink.com/noframes.html
GENSET OLIGOS	Australia, California France, Japan, Singapore	http://www.gensetoligos.com
Integrated DNA Technologies	Iowa	http://www.idtdna.com
Isogen Biosience	The Netherlands	http://www.isogen.nl
Life Technologies	Maryland	http://www.lifetech.com
Midland Certified Reagents	Texas	http://www.mcrc.com
MWG-Biotech	Germany, North Carolina	http://www.mwgbiotech.com
Operon Technologies	California	http://www.operon.com
Oswel (Eurogentec)	Great Britain	http://www.oswel.com
Research Genetics	Alabama	http://www.resgen.com
Sigma-Genosys	Texas	http://www.genosys.com
Synthegen	Texas	http://www.synthegen.com
Synthetic Genetics	California	http://www.syntheticgenetics.com
TIB MOLBIOL	Germany	http://www.lynet.de/TIB-MOLBIOL
TriLink BioTechnologies	California	http://www.trilinkbiotech.com

Real time PCR

The molecular beacons chosen for PCR must have a complement to a sequence in the middle of the expected amplicon, and this sequence must not bind to the used primers (otherwise, high background signals would be observed). Additionally, they must be able to form a stable hybrid with their targets at the annealing temperature of the PCR, whereas the free molecular beacons should stay closed (non-fluorescent) at these temperatures. This is achieved by balancing the length of the probe to the length and GC-content of the beacons arms. The length of the probe is usually between 15 and 30 nucleotides and is designed so that it dissociates from the target at temperatures 7–10 °C higher than the annealing temperature of the PCR. (It is wise to predict this temperature by using the "percent-GC" rule). Sequences of the probe that would interfere with the formation of the hairpin must be avoided; otherwise, the beacon will not be functional. Also the arm sequences should be chosen to open (melt) 7–10 °C higher than the annealing temperature. Because of intramolecular interaction, DNA folding programs must calculate this melting temperature.

One should never trust the calculations completely; a thermal denaturation profile of the beacon with and without target has to be performed (Dissolve 200 nM molecular beacon in 3.5 mM $MgCl_2$ and 10 mM Tris-HCl, pH 8.0) to determin the melting temperatures. Beacons with the correct thermal characteristics will stay non-fluorescent during temperatures below the annealing temperature, fluoresce at the annealing temperature and dissociate from their targets at higher primer extension allowing temperatures. This makes sure that they do not interfere with polymerization.

Protocol 20 Real time PCR using molecular beacons

1. Switch on real time PCR device and prepare necessary steps according to the manual.
2. If necessary, conduct a calibration with pure dyes to be able to separate signals from overlapping fluorophores. Calibration data are required for each fluorophore/filter pair combination. If using multicolor PCRs, the detector must collect data with a unique corresponding filter pair for every fluorophore on the plate. Overlapping spectra of fluorophore combinations should be avoided.
3. Prepare the experimental PCR reactions: Mix 0.34 µM molecular beacon, 1 µM of each primer, 2.5 units of Amplitaq Gold DNA polymerase 0.25 mM dNTP's, 3.5 mM $MgCl_2$, 50 mM KCl, and 10 mM Tris-HCl, pH 8.0 and different amounts of DNA (target) in 6 wells to 50 µl each.
4. Chose filter setup for excitation and emission. Program the thermal cycler to incubate the tubes at 95 °C for 10 min. to activate the Amplitaq Gold DNA polymerase, followed by 40 cycles of 30 s at 95 °C, 60 s at 50 °C, and 30 s at 72 °C (the temperatures may be modified for each experiment).
5. Place the experimental plate in the device, start process, and monitor fluorescence during the 50 °C annealing steps.
6. Analyze data according to the manufacturers' software.

Remarks and troubleshooting: Because of the variety of different cyclers, the procedure does include cycler-specific steps, which cannot be discussed in detail here. Therefore, follow the manufacturers' protocols. High background levels indicate insufficient salt concentration, contamination by either free fluorophores or oligonucleotides that contain the fluorophore but not the quencher, or a principal problem in the design of the beacon. Check if it does not fold into an alternative conformation (if it does change stem and/or probe sequence) or if the probe sequence contributes to the stem (resulting in higher melting temperatures).

Clearly, real time PCR is also possible without using molecular beacons. Protocol 21 describes an alternative to MBs.

Protocol 21 Real time PCR using SYBR Green

1–2. See protocol 20.
3. Prepare the experimental PCR reactions: Mix 5 µl DNA, 2.5 µl of forward and reverse primer (each 20 pmol/µl) 2 µl dNTP's (10 mM), 10 µl 10*PCR-buffer, 1 µl $MgCl_2$ (25 mM), 1 µl SYBR-Green (2000* diluted), 75.5 µl ddH_2O, and 0.5 µl Taq polymerase (5u/µl)
4. Chose "SYBR filters" setup for excitation (490 nm) and emission (530 nm). Program the thermal cycler for cycle step lengths and temperatures.
5–6. See protocol 20.

Remarks and troubleshooting: Because of the variety of different cyclers, the procedure does include cycler-specific steps, which cannot be discussed in detail here. Therefore, follow the manufacturers' protocols.

Figure 20 shows a typical result of such a real time PCR, where the fluorescence is measured in each well (indicated by the well designation A1, A2, C2, C3, D1).

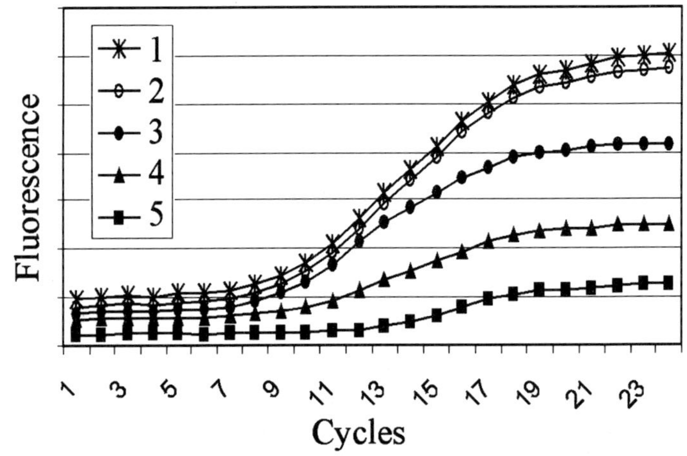

Figure 20

4 Results and discussion

Besides the advantages in sensitivity, selectivity, and others discussed in the introduction, fluorescence has proven to be a technique highly adaptable for a diversity of applications. The number of applications is still growing; additional applications of fluorescence on chip technology are described in other chapters of this book. Because of the complexity of the methods, it is impossible to cover results and discussion here. Advantages, limitations and results of the methods are described, detailed in the subsections included in the text.

5 Troubleshooting

Because of the complexity of the methods and diversity of fluorescence applications, it is impossible to completely cover troubleshooting here. The most important information concerning this topic has been detailed in the subsections included in the text or attached to the protocols. The most effective way of obtaining the necessary information remains discussion with colleagues actually involved in the same laboratory process.

6 Acknowledgements

The authors greatly appreciate contributions from Marieke Overeijnder and Angelika Zotter in the form of helpful discussions and helpful comments on the manuscript from Arminder Gandhum.

References

1 Brand L, Johnson ML (eds) (1997) *Fluorescence spectroscopy* (*Methods in enzymology*, Volume 278), Academic Press, New York
2 Cantor CR, Schimmel PR (1980) *Biophysical chemistry Part 2*. WH Freeman, New York, 433–465
3 Dewey TG, (ed) (1991) *Biophysical and biochemical aspects of fluorescence spectroscopy*. Plenum Publishing, New York
4 Lakowicz JR (ed) (1999) *Principles of fluorescence spectroscopy*, 2nd Ed. Kluwer Academic/Plenum Publishers, New York
5 Lewis GN, Kasha M (1944) Phosphorescence and the triplet state *J Am Chem Soc*, 66: 2100–2116

6 Hochstrasser RM, Weisman RB (1980) In: *Radiationless transitions.* SH Lin (ed), Academic Press, New York, 317
7 Shin DM , Whitten DG (1988) Solvatochromic behavior of intramolecular charge-transfer diphenylpolyenes in homogeneous solution and microheterogeneous media. *J Phys Chem* 92: 2945
8 Rosen CG, Weber G (1969) Dimer formation from 1-anilino-8-naphthalene sulfonate catalyzed by bovine serum albumin – A new fluorescent molecule with exceptional binding properties. *Biochemistry* 8: 3915–3920
9 Hirayama F (1965) Intramolecular excimer formation. I. Diphenyl and Triphenyl Alkanes. *J Chem Phys* 42: 3163–3171
10 MacColl R (1991) Fluorescence studies on r-phycoerythrin and c-phycoerythrin. *J Fluorescence* 1: 135
11 Mathies RA, Stryer L (1986) Single-molecule fluorescence detection: a feasibility study using phycoerythrin. D. L. Taylor, (ed) In: *Applications of fluorescence in the biomedical sciences.* Alan R. Liss, Inc, New York, NY 129–140
12 Panchuk-Voloshina N, Haugland RP, Bishop-Stewart J, et al. (1999) Alexa dyes, a series of new fluorescent dyes that yield exceptionally bright, photostable conjugates. *J Histochem Cytochem* 47: 1179–1188
13 Song L, Hennink EJ, Young IT, Tanke HJ (1995) Photobleaching kinetics of fluorescein in quantitative fluorescence microscopy. *Biophys J* 68: 2588–2600
14 Song L, Varma CA, Verhoeven JW, Tanke HJ (1996) Influence of the triplet excited state on the photobleaching kinetics of fluorescein in microscopy. *Biophys J* 70: 2959–2968
15 Brakenhoff GJ, Visscher K, Gijsbers EJ (1994) Fluorescence bleach rate imaging. *J Microscopy* 175: 154–161
16 Ried T, Baldini A, Rand TC, Ward DC (1992) Simultaneous visualization of seven different DNA Probes by *in situ* hybridization using combinatorial fluorescence and digital imaging microscopy. *Proc Natl Acad Sci* 98: 1388–1392
17 Oefner PJ, Huber CG, Umlauft F et al. (1994) High-resolution liquid chromatography of fluorescent dye-labeled nucleic acids. *Anal Biochem* 223: 39–46
18 Dahan M, Deniz AA, Ha T et al.(1999) Ratiometric measurement and identification of single diffusing molecules. *Chemical Physics* 247: 85–106
19 Grant RL, Acosta D (1997) Ratiometric measurement of intracellular pH of cultured cells with BCECF in a fluorescence multi-well plate reader. *In vitro Cellular & Developmental Biology* 33: 256–260
20 Atsumi T, Sugita K, Kohno M et al. (1996) Simultaneous measurement of Ca2+ and pH by laser cytometry using fluo-3 and SNARF-1. *Cytometry* 24: 99–105
21 Cody SH, Dubbin PN, Beischer AD et al. (1993) Intracellular pH mapping with SNARF-1 and confocal microscopy. I: A quantitative technique for living tissues and isolated cells. *Micron* 24: 573–580
22 Martinez-Zaguilan R, Martinez GM, Lattanzio F, Gillies RJ (1991). Simultaneous measurement of intracellular pH and Ca2+ using the fluorescence of SNARF-1 and fura-2. *Am J Physiol* 260: 297–307
23 Haugland RP, Bhalgat MK (1998) Preperation of avidin conjugates. *Methods Mol Biol* 80: 185–196
24 Haugland RP (1995) Coupling of monoclonal antibodies with enzymes. *Methods Mol Biol* 45: 223
25 Parks DR, Lanier LL, Herzenberg LA. (1986) Flow cytometry and fluorescence activated cell sorting (FACS). In: Weir DM , Herzenberg LA, Blackwell C (eds): *Handbook of experimental immunology.* Blackwell Scientific Publications, Oxford, UK 1986
26 Jackson AL, Warner NL (1986) Preparation, staining, and analysis by flow cytometry of peripheral blood leukocytes. In: Rose NR, Friedman H, Fahey JL, (eds): *Manual of clinical laboratory immunology* 3rd ed. American Society for Microbiology, Washington, DC 226–235
27 Lakowicz JR, (ed) (1997) *Nonlinear and two-photon induced fluorescence*, Vol 5. Plenum Publishing, New York
28 Sako Y (1998) Multi-photon excitation fluorescence microscopy and its application to the studies of intracellular signal transduction, Tanpakushitsu Kakusan

Koso. *Protein, Nucleic Acid, Enzyme* 43: 1927–1930

29 Garrett WR, Yifei Z, Lu D, Payne MG (1996) Multi-photon excitation through a resonant intermediate state: unique separation of coherent and incoherent contributions. *Optics Communications* 128: 66–72

30 López Arbeloa F (1989) Influence of the molecular structure and the nature of the solvent on the absorption and fluorescence characteristics of rhodamines. *Chem Phys* 130: 371

31 Krasnowska EK, Bagatolli LA, Gratton E, Parasassi T (2001) Surface properties of cholesterol-containing membranes detected by Prodan fluorescence. *Biochim Biophys Acta* 1511: 330–340

32 Prendergast FG, Meyer M, Carlson GL et al. (1983) Synthesis, spectral properties, and use of 6-acryloyl-2-dimethylaminonaphthalene (Acrylodan). A thiol-selective, polarity-sensitive fluorescent probe. *J Biol Chem* 258: 7541–7544

33 Macgregor RB, Weber G (1986) Estimation of the polarity of the protein interior by optical spectroscopy. *Nature* 319: 70–73

34 Bushueva TL, Busel EP, Burstein EA (1978) Relationship of thermal quenching of protein fluorescence to intramolecular structural mobility. *Biochim Biophys Acta* 534: 141–152

35 Colucci WJ, Tilstra L, Sattler MC et al. (1990) Conformational studies of a constained tryptophan derivative: implications for the fluorescence quenching mechanism. *J Am Chem Soc* 11: 2

36 Eftink MR (1991) Fluorescence quenching: theory and applications. In: *Topics in fluorescence spectroscopy: volume 2 – Principles*, J.R. Lakowicz (ed), Plenum Press, New York, 53–126

37 F. López Arbeloa (1989) Fluorescence self-quenching of the molecular forms of rhodamine B in aqueous and ethanolic solutions *J Lumin* 44: 105

38 Jones LJ, Upson RH, Haugland RP et al. (1997) Quenched BODIPY dye-labeled casein substrates for the assay of protease activity by direct fluorescence measurement. *Anal Bioch* 251: 144–152

39 Truong K, Ikura M (2001) The use of FRET imaging microscopy to detect protein-protein interactions and protein conformational changes *in vivo*. *Current Opinion in Structural Biology* 11: 573–578

40 Van Der Meer BW, Coker G, Chen, SY (1994) *Resonance energy transfer theory and data*, VCH, New York

41 Perrin F (1926) Polarization de la lumiere de fluorescence. Vie moyenne de molecules dans l'etat excite. *J Phys Radium* 7: 390

42 Weber G (1953) Rotational Brownian motion and polarization of the fluorescence of solutions. *Adv Protein Chem* 8: 415

43 Collett E (1993) *Polarized light: fundamentals and applications.* Marcel Dekker, New York

44 Singh KK, Rücker T, Hanne A et al. (2000) Fluorescence polarization for monitoring ribozyme reactions in real time. *BioTechniques* 29: 344–348, 350–351

45 Dandliker WB, Hsu ML, Levin J, Rao BR (1981) Equilibrium based assays based upon fluorescence polarization. *Meth Enzym* 74: 3–28

46 Murakami A, Nakaura M, Nakatsuji Y et al. (1991) Fluorescent-labeled oligonucleotide probes: detection of hybrid formation in solution by fluorescence polarization spectroscopy. *Nucleic Acids Research* 19: 4097–4102

47 Elson EL, Magde D (1974) Fluorescence correlation spectroscopy I: Conceptual basis and theory. *Biopolymers* 13: 1

48 Magde D, Elson EL, Webb WW (1974) Fluorescence correlation spectroscopy II: An experimental realization. *Biopolymers* 13: 29

49 Schwille P, Oehlenschläger F, Walter N (1996) Analysis of RNA-DNA hybridization kinetics by fluorescence correlation spectroscopy. *Biochemistry* 35: 10182

50 Fries JR, Brand L, Eggeling C et al. (1998) Quantitative identification of different single-molecules by selective time-resolved confocal fluorescence spectroscopy. *J Phys Chem A* 102: 6601–6613

51 Nie S, Chiu DT; Zare RN (1994) Probing individual molecules with confocal fluorescence microscopy. *Science* 266: 1018–1021
52 Cubitt A, Heim R, Adams S et al. (1995) Understanding, improving and using green fluorescent proteins. *TIBS* 20: 448–455
53 Cody CW, Prasher DC, Westlor WM et al (1993) Chemical structure of the hexapeptide chromophore of the Aequorea green-fluorescent protein. *Biochemistry* 32: 1212–1218
54 Ward W, Prentice H, Roth A et al. (1982) Spectral perturbations of the *Aequoria* green fluorescent protein. *Photochem Photobiol* 35: 803–808
55 Chalfie M, Tu Y, Euskirchen G et al. (1994) Green fluorescent protein as a marker for gene expression. *Science*. 263: 802–805
56 Kahana J, Schapp B, Silver P (1995) Kinetics of spindle pole body separation in budding yeast. *Proc Natl Acad Sci USA* 92: 9707–9711
57 Ludin B, Doll T, Meill R et al. (1996) Application of novel vectors for GFP-tagging of proteins to study microtubule-associated proteins. *Gene* 173: 107–111
58 Casper S, Holt C (1996) Expression of the green fluorescent protein-encoding gene from a tobacco mosaic virus-based vector. *Gene* 173: 69–73
59 Wang S, Hazelrigg T (1994) Implications for bcd mRNA localization from spatial distribution of exu protein in *Drosophila oogenesis*. *Nature* 369: 400–403
60 Mitra R, Silva C, Youvan D (1996) Fluorescence resonance energy transfer between blue-emitting and red-shifted excitation derivatives of the green fluorescnet protein. *Gene*. 173: 13–17
61 Heim R, Cubitt A, Tsien R (1995) Improved green fluorescence. *Nature* 373: 663–664
62 Ehrig T, O'Kane D, Prendergast F (1995) Green-fluorescent protein mutants with altered fluorescence excitation spectra. *FEBS Lett* 367: 163–166
63 Tyagi S, Kramer FR (1996) Molecular beacons: probes that fluorescence upon hybridization. *Nat Biotech* 14: 303–308
64 Tyagi S, Bratu DP, Kramer, FR (1998) Multicolor molecular beacons for allele discrimination. *Nat Biotech* 16: 49–53
65 Leone G, van Schijndel H, van Gemen B et al. (1998) Molecular beacon probes combined with amplification by NASBA enable homogeneous, real-time detection of RNA. *Nucleic Acids Res* 26: 2150–2155
66 Matsuo T (1998) *BBA-Gen. Subjects* 1379 (2): 178–184
67 Sokol DL, Zhang XL, Lu PZ, Gewitz AM (1998) Real time detection of DNA RNA hybridization in living cells. *Proc Natl Acad Sci USA* 95: 11538–11543
68 Li J, Fang X, Schuster S, Tan W (2000) Molecular beacons: detecting protein-nucleic acid interactions. *Angew Chem, Int Ed* 39: 1049–1052
69 Liu X, Farmerie W, Schuster S, Tan,W (2000) Molecular beacons for DNA biosensors with micrometer to submicrometer dimensions. *Bioanal Chem* 283: 56–63
70 Steemers FJ, Ferguson JA, Walt DR (2000) Screening unlabeled DNA targets with randomly ordered fiber-optic gene arrays. *Nat Biotech* 18: 91–94
71 Fang X, Liu X, Tan W (1999) Single and multiple molecular beacon probes for DNA hybridization studies on a silica glas surface. *Proc SPIE-Int Soc Opt Eng* 3602: 149–155
72 Marras SA, Kramer FR, Tyagi S (1999) Multiplex detection of single-nucleotide variations using molecular beacons. *Genet Anal* 14: 151–156
73 Piatek AS (1998) Molecular beacon sequence analysis for detecting drug resistance in Mycobacterium tuberculosis. *Nat Biotech* 16: 359–363

3 Immunoanalytical Methods

Franz Gabor, Oskar Hoffmann, Fritz Pittner and Michael Wirth

Contents

Methods and Tools in Biosciences and Medicine
Analytical Biotechnology, ed. by Thomas G.M. Schalkhammer
© 2002 Birkhäuser Verlag Basel/Switzerland

1 Introduction

The roots of immunoanalytical methods are found in the 19th century. By 1890 von Behring and Kitasato demonstrated the neutralising characteristics of bacterial antitoxins. In an effort to identify the origin of the most common bacterial diseases at the turn to the 20th century, the Widal test to verify typhoid (1896) or the Wasserman test to confirm syphilis (1906) antibodies were used as analytical tools [1]. Subsquenty, the rapid progress in biosciences, together with a deeper understanding of the basic principles of the immune system, led to a vast increase in different methods taking advantage of the specific interaction between antibodies and antigens.

It is beyond the scope of this chapter to describe all the laboratory methods, that rely on the antigen-antibody interaction. But a few methods that are currenty used will be described in more detail. Since the analysis of insulin by Berson and Yalow [2], several different immunoassays for determination of a wide variety of drugs, marker substances in blood, and biological proteins have been developed. To date, many assays are commercially available as kits being used routinely in automated clinical laboratories. When a kit cannot be purchased, part of this chapter should help to establish an immunoassay in your lab.

Additionally, immunoanalytical methods become more and more important in our understanding of how cells function. The development of flow cytometers makes multiparametric analysis of single cells possible. Furthermore, confocal laser scanning microscopic imaging of tissues helps us to visualize processes occuring within cells and tissues. Both techniques use immunoflourescence and provide for a powerful tool in life sciences in the future.

2 Structure and characteristics of antibodies

The antibodies are glycoproteins that belong to the group of immunoglobulins. Primarily, they are secreted by plasma cells in response to an immunogen. Each plasma cell secretes about 2000 antibody molecules per second, amounting to about 20% (w/w) of blood. Because blood can be sampled easily, antibodies provide for a powerful analytical tool as a result of their specific interaction with antigens.

Structurally, antibodies are Y-like proteins composed of four polypeptide chains, which stick together by three disulfide-bridges and hydrophobic interactions (Fig. 1). Each molecule contains two identical heavy chains, each composed of about 440 amino acids, and two identical light chains, each containing 210 to 230 amino acids. There are five types of different heavy chains termed as α-chains, γ-chains, δ-chains, ε-chains, or μ-chains. Accordingly, five classes of antibodies, namely IgA, IgG, IgD, IgE, and IgM, are distinguished (Tab. 1). These isotypes of antibodies can contain two identical types of light chains, either κ-chains or λ-chains. The heavy and light chains are made up of domains with sequence homologies comprising about 110 amino acids, which fold into parallel sheets linked by an intradomain disulfide bond. Thus, light chains contain two domains and heavy chains contain four domains.

Table 1 Isotypes of immunglobulins. IgM and IgA additionally contain joining peptides to connect the Y-units. IgA also exhibits a secretory protein.

Isotype	heavy chain	light chain	molecular formula	mass (kD)	valency	function
IgG	γ	κ or λ	$\gamma_2\kappa_2$ or $\gamma_2\lambda_2$	150	2	Secondary response
IgM	μ	κ or λ	$(\mu_2\kappa_2)_5$ or $(\mu_2\lambda_2)_5$	950	10	Primary response
IgA	α	κ or λ	$(\alpha_2\kappa_2)_{1-3}$ or $(\alpha_2\lambda_2)_{1-3}$	180–500	2, 4, 6	Secretory Ig
IgD	δ	κ or λ	$\delta_2\kappa_2$ or $\delta_2\lambda_2$	175	2	Not known yet
IgE	ε	κ or λ	$\varepsilon_2\kappa_2$ or $\varepsilon_2\lambda_2$	200	2	Allergy-Ig

The amino-terminal domains of the heavy and light chains mediate antigen binding and are variable, whereas the remainder are constant regions. Each variable domain is composed of four framework regions and three hypervariable regions. The short loops of the hypervariable region of each heavy and light chain are referred to as complementary determining regions (CDR). Each CDR comprises 5–10 amino acids and protrudes, forming the site of interaction with the antigen. The area of interaction between the epitope, which represents the deteminant on the surface of the antigen, and the paratop which is the complementary structure formed by six CDRs of the antibody, was found to be about 1.6 × 0.6 nm. The interaction between antigen and antibody is reversible and is in equilibrium with the free components. Thus, the precise fitting is a prerequisite for non-covalent interactions such as hydrogen bonds, van der Waals forces, hydrophobic bonds, and sometimes ionic interactions, which stabilize the immune complex.

The high variability of the antigen-binding site derives from about 45,000 different variable domains of the heavy chain, which are genetically encoded by 4 J(joining)-regions, 15 D(diversity) regions, and 250 V(variable) regions, which is extended by three kinds of combination. The variable domain of the light chains is encoded by 250 V-segments and 4 J-segments, resulting in about 3000 different varaiable domains of the light chain. Linking the variability of heavy and light chains results in at least 1.35×10^8 antibodies with different specificity. Additionally, somatic mutations contribute to the diversity of antibody specificity.

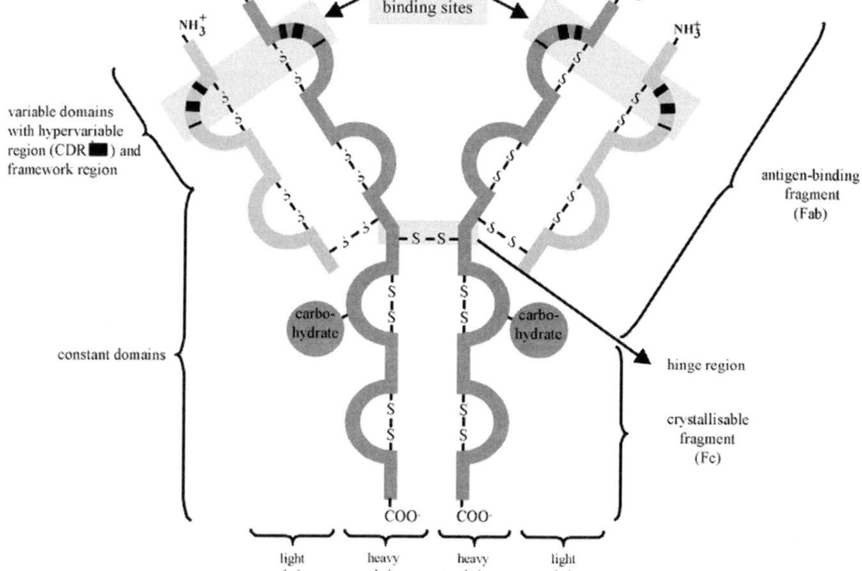

Figure 1 Scheme of IgG-molecule

Exposition to proteolytic enzymes demonstrates the highly folded structure of antibodies. Upon treatment with papain, the heavy chains are cleaved N-terminally to the interconnecting disulfide bond at amino acid 224, yielding three fragments. Two single fragments with paratops at the N-terminus for antigen binding are known as Fab-fragments, corresponding to the arms of the Y. The third fragment that corresponds to the base of the Y was isolated by crystallisation and is known as Fc-fragment. This part of the immunglobulin is not involved in antigen-binding but plays a key role in immune regulation, e. g., activation of the complement system and macrophages. The region between the Fab- and Fc-fragments is known as hinge-region, which provides for flexibility of the Fab-fragments. Lateral and rotational movement facilitates the binding of two determinants of different conformations of a certain antigen. Especially as a reagent in immunoanalytical techniques, the product of pepsin treatment becomes more and more important. Pepsin attacks the immunoglobulin at C-terminal positions below the heavy chain connecting disulfide bond at amino acid 234. Thus, a divalent fragment referred to as $F(ab)_2$-fragment is obtained, which possesses two antigen binding sites, but the Fc-portion is fully digested by pepsin. This $F(ab)_2$-fragment is a useful substitute for intact immunglobulins, as the Fc-fragment is known to interfere in many techniques, resulting in higher background staining or higher signal/noise ratio. There are three terms closely related to each other that describe utility of antibodies in practice. Affinity describes the exactitude of stereochemical fit of the epitope of the antigen to the paratope of the antibody. Thermodynamically, it indicates the strength or energy of the interaction. Deriving from the law of mass action, it is described by an association constant K_a indicating the ratio of bound and unbound antigen (mol^{-1}). Thus, the affinity constant is the reciprocal value of that concentration of a certain antigen, where half of the antigen is bound to the paratopes. In practice, this is the amount of antigen-antibody complex at equilibrium. High affinity antibodies will bind larger amounts of antigen in a shorter period of time than low affinity antibodies.

Consequently this term K_a is independent from the number of paratopes of an antibody molecule. To characterize the overall strength of the antigen-antibody interaction, the term avidity is used. It includes affinity, valency of the antibody, and arrangement of the interacting compounds. Assuming that the affinity of an antigen to the paratope of IgG and IgM is equal, the avidity of IgM is five times higher than that of IgG because of the pentameric IgM. Thus, avidity is also expressed by an association constant K for a particular test system. When an antigen binds to an antibody, this interaction is termed to be specific. But some antigen determinants are shared between molecules, especially closely related ones. Thus, cross-reactivity describes the fact that common epitopes on different antigens are bound by the same antibody. On the other hand, cross-reactivity also can derive from binding of structurally different epitopes by the same antibody, especially in antisera.

3 From antigens to antibodies

3.1 Basic considerations

Antibodies used as reagents in immunoanalytical methods are host proteins that are produced by the immune system in response to foreign large molecules in the body. To provoke formation of antibodies by B-cells and cellular immune response mediated by T-cells, an immunogen is required. According to the clonal selection theory by Sir MacFarlane Burnet, the immunogen binds to well-suited antibodies located at the surface of virgin B-cells, which determine specificity of the antibodies finally obtained. The immunogen is degraded within the virgin B-cell, and its fragments are presented at the surface as a complex with class II proteins. This complex is recognised by T-helper cells, which induce differentiation of B-cells to antibody-secreting plasma cells. For these reasons, epitope binding, degradability, and mediation of the B-cell – T-cell interaction require a minimum molecular weight of the immunogen of about 5 kDa.

In this context, it has to be considered that animals usually used for immunization have developed some self-tolerance, eliminating well-suited B- or T-cells. Thus, the immunogen or carrier protein should be foreign to the immunized species.

3.2 Antigen preparation

Large molecules

According to the requirements of epitope binding, degradability, and mediating interaction of B- and T-cells, high-molecular-weight proteins should elicit an immune response, which usually increases with a higher degree of conformation and structural rigidity. The antigenic sites at the surface of protein molecules can be made up of different domains of amino acid side chains, which are distant in sequence but close in space. In contrast to these conformation-dependent determinants, fibrous proteins possess sequential determinants comprising up to six amino-acid residues. Probably all proteins are immunogenic, but individual proteins may differ markedly in immunogenicity. Self-aggregation of proteins usually is associated with a slight change in its specificity, but with increased immunogenicity. On the other hand, large molecules can be made more immunogenic by denaturation. Heating or heating in the presence of sodium dodecyl sulphate will increase immunogenicity, but, concurrently, specificity will be altered because hidden epitopes will become accessible. Polysaccharides can act as immunogens in purified form, but only in humans and mice, not in rabbits or guinea pigs. Additionally, antibodies against nucleic acids are described, but they are elicited only in humans suffering from the autoimmune disease lupus erythematodes or in animals with similar diseases.

Small molecules

In contrast to large molecules, small molecules including drugs, steroids, or peptides do not elicit an immune response. As first shown by Landsteiner exemplified by 2,4-dinitrophenol, low-molecular-weight compounds termed as haptens have to be coupled to carrier proteins to provide for class II protein-T-cell binding, which results in formation of antibodies reacting specifically with the dinitrophenyl group. To gain better access to virgin B-cell antibodies, antennary exposition of the hapten by covalent linkage via spacer molecules at the surface of the carrier molecule is favored.

The most common carrier proteins are serum albumins of various species, preferably the well-characterized bovine serum albumin or high immunogenic keyhole limpet hemocyanin from the mollusc megathura crenulata, but ovalbumin and thyroglobulin or fibrinogen also are described. For conjugation of hapten and carrier protein, different protocols are described in the literature (see Tab. 2), but the use of heterobifunctional cross-linkers and two-step mechansims is highly recommended to avoid formation of homopolymers and precipitation of the hapten-carrier complex.

Table 2 Reagents for conjugation of haptens to proteins

Hapten with carboxylic group – carrier with amino group
Mixed anhydride method: Activation of the tributylammonium salt of the hapten in anhydrous milieu with isobuytlchloroformate to yield the rather unstable mixed anhydride. Coupling at alkaline pH, adjustment of pH necessary.
Notes: Racemisation omitted; no side products to be removed; isolation of activated hapten not necessary [3].
Carbodiimide method: Activation with DCC in anhydrous or EDAC in aqueous milieu at pH 4–5, coupling at pH 8.
Notes: Works well; probable formation of autopolymers; remove excess cross-linker; use water-miscible anhydrous solvents for activation such as DMF, THF, or dioxane [4, 5].
Active ester method (carbodiimide/N-hydroxy-succinimde method): Activation and coupling is performed at neutral pH.
Notes: Works well; isolation of activated intermediate recommended; thiol-, hydroxyl-, and amino groups can react as well [6].

Hapten with amino group – carrier with amino group
Imidoester method: Homobifunctional cross-linker, activation and coupling at pH 9.
Notes: Isolation of activated intermediate required to omit polymerisation, design of spacer-length possible (dimethyl-adipimidate 0.86 nm, dimethyl-pimelimidate 0.92 nm, dimethyl-suberimidate 1.1 nm) [7].

Glutardialdehyde: Homobifunctional cross-linker, activation and coupling at alkaline pH.
Notes: Isolation of intermediate is recommended; reagent is a polymer; sometimes reduction by sodium-borohydrid is required for stability [8].

SPDP (N-succinimidyl-3-(2-propyldithio)-propionate: Heterobifunctional, hapten activated at pH 7–9, coupling to thiol-containing carrier after conversion of amino- to thiol-residues with 2-iminothiolan or cleavage of carrier disulfide bonds with dithiothreitol.
Notes: Mild conditions; degree of substitution measured by appearance of 2-thiopyridon [9].

Hapten with hydroxyl-group – carrier with amino group

Hemisuccinoylation: Hapten converted to carboxylderivative by reaction with succinic anhydride in anhydrous milieu followed by use as carboxylated hapten.
Notes: Spacer is introduced [10].

Divinylsulfone: Derivatisation of hydroxyls at pH 11, coupling at pH 9.
Notes: Inactivation of excess reagent by addition of glycine [11].

Periodate oxidation: Mildly alkaline conditions required for cleavage and oxidation of cis-diol partial structures, coupling of aldehyde at pH 9.
Notes: Stability is increased by reducing the aldimin with sodium borohydride [12].

The amount of hapten coupled per molecule of carrier protein can be determined only after rigorous purification of the artificial antigen by dialysis, gel permeation chromatography, or precipitation with ammonium sulphate or any other useful method. Most simply, the degree of substitution can be determined by UV-difference spectroscopy of the carrier protein and the conjugate. Appearance of an additional peak at the maximum absorbance of the hapten in the spectrum of the conjugate in comparison to the carrier protein is a qualitative confirmation of successful coupling. When the solutions are carefully prepared by weighing accurate amounts of the lyophilised conjugate and the carrier protein and the solutions are adjusted to the same concentration, the hapten content of the conjugate can be estimated quantitatively. An alternative is an indirect assay by determining the number of free amino groups of the carrier protein prior to and after coupling of the hapten. The difference points to the number of modified amino residues, which corresponds to the number of hapten molecules attached. A rather time-consuming method is the 2,4-dinitro-2-fluorobenzene test, originally applied by Sanger for determination of insulin. A simple and rapid alternative is the trinitro-benzene sulfonic acid test, which avoids total hydrolysis and extraction steps of dinitrophenylated amino acids [13]. But it should be considered that non-covalent interactions also can contribute to hapten-fixation on the carrier protein, which amounts in the case of bovine serum albumin to about 2 to 3 hapten molecules. While bovine serum albumin has 59 lysine residues, only 35 are accessible at the surface for coupling at neutral pH. A conjugation rate of about 10 hapten molecules per carrier molecule is sufficient to obtain not only serum albumin-specific antibodies but also hapten-specific ones [14].

3.3 Immunization

Before an immunization protocol is started, the legal responsibilities should be considered. In the countries of the European Union, as well as in the USA, a personal and a project license is required for all immunization and bleeding procedures.

The choice of the animal species to be immunized greatly depends on the facilities. Most commonly, rabbits of 4 to 6 months of age are used, as about 25 ml of serum is yielded from a single bleed. Guinea pigs, which are often hard to bleed, rats, or hamsters yield 1 to 2 ml, while mice yield only 200 µl. For rabbits, two animals should be used as a minimum, whereas in the case of rodents three to six animals are recommended. It is self-evident that the animals should be held under proper environmental conditions preventing any stress.

Prior to immunization of an animal, the antigen is mixed with adjuvants. Because adjuvants enhance the humoral immune response and activate unknown sectors of the immune system, immunization never should be done without adjuvants. Heat-killed bacteria containing lipopolysaccharides (*Bordetella pertussis*) or muramyldipeptides (*Mycobacterium tuberculosis*) stimulate lymphokine production resulting in strong local inflammatory reaction, which mobilizes macrophages. On the other hand, aluminium salts precipitate the antigen and mineral oils form water-in-oil emulsions. Both result in formation of a depot, which prevents rapid degradation of the antigen and guarantees sustained nonspecific stimulation of the immune response. Most commonly, the primary injection is given as an emulsion of the immunogen in complete Freund's adjuvant containing killed *Mycobacterium tuberculosis* and mineral oil, which provokes aggressive and persistent granulomas [15]. But the boosts are administered as a mixture with incomplete Freund's adjuvant containing mineral oil and emulsifier only.

As a rough basic rule, in rabbits about 0.5 to 1 mg antigen in adjuvant is administered, preferably subcutaneousely, on multiple sites at low volumes for the primary immunization, which corresponds to 50 to 100 µg antigen for mice. When only small amounts of poor immunogens are available, the antigen might be injected in rabbits in the poplietal lymph node at the lower leg to enhance the probability of an appropriate immune response. In this latter case, however any adjuvant must be omitted. The dose of the following boost injections about three to four weeks after the first injection might be one-third to one-half that of the first immunization. For deeper insight into immunization procedures, there is some useful literature available [16].

UNIVERSITÄTSBIBLIOTHEK
BODENKULTUR
Wien
Peter-Jordanstraße 82
A-1190 Wien

3.4 Sampling serum and storage

After the first injection of the immunogen, the primary response results in exponential increase of antibody secretion occuring within 30 days. In the majority of cases, the antibody concentration in the blood, termed as titer, reaches a maximum within 9 to 11 days so that the first test bleed can be collected.

When using rabbits, the animal is placed in a small box and and a small distal area of the rear ear is rinsed with a small amount of 70% ethanol. When the marginal ear vein is not apparent, it might be dilated by gently rubbing or warming with a lamp, but avoid use of toluene or xylene. The area, which should be incised or opened by insertion of a thick injection needle, is covered with Vaseline®. The blood is collected by droping in a clean test tube. When the blood flow stops by clotting, the incision area should be wiped with sterile warm water. After collecting usually 5 ml, but 20 ml at the maximum, the needle is removed and gentle pressure is applied to the cut for half a minute. Check to be sure that the blood flow has stopped.

Before screening this first test bleed for presence of specific antibodies, the blood is allowed to coagulate for one hour at 37 °C. The clot is removed with a Pasteur pipette or by simply rotating a skewer made from wood followed by centrifugation for 10 min. at 13,000 rpm to remove any remaining insoluble material. The clot might be allowed to retract overnight to yield an additional few milliliters of serum. Additionally, the serum can be heated to exactly 56 °C for 10 to 20 min. to inactivate the complement proteins. These proteins can interfere with some immunoanalytical techniques by binding to the antigen-antibody complex. The serum can be stored for many years at –20 °C or below. When stored at 4 °C for weeks, it is recommended to add 0.02% sodium azide or thiomersal at 1:10,000. But be aware of the fact that sodium azide is highly poisonous and can interfere with some immunoanalytical techniques.

Test bleeds can be taken from mice in a similar way or from rats by incision of the tail vein. The test bleed usually yields a volume of 200 to 400 µl and can be treated as described above.

The first test bleed predominantly contains IgM and only few IgG, but antibody formation declines with time.

About one month after the first immunization, another exposure to the immunogen provokes the secondary immune response, which yields a stronger response by activating memory cells. Moreover the antibodies of the secondary response belong predominantly to the IgG-type and often exhibit higher affinity, which might even be increased by the number of boosters. The titer should be monitored by appropriate methods such as enzyme immunoassays or simple immunodiffusion assays [17]. When a good titer has developed, regular boosts are preformed every 6 weeks, and up to 50 ml blood (rabbit) is collected 10 days after the booster injection.

Through the course of time, affinity, specificity and cross-reactivity are altered. The concentration of the antigen in the animal decreases with time so that lymphocytes producing low-affinity antibodies are not stimulated further, whereas those secreting high-affinity antibodies are still stimulated. By boost injections with low doses of antigen, the threshold value for stimulation remains high, and only those lymphocytes remain to be stimulated that produce high-affinity antibodies. The decrease in specificity with time probably results from early formation of antibodies with low affinity. Each determinant of the antigen provokes formation of specific antibodies with different lag times. Thus, shortly after administration of the immunogen the number of antibodies with different specificity is low, but it increases with time. Additionally, cross-reactivity increases by time. In response to different determinants located at the surface of an antigen, a higher number of different antibodies is formed. Thus, it is likely that certain antibodies also recognize antigens with similar conformation.

3.5 Polyclonal antibodies

According to the clonal selection theory, a clone of B-cells produces antibodies of only one specificity. When an animal is immunized with an antigen, a random number of clones is activated and, consequently, a polyclonal antiserum is obtained. This strategy of the immune system to produce polyclonal antibodies yields important advantages in terms of affinity and specificity. A major advantage of polyclonal antibodies is the formation of large insoluble immune complexes with polyvalent antigens. At a molar ratio of 4 Mol antigen/mol antibody, the multivalent antigens and the bivalent antibodies forms a three-dimensional network, which is opaque and can be determined quantitatively. Additionally, polyclonal antibodies are usually well-suited for cell staining, immunoblotting, and immunoassays with labeled antigens, but it is difficult to perform immunoassays with labeled antibodies. In the latter case, use of monoclonal antibodies is recommended.

When using antibodies as a reagent for immunoanalytical techniques, it should be considered that there is a batch-to-batch variation of polyclonal antibodies, even those deriving from the same animal. The antibodies are heterogenous with regard to specificity, isotype composition, and optimum binding conditions so that each assay has to be adapted. Additionally – with exception of specific pathogen-free animals – the antiserum contains antibodies against all antigens to which the animal was ever exposed. Thus, the antiserum contains at most 20% to 30% immunglobulins, of which about 10% are specific antibodies against the desired immunogen.

3.6 Monoclonal antibodies

In 1984 G. Köhler and C. Milstein were awarded with the Nobel Prize for development of the hybridoma technique, which yields monoclonal and, consequently, monospecific antibodies [18]. The rationale for this technique is to fuse antibody-secreting lymphocytes from the spleen, which produce polyclonal antibodies, with immortal myeloma cells. Thus, the capability of antibody production deriving from the B-cells and the immortality deriving from the B-lymphocyte tumor cells are joined [19]. In order to render the polyclonal antibodies in the supernatant of the fused cells monoclonal, certain antibody-secreting cells are selected by cloning. The growing hybridomas are diluted and grown in microplate wells as far as the antibodies secreted might stem from the same fused, antibody-secreting cell.

Production of monoclonal antibodies requires mice, usually the inbred strain BALB/c, for immunization and plasmacytoma cells such as the non-secreting, HAT-sensitive P3-X63-Ag8.653 cells. Additionally, good tissue culture facilities including a 5% CO_2/95% air incubator, a sterile workbench, and an inverted microscope for visual control of the cultures are needed.

In the following some basic items for production of monoclonal antibodies are given. There are many different protocols for production of monoclonal antibodies; for detailed information; the reader is referred to the literature [20, 21].

Per immunogen, a minimum of three BALB/c mice is immunised by subcutaneous injection of about 25 μg immunogen in complete Freund's adjuvant at multiple sites at the back. Four weeks later, half of the amount of immunogen mixed with incomplete Freund's adjuvant is administered intraperitoneally. About one week after the booster injection, the animals are sacrified and the spleen is removed under aseptic conditions. Using a stainless sieve, the spleen is homogenised, and clotted cells of the connective tissue are removed by centrifugation. Erythrocytes are lysed by osmotic shock, and the spleen cells are recovered in the pellet after centrifugation.

The myeloma cells P3-X63-Ag8.653 represent a variant of BALB/c plasmacytoma cells that do not secrete antibodies and that are sensitive to HAT medium. They should be in the logarithmic phase of cell growth.

While fusion of the cells in early work was achieved by inactivated Sendai virus, PEG 4000 is commonly used today. Fusion of the lymphocytes with myeloma cells by PEG requires strict compliance with the selected procedure with regard to time and amounts of reagents added. After fusion, the cell suspension contains hybridoma cells, non-fused lymphocytes, and myeloma cells. Whereas lymphocytes are mortal and lose their viability during further propagation, the myeloma cells are immortal. In order to eliminate the myeloma cells, the cell suspension is cultured in a selective culture medium (HAT medium) containing hypoxanthin, aminopterin, and thymidin. The myeloma cells suffer from a deficiency of the enzyme hypoxanthin-guanine-phosphoribosyl transferase (HPGRT). Thus, they are not able to synthesize DNA in presence of the HAT medium and lose viability. In contrast, in hybridoma cells the enzyme deficiency is compensated by the parent lymphocytes and they still grow.

About two weeks later, the cells are cultured in HAT medium and cells from dense cultures are propagated further on feeder layers containing macrophages from the peritoneum of mice. They are obtained by rinsing the abdominal cavity of adult mice with sterile saccharose-solution. Upon propagation of the cell suspension containing hybridomas on feeder-layers, non-growing hybridomas (99%) are degraded by the macrophages, whereas the growing hybridomas (1%) take advantage of the growth factors provided by the macrophages.

Different methods are available to select the hybridomas secreting relevant antibodies [22]. Most commonly, enzyme-linked immunosorbent assays are used. After coating the wells of a microplate with antigen and washing, an aliquot of the supernatant of the hybridomas is added and washed after incubation. The antigen-bound antibodies are detected with goat-anti-mouse IgG/IgM labeled with horseradish peroxidase. After removal of excess second antibody by washing and addition of substrate solution the absorption of the cleaved, stained substrate is determined photometrically after stopping the enzyme reaction, by addition of acid.

According to the results of screening the hybridoma-containing wells for antibody secretion, relevant hybridomas are cloned by limited dilution rather than by the agar technique. Usually limiting dilution is done in three subsequent series, ensuring that the antibodies in the supernatant derive from fusion of a single B-cell and a myeloma cell.

A problem sometimes encountered with production of monoclonal antibodies is loosening of chromosomes [23]. Thus, a significant number of initially positive clones can be lost. This can be prevented by omitting overgrowth of the hybridomas in the wells, but this phenomenon occurs as the hybridomas become monoclonal.

The hybridoma cells can be propagated continuosly *in vitro*, and the supernatant contains homogenous antibodies, which recognize only one or a few closely related antigens. In contrast to polyclonal antibodies, monoclonal antibodies are homogenous molecules with regard to specificity, affinity, and isotype (see Tab. 3). Additionally, in contrast to polyclonal antisera, the predominant protein in the supernatant of hybridomas is the specific antibody that might be used as a homogenous reagent for nearly all immunoanalytical techniques especially for antibody-labeled techniques. Because of their monospecificity, monoclonal antibodies exhibit only poor antigen-precipitating properties as single reagents. Thus, indirect measurement of antigen-antibody-binding by use of a label is possible, which in turn represents the more sensitive assays.

Table 3 Characteristics of polyclonal antisera and monoclonal antibodies (adapted with modifications from [24, 25]

Characteristics	**Polyclonal antisera**	**Monclonal antibodies**
Purity of immunogen	Important	Not important
Time and expense	Low	Initially high
Specific antibodies	0.1–1.0 mg/mL	5–25 µg/mL in culture supernatants mg range in bioreactors
Nonspecific antibodies	About 10 mg/mL (> 80%)	None in serumfree supernatants
Specificity	Recognizing all determinants of immunogens the animal	Recognizing a single determinant of the imunogen
Affinity	Mixture of high and low Affinity antibodies	High or low according to the screening assay
Avidity	High	Low
Cross-reactivity	Yes	Usually not, but cannot be excluded
Antigen-precipitation	Yes	No
Isotypes and subtypes of immunoglubulins	Typical spectrum	One isotype and subtype
Physical stability	Usually high	Differs markedly
Utility for:		
Cell staining	Good	Antibody-dependent, but excellent with pooled antibodies
Immunopreciptation	Good	Antibody-dependent, but excellent with pooled antibodies
Immunoblots	Good	Antibody-dependent, but excellent with pooled antibodies
Immunoaffinity Purification	Poor	Antibody-dependent, poor with pooled antibodies
Immunoassays with		
Labeled antibody	Difficult	Good, excellent with with pooled antibodies
Labeled antigen	Good	Antibody-dependent, but excellent with pooled antibodies

4 Immunoassays

Immunoassays take advantage of the specific interaction between antigen and antibody. They obey the law of mass action:

$$\text{antibody} + \text{antigen} \leftrightarrow \text{antigen} - \text{antibody complex.}$$

At equilibrium, the antigen-antibody complex is formed as described by the association rate constant K1 and concurrently dissociated to yield again its free components according to the dissociation rate constant K2. But when at a constant amount of one free component the amount of the second free component is increased, the equilibrium of the interaction is driven toward complex formation. Thus, immunoassays enable detection and quantification of antigens and antibodies. Formerly, non-labelled techniques such as immunoprecipitation, agglutination, or light scattering were used, but they still suffered from low sensitivity in the micromolar range. Today labelled immunoassays are used that exhibit at least 1000 fold higher sensitivity. Antibodies or antigens are decorated with labels that impart measurable signals such as radioactivity, fluorescence emission, or enzymic activity. The extent of complex formation is then determined by measuring the amount of labelled antigen or antibody. Depending on the assay design, the signal intensity quoted is directly or inversely proportional to the analyte of interest.

In order to describe the quality of an immunoassay, the terms sensitivity and accuracy are used.

In a narrower sense, sensitivity corresponds to the change in response per amount of reactant, whereas sometimes the term is used to characterize the detection limit of the assay. Accuracy relates the amount of substance determined by the assay to the true value.

4.1 Design of an immunoassay

The criteria to classify immunoassays can be as follows: (1) determination of the antigen or antibody, (2) labelling of antigen or antibody, (3) homogenous or heterogenous assay, and (4) competitive or non-competitive assays.

Homogeneous assays

Homogenous methods are based on changes in enzyme activity, which is mediated by formation of an immune complex. Consequently, modulation of enzyme activity reflects the degree of the immunochemical reaction resulting in enhancement or inhibition. Thus, measurement of the degree of complex formation is carried out without any physical separation of bound and free compounds. Enzymes used as labels in homogenous assays are lysozyme, malate dehydrogenase, and β-galactosidase.

An assay design using enzyme-labelled antigen or analyte is referred to as enzyme multiplied immunoassay technique (EMIT) and was developped for determination of antiepileptic drugs, cardioactive drugs, anti-asthmatic drugs, and drugs of abuse (see Fig. 2). The EMIT assay system requires an active enzyme-hapten-conjugate. Upon binding of hapten-specific antibodies, the activity of the conjugate is inhibited by inducing or preventing changes in conformation of the enzyme as necessary for enzyme activity [26]. Thus, free hapten as analyte and hapten-enzyme conjugate compete for antibody binding.

In the presence of excess conjugate, the enzyme activity is directly proportional to the concentration of the hapten.

Figure 2 Scheme of EMIT using enzyme-hapten conjugate, free hapten as analyte, and hapten-specific antibodies for competitive binding.

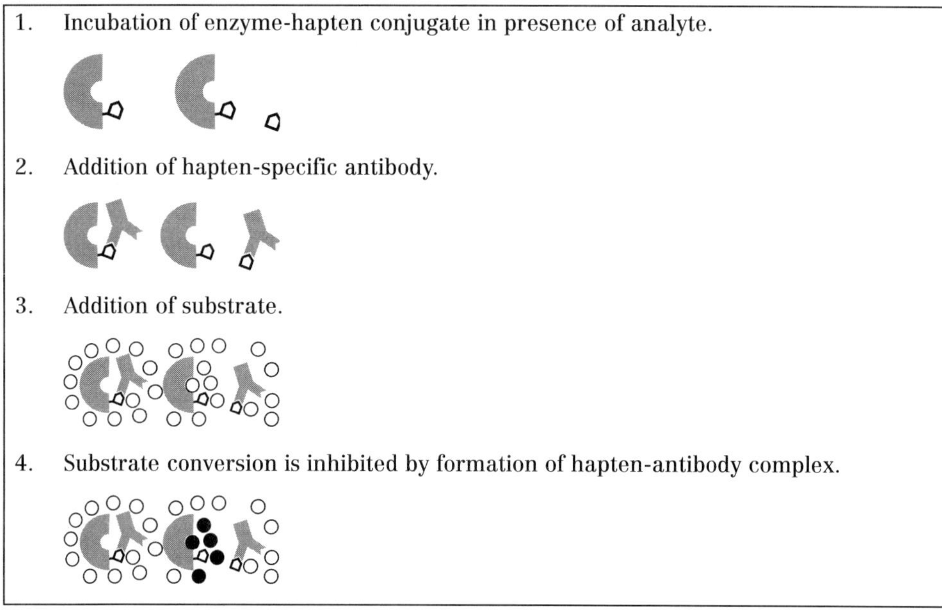

Generally, homogeneous methods for detection of small or large analytes are less sensitive than heterogeneous assays. A major drawback of homogeneous assays is reproducible and time-consuming preparation of well-characterised conjugates. In order to modify the enzyme activity by complex formation, the hapten needs to be exposed to guarantee unrestricted interaction with the antibody.

On the other hand, a separation step as a source of imprecision is omitted. When a homogeneous method is established, the assay still consists of mixing and measuring facilitating automatisation.

Heterogeneous assays

In contrast to homogeneous methods, in heterogeneous systems the activity of the labelled ligand is not altered upon formation of the antigen-antibody complex. Thus, most radioimmuno- and enzymeimmuno-techniques require separation of bound and free-labelled antibody or antigen in order to be able to discriminate between bound and free-labelled ligand, which in turn allows quantification of the analyte of interest.

Usually the activity of radiolabelled reactants is not affected by complex formation, whereas the activity of enzyme-labelled immunoreactants can be altered upon binding to the complementary antigen or antibody depending on the site of coupling the enzyme. In heterogenous enzyme immunoassays, usually horse radish peroxidase, alkaline phosphatase, β-galactosidase, or glucose-oxidase are used as labels.

Methods for separation include use of solid-phase-bound antigen or antibody as well as second antibody precipitation. When enzyme-labelled antigen or antibody is used for detection of the reciprocal solid-phase-bound reactant, this is referred to as enzyme-linked immunosorbent assay (ELISA) [27]. The amount of the bound or free fraction is determined by enzyme-driven conversion of a colourless or non-fluorescent substrate to a intense coloured or fluorescent substrate, which is quantitated by photometry or fluorimetry.

Competitive assays

In competitive assays, the competition in binding to a fixed antibody between a constant amount of labelled antigen and unknown amounts of sample antigen is measured. Of course, this assay can be reversed for measuring the antibody. As the labelled analyte competes with unlabelled analyte for solid-phase binding, the reaction product measured is inversly proportional to the test antigen or antibody (see Fig. 3).

Competitive assays are easy to perform but require high accuracy in dispensing. Additionally, high purity of the labelled ligand is required, but the result is less influenced by contaminants such as cross-reactive and non-specific binding compounds. However, the measurement of extremely low concentrations of analytes can be affected by non-specific adsorption. Furthermore, optimum conditions must be established for each step of the assay. This includes optimisation of the concentration of the solid-phase ligand and the labelled ligand, the type as well as amount of substrate, and incubation time and temperature. The main advantage of competitive assays is that the detection limit is found in the picogram-range and they are easy to perform once established. Therefore, in most laboratories this type of assay is the first design that is applied upon establishing an assay for a new compound to be quantified.

Non-competitive assays

These are often called immunoradiometric assay (IRMA) in the case of radioactive tracers or immunoenzymometric assays. In this assay excess of immobilized antibody is incubated with standard antigen or sample antigen. Formation of the antigen-antibody complex is measured in a second step. When multivalent antigens are measured, the antigen can be detected by a second, labelled antibody. Because at least three layers of immunoreactants corresponding to antibody-antigen-antibody are built up, this technique is called sandwich immunoassay. Since the measured signal derives from the amount of antigen-

Figure 3 Competitive ELISA with solid-phase antibody using enzyme-labelled antigen for quantification of test antigen.

1. Adsorb specific antibody to the solid phase.
2. Remove excess of unbound antibody by washing.

3. Saturate remaining binding sites by incubation with blocking buffer.
4. Remove excessive blocking agents by washing.

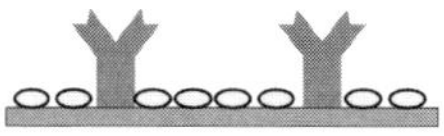

5. Add test antigen in the presence of constant amounts of enzyme-labelled antigen.

6. Remove excess reactants by washing.
7. Add substrate solution and incubate for a certain period of time at appropriate temperature.

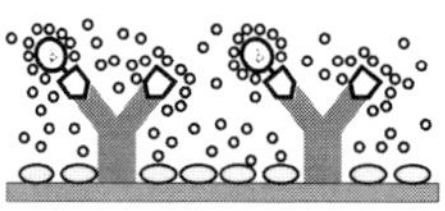

8. Stop enzyme activity and quantify converted substrate, which is inversely proportional to the test antigen.

bound, labelled second antibody, there is a direct proportionality between analyte and signal quoted (see Fig. 4).

Conversely, this design is also applicable to immobilized antigen, and the antigen-antibody complex can be detected by a second, heterologous-labelled antibody, which recognizes the Fc-part of the first antibody.

The results obtained by sandwich ELISA are comparable to those of radioimmunoassays in terms of sensitivity. Additionally, optimisation steps are required as described above.

Figure 4 Sandwich ELISA for determination of polyvalent antigens with second enzyme-labelled, antigen-specific antibody.

1. Adsorb specific first antibody to the solid phase.
2. Remove excess of unbound antibody by washing.

3. Saturate remaining binding sites by incubation with blocking buffer.
4. Remove excessive blocking agents by washing.

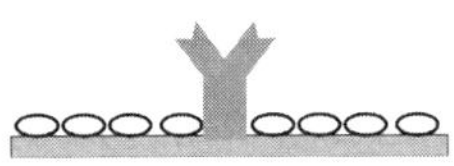

5. Incubate with multivalent, unlabelled antigen.
6. Remove unbound antigen by washing.

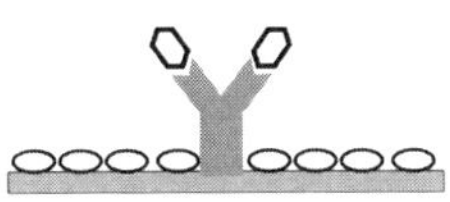

7. Add excess of second, enzyme-labelled antibody.

8. Remove unbound reporter antibody by washing.
9. Add substrate solution and incubate under appropriate conditions.

10. Stop enzyme activity and quantify converted substrate, which is directly proportional to the concentration of the antigen.

4.2 Practical considerations

Antigen and antibody

It is highly desirable to have pure antigen at one's disposal. When two monoclonal antibodies or affinity-purified antibodies are available, which recognize two independent epitopes of the antigen, the sandwich assay is appropriate to detect even impure antigen preparations. To perform competitive assays for detection of antigen, pure or partially purified labelled antigen is required. On the one hand, the solid support is coated with antiserum or culture supernatant, and subsaturating amounts of a mixture containing labelled and non-labelled antigen is allowed to compete for binding to the antibody. On the other hand, the antigen is immobilsed and a mixture of antigen and labelled antibody is added. Consequently, high amounts of antigen inhibit binding of the labelled antibody to the immobilized antigen. If pure antigen is not available, the wells are coated with crude antigen preparation followed by detection with saturating amounts of labelled antibody.

For this reason and because of possible interference with compounds of the antiserum or the biological matrix containing the analyte, it is sometimes necessary to purify and concentrate the antibodies, especially polyclonal antisera. This can be done most simply by ammonium sulphate precipitation. IgG usually precipitates at 33% to 40% ammonium sulphate saturation. This can be performed simply by mixing equal volumes of antiserum and PBS-buffer, followed by addition of two volumes of saturated ammonium sulphate solution. After stirring for 30 min. the pellet is collected by centrifugation and washed with half-saturated ammonium sulphate solution. Finally, the pellet is dissolved in PBS (one volume or less) and dialysed against PBS. Thus, the antibody preparation is purified and concentrated.

An alternative method is to adsorb serumproteins on a cation-exchange matrix. The IgG is neutral or slightly positively charged at pH 6.5, whereas most of the serum proteins are negatively charged. Consequently, the serum proteins are bound to the DEAE-matrix, whereas IgG passes the column or is detected in the supernatant of the batch. Large amounts of ion exchanger are needed and the preparation is diluted. Furthermore, the pH should be adjusted to 8 as soon as possible to prevent denaturation of the immunoglobulin.

Much more comfortable is isolation of IgG by purification with protein A immobilised on beads that are commercially available [28]. Because the Fc-proportion of IgG is bound to the surface protein of *Staphylococcus aureus* at pH 8, all contaminants are removed by washing the column. The IgG is dissociated from the beaded carrier by elution at pH 2. Thus, pure and concentrated IgG is yielded by elution in counter-current flow, but the pH of the eluate should be adjusted to 8.5 immediately. Only minimum amounts of the commercially available matrix are necessary, as the IgG-binding capacity of protein A is high (about 20 mg IgG/mg protein A).

For preparation of monospecfic antibodies from polyclonal antisera, affinity chromatography on immobilzed antigen is performed. This rather time-consuming method requires synthesis of an affinity matrix containing covalently bound antigens via 6C-12C spacers. The antiserum is bound to the matrix at physiological pH and, after removal of impurities by washing the column, the monospecific antibodies are recovered by lowering the pH, enhancing the ionic strength of the buffer, or addition of chaotropic agents. Again, physiological conditions in the eluate should be restored as soon as possible.

In general, the sensitivity of immunoassays increases with purity of antibody and antigen preparation.

Diluents

The diluent should provide optimum conditions for formation of the antigen-antibody complex. Furthermore, extremely low concentrations of analytes are applied, as they are susceptible to surface adsorption on the reaction vessels and denaturation. Whereas monoclonal antibodies and enzyme labels are more sensitive to ionic conditions and temperature, for polyclonal antibodies a pH near neutrality and an ionic strength of 0.15 is recommended, usually PBS pH 7.4. Addition of bovine serum albumin up to 1% minimizes adsorption to the surface of the vessel and stabilizes antigens and antibodies, especially in highly diluted solutions. However, in assays for haptens addition of albumin should be avoided, as it binds many drugs and hormones. Washing buffers can contain up to 0.5% Tween 20 to enhance solubilisation of non-specifically adsorbed compounds.

Inclusion of bacteriostatic agents, e. g., 0.02% sodium azide, enhances storage stability of antibody solutions but sometimes hardly interferes with enzymic detection. To improve storage stability of concentrated antibody solutions upon freezing, additon of glycerol up to 50% is recommended. Prior to the assay, the solution should be highly diluted with working buffer.

Solid supports

A solid matrix with immobilised antigen or antibody greatly facilitates manipulation during immunoassays. Formerly, beads were used as carriers, but today 96-well microplates are used in labs routinely for many purposes. Sometimes the reactants are immobilised covalently, but non-covalent interactions are used more frequently. In the latter case, the solid support can be made from polyvinylchloride or polystyrene, each exhibiting a protein-binding capacity of up to 300 ng/cm^2. Because polyvinylchloride microplates are not translucent, they are not suited for enzyme immunoassays measuring conversion to coloured substrate. Translucent polystyrene microplates with flat bottoms are recommended for enzyme immunoassays, as they are commercially available with high and low protein-binding capacity.

Infrequently, protein-A-coated beads serve as solid support. As protein A interacts with the Fc-fragment of the IgG-molecule, the site-directed immobilisation of the antibody renders the antigen-combining sites freely accessible.

Furthermore, so-called "magneto-beads" are commercially available and contain a core made from iron, which facilitates separation by magnetic fields omitting centrifugation.

Detection

There are three ways of detection in immunoassays: (1) direct detection using labelled antigen or antibody; (2) indirect detection using a secondary labelled analyte, most frequently a second antibody recognizing the Fc-fragment of the first antibody; and (3) use of the biotin-avidin system comprising detection of biotinylated antigen or antibodies with enzyme-tagged or radiolabelled avidin. Direct detection requires fewer steps in the assay procedure and background problems are minimal. However, pure analyte is required and needs to be conjugated covalently with the label. In the case of radioimmunoassays, labelling procedures are well established, but for enzyme immunoassays this can be crucial to do in the lab. Additionally, direct detection is less sensitive than the indirect way. The main advantage of indirect detection is commercial availability of labelled antibodies, which might be used to detect a wide range of antibodies with different specificity. The interaction between the glycoprotein streptavidin and the vitamin biotin is characterised by a dissociation constant of 10^{-15} mol/L, which corresponds to a half life of 160 days and approaches covalent interactions. Streptavidin consists of four identical subunits, each binding one molecule of biotin. Thus, an enhancement effect can result from the multivalency of streptavidin, which might be amplified further by a multilayer design using avidin-biotin complexes. Reagents for biotinylation of proteins containing spacer molecules are commercially available, and avidin is easily coupled to enzymes.

The type of the label needs to be carefully selected prior to the experiment. Radiolabelling is easy to perform by the chloramin-T iodination or by the lactoperoxidase method [29], but safety regulations concerning use, storage, and disposal need to be fulfilled. In radioimmunoassays low amounts of radioactivity are necessary, about 10 to 100 µCi/µg in the case of ^{125}I exhibiting a half life of 60 days, and about 50,000 cpm/well are required for reliable results. Consequently, danger to health is minimal if handled correctly, but special equipment is a prerequisite. In comparison, labelling with enzymes is safe, but it is emphasized that some cross-linking agents and substrates are hazardous. The equipment necessary for enzyme immunoassays is inexpensive and the reagents exhibit a long shelf-life. Additionally, enzyme immunoassays allow a wide variety of assay designs, and detectability is comparable to that of radioimmunoassays, which is supported by the amplification effect of enzymes and by use of fluorogenic substrates. Usually 5 to 50 ng enzyme-labelled antibody/well yields a colour development by substrate conversion corresponding to an absorption of 0.2 to 2.0. In practical terms, direct visualisation of the result on the microplate is possible. For further enhancement of detectability, fluorogenic substrates might be used. A high quantum yield provides for the best detection limit, a large Stoke's shift should minimise the interaction between excitation and emission, and naturally occuring fluorophores in the sample should not interact. Quenching effects need to be characterised for each individual assay design. A major disadvantage of enzyme immunoassays is the possibility of interference of enzyme activity with compounds of the biological matrix of the sample, which might be reduced by selection of an appropriate assay design.

4.3 Protocols

Quantification of cortisol by ELISA

Protocol 1 Antigen preparation

1. A solution containing 5.5 mmol cortisol and 22 mmol succinic anhydride in anhydrous pyridine is refluxed under protection from light and humidity for 4 h.
2. The solvent is completely removed by evaporation under vacuum, and the oily residue is dissolved in chloroform followed by extraction with distilled water. The extraction is repeated three times and the organic phase is dried with anhydrous sodium sulphate.
   Formation of cortisol-21-hemisuccinate is confirmed by thin-layer chromatography on KGF254 in chloroform/methanol/water (65+30+6, v/v/v) indicated by hRf 49 as compared to cortisol with hRf 65. IR-spectroscopy and mass spectroscopy confirm these results.
3. The antigen is prepared by the mixed anhydride method.
   Under a hood, a solution containing 1.6 mmol cortisol-21-hemisuccinate in 32 ml anhydrous dioxane is cooled to 13 °C and 1.6 mmol tri-N-butylamine is added under stirring. After addition of 1.6 mmol isobutylchloroformate, the solution is incubated for 30 min. followed by addition of a solution containing 880 mg BSA in a mixture of 26 ml distilled water pH 10 and 17.2 ml dioxane. Immediately, the pH is adjusted to 7.5–8.5 by addition of 1 M NaOH and maintained for the next 2 h.
4. The BSA-cortisol conjugate is dialysed extensively against distilled water followed by lyophilisation. For immunization of mice, the conjugate is further purified by chromatography on Sephadex G-25 using TRIS-buffer pH 8.6 for elution to remove even traces of the free immunosuppressive drug.

The degree of substitution is determined by UV-difference spectroscopy. At the absorption maximum of cortisol at 246 nm, the difference in absorption of accurately equimolar solutions of conjugate and carrier protein is determined and calculated from the calibration curve of cortisol to be 19.4 mol cortisol/mol carrier protein.

Protocol 2 Immunization and preparation of monoclonal antibodies

1. Three Balb/c mice were immunised with 100 µg antigen emulsified in 1 ml complete Freund's adjuvant for the first immunization and 50 µg antigen in 1 ml incomplete Freund's adjuvant for the subsequent two boosters.
2. Fusion of isolated spleen cells and $2x10^8$ myeloma cells was performed with PEG as described above, following a protocol reported in the literature [20, 21]. The primary hybridoma cultures (one drop of fused cells in 2 ml medium) were grown in 24-well culture dishes as described above, yielding 75 growing clones.
3. After reaching confluency, the supernatant was screened for cortisol-specific antibodies yielding six positive clones, which were propagated further.
4. Cloning was done by limited dilution. The primary dilution series was done in 96-well microplates at a density of 10^4 cells/well in 200 µl HAT medium. The secondary dilution series was done accordingly from positive clones.
5. Screening of the supernatants resulted in eight cortisol-antibody secreting clones. These clones were further propagated in 24-well microplates.

Protocol 3 ELISA protocol

1. The wells of a 96-well microplate were coated with 100 µl antigen solution (0.003% in PBS containing calcium and magnesium) overnight at 4 °C.
2. After washing three times with 100 µl PBS each, the surface of the wells was saturated by incubation with 100 µl 1% BSA-solution in PBS for 2 h at room temperature.
3. After washing three times with 100 µl PBS each, a mixture of 50 µl hybridoma-supernatant and 50 µl cortisol in PBS (serial dilution, 5–200 ng) was added and incubated for 2 h at 37 °C.
4. The wells were washed three times with PBS. After addition of 100 µl goat-anti-mouse-IgG-conjugated peroxidase, the microplate was incubated for 2 h at 37 °C.
5. The supernatant was removed and the wells were washed again three times with PBS. 100 µl of freshly prepared substrate solution consisting of 10 mg o-phenylendiamin hydrochloride, 9 ml 0.11 M aqeous disodium-hydrogen-phosphate, 1 ml 0.5 M aqueous citric acid, and 10 µl 30% H_2O_2 were added.
6. After incubation for 30 min. under protection from light substrate, conversion was stopped by addition of 100 µl 3 M sulfuric acid. Immediately, the absorption was read at 450 nm using a microplate reader.

The samples were assayed in triplicate at the minimum, including blanks (well-filled with PBS), negative controls (as above, but without addition of cortisol-antibody and cortisol), and positive controls (as above, but without addition of cortisol).

Cross-reactivity of the antibody with structurally related steroids can be determined by the same assay, but add the steroid of interest instead of cortisol.

5 Immunofluorescent analysis by flow cytometry

5.1 Introduction

Because of the fascinating facilities concerning the characterisation of cells, immunofluorescence techniques in combination with flow cytometric analysis have gained more and more interest over the past years [30]. Application of fluorescent-labelled monoclonal antibodies directed against specific proteins or polysaccharides located at the cell surface or – with some limitations – selected cytoplasmic, as well as nuclear, ligands offers the possibility to obtain direct information on the state of activation, proliferation, and differentiation of the cells. In comparison to microscopical methods, the advantage of flow cytometry is the rapid quantification of the amount of a particular stain per cell, which offers a good statistical evaluation because of the large number of cells usually assessed.

The objective of this chapter is to give an overview of the basic principles of flow cytometric measurements and to provide readers with basic considerations concerning immunofluorescence techniques. Upon understanding the important parameters, the specificity and sensitivity of the staining methodology can be optimized for a particular application.

5.2 Principles of flow cytometry

The basic components of a flow cytometer consist of a fluidic system to transport the cells across the microscopic field; an optical system for illumination and detection; and the electronics for light collection, data management and control [31–33].

The most important part of the **flow system** is the flow chamber, where the microscopic observation takes place. An important parameter for successful flow cytometric analysis is the hydrodynamic focusing of the cells in this flow chamber providing for individualisation and correct positioning of the cells under investigation. Furthermore, the sample suspension is injected into a particle-free “sheath fluid”, and the diameter of the sample flow is reduced dramatically at the injection point of the flow chamber to guarantee serial alignment of the individual cells during optical analysis [34].

The optical system consists of three parts, the first is the **illumination optics,** which represents the light source for the flow cytometry and is made up of either a conventional lamp or a laser, generally an argon ion laser. Via prisms and mirrors, and if necessary monochromatic filters, the illuminating light is directed to the liquid stream and focused by several lenses to the point of microscopic observation.

The other two components belong to the detection system of the flow cytometer. The **forward scatter collection optics** is a microscope with low numerical aperture observing the liquid stream from opposite the illuminating light. The "forward scatter" (FS) collects the light scattered by the cells in the flow stream in the range of 2° to 20° off the axis of the illuminating light. The scattering signal measured by the FS corresponds to the diameter and, therefore, the size of the cell under investigation. The second detection component includes the **fluorescence and side scatter collection optics,** comprising a second microscope with a long working distance and high numerical aperture, which analyses the light scattered perpendicular to the illumination beam as well as the liquid stream. The "side scatter" (SS) signal is dependent on the surface characteristics and the granularity of the measured cells, whereas the fluorescence detectors collect the light emitted by fluorescent compounds associated with the analysed cells.

The electronic system is necessary for the correct collection and processing of the light signals gained. Conversion of optical into electrical signals is done by semiconductor photodiodes or/and photomultiplier tubes. To analyse two or more fluorescent compounds with overlapping emission spectrum (e. g., fluorescein isothiocyanate and phycoerythrin) simultaneously, the flow cytometer should be equipped with an electronic fluorescence compensation network for correction of fluorescence overspill and proper calculation of the different fluorescence signals. For data calculation, either a linear or a logarithmic amplifier can be used, depending on the dynamic range of the parameter under investigation. In addition, a trigger circuit provides for correct separation between signals and electronic noise or debris by setting a threshold level for one parameter (e. g., forward scatter). Each analog pulse acquired by the photomultiplier tubes that exceeds this trigger threshold is converted into a digital signal and transmitted to a computer for further storage and data evaluation.

5.3 Data evaluation

Today, state-of-the-art flow cytometers offer the possibility to analyse five parameters per cell simultaneously: forward and side scatter as well as up to three fluorescence light parameters. Normally, the intensity of the scattered light is analysed on the basis of linear amplification, but the intensity of immunofluorescence is analysed upon logarithmic amplification over four decades. Thus, a single gain setting can be used for data acquisition allowing direct comparison between relative fluorescence intensities of various cells.

For evaluation a multitude of data plots is available. Usually a two-dimensional dot plot is used to characterize the scattering properties of a cell population by plotting the FS signal *versus* the SS signal. One-dimensional histograms are best suited for illustration of the intensity of fluorescence staining emphasising the quantitative aspects of flow cytometry.

In order to exclude non-specifically stained particles (e. g., dead cells, debris) or conglomerates of cells from analysis, cells for data evaluation can be preselected by setting lower and upper thresholds for certain parameters ("gating") [35]. Statistical calculation of data is done by the cytometer software, making the evaluation of immunofluorescence assays rather simple. In this statistical evaluation, the cell populations are usually described by the relative mean and the coefficient of variation (CV) of a particular fluorescence parameter as well as the number of cells in the respective population. The mean is defined as the arithmetic mean in the case of linear scale and geometric mean for logarithmic scale. The CV is calculated as half the width of the distribution at 0.6 times maximum height. Usually the distribution of fluorescence intensities on a logarithmic scale varies symmetrically around the mean because of both biological variation (e. g., size of cells, density of target antigen and staining antibody) and variation of measurement (e. g., focusing, orientation). Absolute quantification of the amount of antigen molecules per cell is difficult because of the necessity to include a reference method for calibration of the relative immunofluorescence [36].

5.4 Basic considerations on immunofluorescence techniques

The area of immunofluorescence covers the staining of cellular structures with antibodies and other specific ligands directly or indirectly labelled with fluorescent dyes [37]. During an immunofluorescence analysis, all the steps prior to the flow cytometric measurement, including preparation of cells, staining, and fixation, are most critical.

Preparation of cells

In order to obtain appropriate data from flow cytometric analysis, attached or connected cells have to be individualized prior to staining and measurement. Preparation of a single-cell suspension is done easily by repeated re-suspension in the case of natural single cells derived from blood, bone marrow, or non-adherent cultured cells. Adherent growing cells or cells from tissues can be separated mechanically or enzymatically either prior to or after the staining procedure. Upon use of enzymes for disintegration of cells, care has to be taken in order to ensure that the antigen of interest or the attachment of the immune label is not affected by the dissociation procedure.

Recent protocols often include pre-enrichment steps, especially in the case of blood samples, e. g., the depletion of erythrocytes for analysis of leukocytes. Among these procedures, density gradient centrifugation and lysis are used frequently. Pre-enrichment of rare cells is advantageous because of the reduction of staining volume and elimination of cells that could be stained unspecifically, such as dead cells, resulting in improved discrimination of the cells of interest.

Staining of cells
The staining procedure can be performed either directly or indirectly. In the case of direct staining, the fluorescence-label is attached covalently to the antibody against the specific cellular antigen under investigation and used for analysis. Indirect staining is done by incubating the cells with an unconjugated primary antibody followed by detection of the antibody with a secondary, fluorescent-labelled, usually polyclonal anti-antibody. Direct staining is preferred because of easier handling and better control of the staining parameters. Especially in routine use and for muliparametric analysis using a combination of various markers in one staining step, control and handling are important to consider [38]. Prior to use the labelled primary antibodies of interest have to be purified. Indirect staining methods facilitate application of the primary antibody without any previous labelling and purification. A further advantage of indirect staining protocols is the possibility to improve the sensitivity of detection by increasing the number of secondary antibodies bound to one primary antibody, thus increasing the number of fluorescence markers per specific antigenic determinant. But particularly in multiparameter analysis, control of proper indirect staining is not trivial and requires highly purified isotype-specific secondary antibodies or complete blocking of free binding sites of the secondary antibody prior to staining of the next parameter to avoid any cross-reactions. To combine high specificity and high sensitivity, an indirect staining protocol using primary antibodies conjugated to haptens such as digoxigenin or biotin can be followed. In these systems the detection is pursued very specifically by fluorescence-labelled secondary antibodies directed against the hapten or streptavidin, while the sensitivity is controlled by the number of marker molecules per mol secondary antibody/streptavidin.

In order to avoid any loss of or damage to cells, the number of steps in the staining protocol should be kept to a minimum. Additionally, reproducible results can be expected only if the concentrations of cells and antibodies used for detection are standardized. In so doing it is important to apply concentrations that are as low as necessary for predominance of specific high-affinity antigen-antibody interactions over low-affinity cross-reactions but are high enough to ensure that staining is as bright as possible to obtain good quality of data.

The optimum concentration of cells depends mainly on the nature of cells and the antibodies used for detection and should be determined in preliminary studies. The lower limit amounts to approximately 10^5 cells in 10 µl providing for flow cytometric analysis of 10^4 cells. Thus, 10 µl is the minimum volume necessary for proper staining in order to avoid significant changes in the concentration of the staining antibody during the washing procedure. Usually staining volumes of 20 to 100 µl containing 0.7 to 5 × 10^6 cells are used. All these concentrations refer to viable positive cells; additional non-stainable cells can be neglected. However, for flow cytometric measurement of rare cells the number of negative cells has to be considered upon optimization of the staining protocol. To stain larger numbers of cells, it is better to increase the volume of the cell suspension than the concentration of cells.

To obtain results that guarantee adequate statistical evaluation, the optimum concentration of staining reagents should be determined by titration. For that matter, it should be kept in mind that too-low concentrations resulting in poor discrimination between positive and negative cells are as problematic as too-high concentrations resulting in non-specific staining of negative cells or subpopulations due to low affinity cross-reactions. Titration can be performed by flow cytometric analysis of a mixture of positive and negative cells stained with varying concentrations of the fluorescent-labelled antibody using a standard protocol. By plotting the mean cell-associated fluorescence intensity against the concentration of the staining reagent, optimum conditions in an antibody concentration exist where the difference in cell binding between positive and negative cells is highest. If there are any problems with cross-reactions, it might be useful to apply lower concentrations. Higher concentrations are recommended in the case of low antigen density.

The incubation with the antibody should be kept as short as possible because high-affinity antigen-antibody reactins occur very fast. Extending the incubation time may lead to enhanced low-affinity cross-reactions and non-specific adsorption. Since with most staining reagents about 90% of the maximum staining is achieved within 5 min., doubling the time for safety reasons results in a standard staining time of 10 min. For immunofluorescent staining of intracellular components, a prolonged staining period after fixation and permeabilisation of the cells is necessary. To guarantee appropriate diffusion of the staining antibodies into the cytoplasm to reach the antigens of interest, staining times of about 1 h are used.

To avoid any problems deriving from cell physiology, staining of live cells is done at an incubation temperature of 4 °C. At this temperature, the metabolism of the cells is minimized and the fluidity of the cell membrane is reduced, resulting in a preferred binding of the staining reagents to the cell surface. In the case of staining cytoplasmatic parameters of fixed and permeabilized cells, the temperature during staining should be raised to 37 °C in order to enhance diffusion of the staining antibodies. Staining of live cells at higher temperatures may lead to altered results that are due to modulated expression of the antigenic receptors (e.g., lymphocytes can "cap" their receptors with the staining antibodies, throw off the cap into the medium, and then appear "negative"). Blocking cell physiology by addition of 0.003% sodium azide in the staining medium may reduce these problems.

Washing and fixation of cells

Prior to flow cytometric analysis, unbound staining antibodies have to be removed by washing. Therefore, the cell suspension is spun down at 1000 rpm followed by removal of the supernatant and resuspending the cells in fresh medium or isotone buffer. Because under even optimum conditions 10% of the cells are lost per washing step, washing should be reduced to a minimum. Applying an adequate low concentration of reagents during staining can help to restrict washing of the cells to one or two steps, particularly in direct staining protocols. In the case of indirect staining, proper washing of the cells is very important to improve staining results and to avoid any reaction between primary and secondary antibody in solution. In these protocols, the specificity of the antigen-antibody interaction often is improved by addition of 1% BSA to the washing solution for blocking of residual non-specific binding sites.

For routine analysis cells usually are fixed prior to flow cytometry using paraformaldehyde (2% in PBS) to avoid time-dependent alterations of the cell staining as well as the risk of infections, thus standardizing analysis parameters. However, like all other manipulations, fixation of cells may lead to unpleasant side effects. Especially for analysis of rare cells, discrimination between rare positive cells and non-specifically stained dead cells is crucial. After fixation it is very difficult to distinguish whether a cell was dead before staining or not. In such cases it is more convenient to analyse live cells without fixation immediately after the staining procedure in the presence of propidiumiodide, gating out dead cells according to the scatter and the propidiumiodide fluorescence. In contrast, staining of cytoplasmatic or nucleic substructures requires fixation and permeabilisation of cells because of the inability of antibodies to penetrate the cell membrane of live cells [39]. For this reason, cells are fixed either by addition of ice-cold methanol (10 min. at –20 °C) and subsequent rehydratisation in PBS or with para-formaldehyde (2% in PBS, 10 min). In the case of para-formaldehyde, fixation has to be followed by an extra permeabilisation step using Triton X-100 (0.1% in PBS, 10 min.).

Measurement and data evaluation

Prior to flow cytometric analysis, the sample usually amounting to 50 to 200 µl has to be resuspended in 1 ml Cell Pack. Amplification of the fluorescence signals during flow cytometry should be adjusted in such a way that the autofluorescence signal of unlabelled cells is put in the first decade of the 4-decade log range. For each measurement between 5000 and 10,000 cells of interest are accumulated. Generally data evaluation is done by using a forward *versus* side scatter plot for gating and the relative mean of the particular fluorescence parameter for further calculation, as described above in detail.

5.5 Standard protocols – immunofluorescence staining and flow cytometric analysis

Protocol 4 Direct staining of surface antigens

1. Prepare 50 µl of a single-cell suspension containing 1 × 10^6 cells.
2. Add 50 µl of the specified fluorescence-labelled antibody solution and incubate for 10 min. at 4 °C. The adequate concentration of antibody depends on antigen density on target cell, fluorescein/protein ratio of the labelled antibody, concentration and nature of cells, etc., and has to be determined by titration in a preliminary study.
3. Add 100 µl PBS and spin down the cell suspension at 4 °C and 1000 rpm. Discard 150 µl of the supernatant and resuspend the cells using 150 µl of PBS. This washing step can be repeated once. Finally the volume is adjusted to 100 µl.
4. If fixation is required, add 100 µl of a para-formaldehyde solution (4% in PBS) and incubate for 10 min. at r.t.
5. Dilute the sample with 1 ml Cell Pack and analyse by flow cytometry, accumulating 5000–10,000 cells per measurement.

Protocol 5 Indirect staining of surface antigens

1. Prepare 50 µl of a single cell suspension containing 1 ×10^6 cells.
2. Add 50 µl of the specified primary antibody solution followed by an incubation for 10 min. at 4 °C. The adequate concentration of antibody depends on antigen density on target cell, fluorescein/protein ratio of the labelled antibody, concentration and nature of cells, etc., and has to be determined by titration in a preliminary study.
3. Add 100 µl 1% BSA in PBS and spin down the cell suspension at 4 °C and 1000 rpm. Discard 150 µl of the supernatant and resuspend the cells in 150 µl of 1% BSA/PBS. This washing step can be repeated once. The volume should be adjusted to 50 µl.
4. Add 50 µl of the corresponding fluorescence-labelled secondary anti-antibody solution and incubate again at 4 °C for 10 min. For considerations on the concentration of antibody, see above.
5. Repeate the washing procedure as described above.
6. If fixation is required, add 100 µl of a para-formaldehyde solution (4% in PBS) and incubate for 10 min. at r.t.
7. Dilute the sample with 1 ml Cell Pack and analyse by flow cytometry accumulating 5000–10000 cells per measurement.

Protocol 6 Staining of cytoplasmatic or nucleic structures

1. Prior to staining, the single-cell suspension has to be fixed and permeabilized:
2. By addition of ice-cold methanol (70%) and incubation for 10 min. at –20 °C followed by rehydratisation at 4 °C
   A. By addition of para-formaldehyde in PBS (final concentration 2%) and incubation for 10 min. at r.t.
   B. Use of para-formaldehyde for fixation requires additional permeabilisation of the cells using 0.1% Triton X-100 in PBS and incubation for 10 min. at r.t.
3. Staining is performed as described above, but incubation time is extended up to 1 h.
4. During washing the cells should be incubated after resuspension for about 15 min. to guarantee proper removement of antibody and blocking of residual binding sites.

5.6 Conclusions

All in all, immunofluorescence techniques in combination with flow cytometric analysation offers exciting possibilities to determine specific antigens at the surface of cells or even cytoplasmatic and nucleic structures. The advantages of flow cytometric analysis are based mainly on the ability to determine cell-associated fluorescence intensities at the single-cell level and the rapid assessment of a large number of cells leading to excellent statistical data. Thus, flow cytometry is best qualified to improve characterisation of cells in all areas, ranging from basic immunology and molecular biology to clinical medicine and development of new strategies for pharmaceutics, as well as discrimination between even very close subsets of mammalian cells as applied to detect and monitor certain diseases [40, 41].

6 Immunofluorescence studies using confocal laser scanning microscopy

6.1 Introduction

Confocal laser scanning microscopy has been established within the last 15 years as a valuable tool for obtaining high-resolution images and three-dimensional reconstructions for a variety of biological specimens [42, 43]. Convential light microscopy creates images with a depth of field of 2 to 3 μm, and the real information in the focal plane is overlayed with "out-of-focus" information from optical planes above and below. This is especially a problem in fluorescence miroscopy where "out-of-focus" fluorescence creates a diffuse haze around the objects of interest. The aim of confocal light microscopy is to use only the information of the plane that is in focus and to divide three-dimensional objects into multiple optical slices. These images can then be digitized and reconstructed into three-dimensional objects using modern computer technology.

Components of a confocal laser scanning microscope
Conventional epifluorescence microscope: Objective (oil or water immersion), stage, focus motor (for optical slicing).

Confocal unit: laser, excitation filter, illumination pinhole, scanning unit (oscillating mirrors), dichroic mirror, detection pinhole, reflection filter, photomultiplier.

6.2 Optical principle of a confocal laser scanning microscope

The principle of a confocal laser scanning microscope is shown in Figure 5. A laser light beam from a strong laser source is expanded to make optimal use of the optics in the objective. This improves spatial resolution by the objective compared to that of conventional light sources. To generate a two-dimensional image, the laser light beam is scanned across the specimen. This is achieved by an x-y deflection mechanism (reflection from galvanometer-driven oscillating mirrors [44]). This enables the scanning laser beam to focus by an objective lens on a very small area of the fluorescent specimen termed the excitation light. Using a high numerical aperture objective lens, the crossover of the conical beam of laser light is approximately 0.2 μm in diameter. The fluorescent specimen emits light of a higher wavelength that has to pass the objective lens, the oscillating mirrors, the reflecting dichroic mirror, an excitation light filter, and a confocal pinhole and ultimately is focused onto a photodetector.

The emitted light from the fluorescent specimen is "de-scanned" by the same oscillating mirrors scanning unit. The reflecting dichroic mirror or beamsplitter allows only the emitted light up to a defined wavelength to pass through and

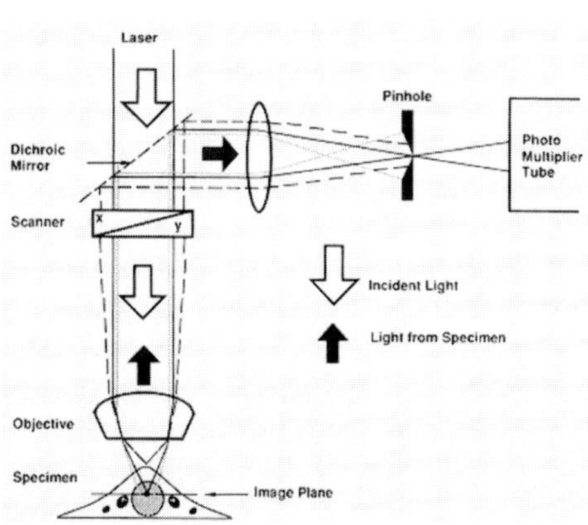

Figure 5 Confocal laser scanning microscope (cf. to [44])

reflects light with higher wavelengths (which includes all the excitation light coming back from the specimen). The confocal pinhole is constructed so that it eliminates all light from "out-of-focus" planes. This is especially important when evaluating thick specimens. A "confocal spot" is the term used for the area that is detected by the laser beam. To generate a two-dimensional image of a small area of the specimen or "optical section", a raster sweep of the specimen is done at one particular focal plane. As the laser scans across the specimen, the analog light signal is detected by the photomultiplier and converted into a digital signal. The combination of scans creates a pixel-based image (256 × 256 pixels, 512 × 512 pixels, or 1056 × 1056 pixels) that is displayed on a computer monitor. The relative intensity of the fluorescent light emitted from the laser hit point on the fluorescent specimen corresponds to the intensity of the resulting pixel in the image (8-bit grayscale). Colour can be added to the 8-bit grayscale images to define reacting proteins in a double or triple labeling and prior to three-dimensional reconstruction to highlight regions of interest. Most frequently, assigning colours by thresholding is used to make images look clearer. This is done by assigning various intensity ranges to different colours. The fluorescence intensity of each assigned colour is ramped from dark to light to reflect fluorescence intensity, termed pseudocolor. The colours are usually shown in an 8-bit colour look-up table (LUT), which is included in the image legend.

6.3 Pinhole and confocality

The confocal microscope produces an optical slice through a specimen of defined thickness as a consequence of the pinhole [45–47]. The numerical aperture of the lens influences the thickness of the optical slice: Oil- and water-immersion lenses are most suitable and can generate slices as thin as 1 μm. The diameter of the pinhole inversely correlates with confocality. For example, a large pinhole increases the signal but decreases the axial resolution of a plannar object and can result in a loss of confocality resulting in conventional microscopy. A small pinhole limits brightness. In practice, microscope users often open up the size of the pinhole to increase the signal from a weekly fluorescent or scattering object. For each confocal microscopy specimen, it is necessary to determine an optimal pinhole diameter for slice thickness and brightness.

6.4 Commonly used laser sources and fluorophores

Specimens are stained using antibodies labeled with fluophores for imaging with CLSM. This technique is called immunofluorescence histochemistry. However, other techniques using direct staining with specifically reacting substances like phalloidins also can be used. Fluorescent probes are available for a broad range of biomolecules, permitting the use of the CLSM to image macromolecular structures (proteins, lipids, carbohydrates, and nucleic acids) and physiological ions (e. g., calcium, protons) for static fluorescent specimens, as well as for dynamic quantitative imaging and measurements [48, 49]. The choice of fluorophores used for immunofluorescence is critical because auto-fluorescence intrinsic from the tissue can mimic the appearance of the fluorophores. To exclude auto-fluorescence interfering with the fluorophore emission, the unstained specimens are examined with the same set of filters for excitation that are used with the specific fluorophore. Auto-fluorescence is usually the highest, with excitation between 400 and 450 nm, and decreases with increasing wavelengths [50]. Additionally, the choice of fluorophores depends on their excitation/emission wavelength, the laser source, and the filter sets available.

Immunofluorescence can be performed by either direct or indirect methods [45]. Direct immunofluorescence involves the conjugation of the primary antibody with a fluorophore such as fluorescein or rhodamine. For indirect immunofluorescence, the primary antibody is detected by a secondary fluorescence-labeled antibody that is directed against the first antibody. This method is more commonly used because it is easier to label secondary antibodies than primaries and it is more sensitive, i. e., more than one molecule of the secondary antibody can bind to the primary antibody. The specificity of the staining antibodies for an indirect staining protocol is especially important for multi-colour immunofluorescence because of cross-reactivity. Each individual secondary antibody must recognize only one primary antibody and must not cross-react with intrinsic proteins in the specimen or with other secondary antibodies. Unspecific staining of the tissue by a secondary antibody can be determined by staining the tissue in the absence of the primary antibody. To ensure that a secondary antibody recognizes only the primary antibody, it is necessary to determine whether the secondary antibody binds to the primary antibody. To avoid cross-reactivity, it is ideal to use secondary antibodies that have been raised in species different from the species in which the primary antibodies have been derived.

Some of the important considerations for choosing a laser for a CLSM system are cost, wavelength of emission, output power at each wavelength, efficiency, and stability. Laser lines below 400 nm are expensive to achieve, therefore, only a few CLSM systems with such lasers are available commercially.

Table 4 Commonly used laser sources [44]

Source		Excitation lines	
He Ne	543 nm	633 nm	
Argon Ion 1	488 nm	514 nm	
Argon Ion 2	458 nm	488 nm	529 nm
Argon Krypton	488 nm	568 nm	647 nm

Use of filters in a CLSM system: Gray filters are used for the attenuation of the laser power, and single lines are selected by excitation filters. Ideally, these filters should have minimal loss of laser intensity and should permit the detection of the fluorophores around their maximal emission wavelength. For multi-labelling experiments, fluorophores should have only minimal overlap in their excitation/emission wavelengths. To eliminate channel "crosstalk", it is necessary to use sequential excitation and imaging of each fluorophore to spectrally isolate their fluorescence emission. This method requires special filter combinations based on each of the fluorophore's excitation and emission wavelengths and is time-consuming because the specimen must be scanned several times with different excitation and emission wavelengths.

Anti-fade reagents
Because of the high intensity and focus of the laser beam, the labeled specimens undergo significant bleaching/fading. The bleaching effect is more severe than in conventional epifluorescent microscopy, where the entire specimen is observed under lower-power excitatory light. This results in accelerated degradation of fluorophores and makes longer observation periods impossible. Several factors influence fading, among them the fluorescence intensity, pH, and base solution of the embedding medium. Antifade reagents can slow down the degradation process of the fluorophore, resulting in longer periods of observation, fluorimetry, and pattern recognition [50].

P-phenylenediamine (PPD): This is one of the most effective antifade reagents, but it is photo/thermosensitive and cannot be used for *in vivo* studies because of its toxic effects. The optimal PPD antifade mixture consists of 90% glycerol, 10 PBS, PPD concentrations between 2 mM and 7 mM, pH 8.5–9.0 [51].

N-propylgallate (NPG): NPG is non toxic and photo- and thermostable. Though less effective as an antifade, it can be used for *in vivo* studies. Optimal working concentrations: glycerol base, 3 mM–9 mM NPG [51].

Vectashield, Vector Laboratories, Inc.,
30 Ingold Road, Burlingame, CA 94010, USA, Phone: (415) 697–3600, Fax: (415) 697–0339.

Slow Fade, Molecular Probes, (www.probes.com)
PO Box 22010, Eugene, OR 97402–0469, USA, Phone: (541) 465–8300, Fax: (541) 344–6504.

FluoroSave™: embedding (without glycerol) and antifade, Calbiochem, La Jolla, CA, USA.

6.5 Standard protocols for labeling

Direct labeling: Staining F-Actin with fluorescent phallotoxins (protocol from Molecular Probes [52], modified).

Preparation of stock solution. Fluorescent phallotoxins are available from Molecular Probes, The Netherlands or Eugene, Oregon. The content (5 µg) is dissolved in 1.5 ml methanol (200 units/ml). Molecular Probe defines one unit as the amount of material used to stain one microscopic slide of fixed cells. This stock solution can be stored at 4 °C. Prior to staining, the necessary volume is taken from the stock solution and transferred to a test tube, evaporated, and then dissolved again in the appropriate buffer immediately before use.

Protocol 7 Procedures for staining slides

This procedure has yielded consistent results in most instances. Modifications may be necessary for particular specimens. For cells grown on glass coverslips (for each coverslip):

1. Fix coverslips or slides in 4% paraformaldehyde in PBS pH 7.3 for 3 min. and then 10 min. at room temperature.
2. Wash several times with PBS.
3. Place the coverslip in a glass petri dish and permeabilize the cells with ice-cold acetone.
4. Air dry sample.
5. Evaporate 5 µl of the stock solution in a small test tube and dissolve it again in 150 µl of PBS. Place on the coverslip for 20 min. at room temperature.
6. Wash twice rapidly with PBS.
7. Mount the coverslip on a slide with the cell side down in FluoroSave.
8. Seal the edges of the coverslip with rubber cement.
9. Specimens prepared in this manner retain actin staining for at least 5 to 6 months when stored in the dark at 4 °C.

Indirect labeling

Protocol 8 Method for immunofluorescence on coverslips or slides

1. Fix coverslips or slides in 4% paraformaldehyde in PBS pH 7.3 for 3 min. and then for 10 min. at room temperature.
2. Wash several times with PBS.
3. To stain intracellular proteins, the membrane must be permeabilized prior to the labeling by rapid dehydration/rehydration using different alcohol concentrations (70%, 80%, 90%, 100%, 90%, 80%, 70%)
4. Wash once with PBS.

5. Block nonspecific staining by incubating in PBS + 0.1% bovine serum albumin (BSA) and normal serum from the host species of the secondary antibody, for example 5% natural goat serum (NGS) for 1 h at room temperature. To detect intracellular antigens, it is necessary to add 0.05% Saponin to this buffer and keep the BSA and Saponin through the procedure.
6. Incubate the primary antibody at room temperature between 1 and 3 h (depending on the characteristics of the primary antibody). Use approximately 100 ml per coverslip, more for slides. Wash several times with PBS, BSA, and Saponin if permeabilized.
7. Incubate with fluorescent-labeled second antibody for 1 h at room temperature in the dark to protect fluorescence (add 5% NGS to the secondary antibody solution).
8. Wash several times with PBS and once with distilled water.
9. Mount in Fluorosave™ (embedding and antifade reagent) or other embedding media containing antifade agents.

Preparation of solutions

- **4% paraformaldehyde in PBS**: Heat 8 g paraformaldehyde in 100 ml PBS until dissolved (must be prepared in a fume hood). Remove particles using a paper filter in a glass funnel. Adjust to the original volume (100 ml), then mix 1:1 with 2x PBS to obtain a 4% paraformaldehyde solution in PBS. (This fixative works best when used fresh.)
- **Phosphate buffered Saline (PBS), 10x**: Dissolve 2.0 g KCl, 2.0 g KH_2PO_4, 1.0 g $MgCl_2.6H_2O$, 80,0 g NaCl, 14.34 g $Na_2HPO_4.2H_2O$, and 1.0 g $CaCl_2$ (add last) in 1000 ml distilled water, adjust pH to 3. Before diluting PBS to use as 2x or 1x solution, adjust pH to 7.2 with 1M NaOH.

Multi-labeling

For multi-labeling of fixed cells with antibodies, the basic protocol must be performed twice. During the labeling, the coverslips must stay in PBS + 0.1% BSA + 5% NGS. It is important to use the right combination of primary and secondary antibodies. If primary antibodies of similar origin (e. g., mouse) are used, cross-reactions can occur during the detection with secondary antibodies. This can be avoided with appropriate blocking steps. If possible, use direct staining with directly labeled probes, such as Texas red-X-Phalloidin or DAPI.

List of primary and secondary antibodies used in the multi-labeling experiment to stain the activation of cytoskeletal proteins:

1. Labeling: Anti-Phosphotyrosine, polyclonal, rabbit IgG (#06–427, Upstate Biotechnology); secondary antibody: FITC (Fluorescein isothiocyanate) anti rabbit IgG (Ex = 496/Em = 516);
2. Labeling: Anti-paxillin, monoclonal, mouse IgG (#P17920, Transduction Laboratories); secondary antibody: TRITC (Tetramethylrhodamine isothiocyanate) anti-mouse IgG (Ex = 543/Em = 571).
3. Labeling: direct, for cytoskeletal F-actin: Texas red-X-Phalloidin, (Ex = 594/ Em = 608).

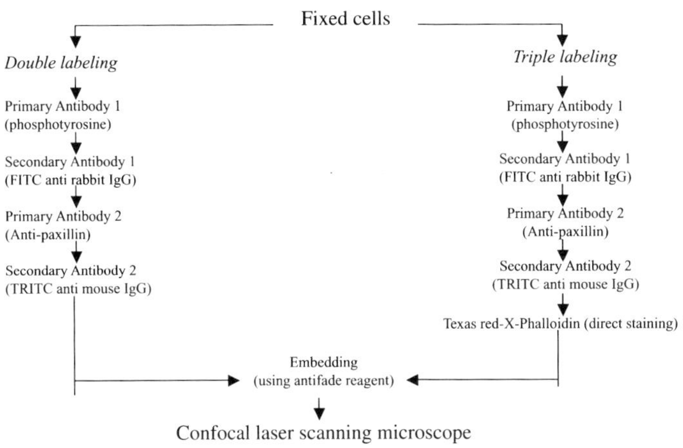

Figure 6

6.6 Conclusions

CLSM for multicolour immunofluorescence is a complex task involving detailed knowledge of immunohistochemistry, confocal microscopy operation and instrumentation, and image presentation and analysis. The development of CLSM systems with multi-wavelength excitation and powerful microcomputer systems allows confocal imaging to be used with triple-labeled specimens.

References

1 Catty D. (1989) Introduction. In: *Antibodies volume I – A practical approach.* The practical approach series, Rickwood D. Hames BD (eds.) IRL Press, Oxford Washington, 1–5

2 Berson SA, Yalow RS (1959) *J Clin Invest* 38: 1996

3 Gabor F, Pittner F, Spiegl P (1995) Drug-protein conjugates: Preparation of triamcinolone-acetonide containing bovine serum albumin/keyhole limpet hemocyanin-conjugates and polyclonal antibodies. *Arch Pharm* 328: 775–780

4 Gabor F, Hamilton G, Pittner F (1995) Drug-protein conjugates: Haptenation of 1-methyl-10α-methoxy-dihydrolysergol and 5-bromo nicotinic acid to albumin for the production of epitope-specific monoclonal antibodies against nicergoline. *J Pharm Sci* 84: 1120–1125

5 Grabarek Z, Gergely J (1990) Zero-length cross-linking procedure with the use of active esters. *Anal Biochem* 185: 131–135

6 Staros JV, Wright RW, Swingle DM (1986) Enhancement by N-hydroxysulfosuccinimide of water soluble carbodiimide-mediated coupling reactions. *Anal Biochem* 156: 220–222

7 Al-Bassam MN, O'Sullivan MJ, Gnemmi E et al. (1978) Double antibody enzyme immunoassay for nortriptyline. *Clin Chem* 24: 1590–1594

8 Riceberg LJ, Van Vunakis H, Levine L (1974) Radioimmunoassays of 3, 4, 5-trimethoxyphenethylamine (mescaline) and 2,5-dimethoxy-4-methylphenylisopropylamine (DOM). *Anal Biochem* 60: 551–559

9 Carlsson J, Drevin H, Axen R (1978) Protein thiolation and reversible protein-protein conjugation. N-succinimidyl-3-(2-propyldithio)-propionate, a new heterobifunctional reagent. *Biochem J* 173: 723–737
10 Erlanger BF, Borek F, Beiser SM, Liebermann S (1959) Steroid-protein conjugates: II. Preparation and characterisation of conjugates of bovine serum albumin with progesterone, deoxycorticosterone and estrone. *J Biol Chem* 234: 1090–1094
11 Houen G, Jensen OM (1995) Conjugation of preactivated proteins using divinylsulfone and iodoacetic acid. *J Immunol Meth* 181: 187–200
12 Butler VP, Chen JP (1967) Digoxin-specific antibodies. *Proc Nat Acad Sci USA* 57: 71–78
13 Habeeb AF (1966) Determination of free amino groups in proteins by trinitrobenzene sulfonic acid. *Anal Biochem* 14: 328–336
14 Butler VP (1977) The immunological assay of drugs. *Pharm Rev* 29: 103–184
15 Freund J (1956) The mode of action of immunologic adjuvants. *Adv Tub Res* 7: 130–148
16 Herbert WJ, Kristensen F (1986) Laboratory techniques for immunology. In: Weir, D.M. et al. (eds.) Handbook of experimental immunology. Blackwell Scientific, Oxford, 133.1–133.36
17 Cooper TG (1977) *The tools in biochemistry.* Wiley & Sons, New York, 259–263
18 Köhler G, Milstein C (1975) Continuos cultures of fused cells secreting antibody of predefined specificity. *Nature* 256: 495–497
19 Köhler G, Howe SC, Milstein C (1976) Fusion between immunoglobulin-secreting and onsecreting myeloma cell lines. *Eur J Immunol* 6: 292–295
20 de St Groth SF, Scheidegger D (1980) Production of monoclonal antibodies: strategy and tactics. *J Immunol Meth* 35: 1–21
21 Köhler G, Milstein C (1976) Derivation of specific antibody-producing tissue culture and tumor lines by cell fusion. *Eur J Immunol* 6: 511–519
22 Lane DP, Lane EB (1981) A rapid antibody assay system for screening hybridoma cultures. *J Immunol Methods* 47: 303–310
23 Goding JW (1980) Antibody production by hybridomas. *J Immunol Methods* 39: 285–308
24 Tijssen P (1985) Practice and theory of enzyme immunoassays. In: *Laboratory techniques in biochemistry and molecular biology* (RH Burdon, PH van Knippenberg (eds.) Volume 15, Elsevier, Amsterdam, Oxford, 59–62
25 Harlow E, Lane D (1988) *Antibodies – a laboratory manual.* Cold Spring Harbor, New York
26 Rowley GL, Rubenstein KE, Huisjen J, Ullman EF (1975) Mechanism by which antibodies inhibit hapten-malate dehydrogenase conjugates. *J Biol Chem* 250: 3759–3762
27 Engvall E, Perlmann P (1971) Enzyme linked immunosorbent assay (ELISA). Quantitative assay of immunoglobulin G. *Immunochemistry* 8: 871–877
28 Sugiura T, Imagawa H, Kondo T (2000) Purification of horse immunoglobulin isotypes based on differential elution properties of isotypes from protein A and protein G columns. *J Chromatogr B Biomed Sci Appl* 742: 327–334
29 Bolton AE, Lee-Own V, McLean RK, Challand GS (1979) Three different radioiodination methods for human spleen ferritin compared. *Clin Chem* 25: 1826–1830
30 Melamed MR, Mullaney PF, Shapiro HM (1990) An historical review of the development of flow cytometry and sorters. In: MR Melamed, T Lindmo, ML Mendelsohn (eds): *Flow cytometry and sorting.* John Wiley & Sons, Inc. New York, 1–9
31 Shapiro HM (1988) *Practical flow cytometry.* Alan R. Liss, Inc., New York
32 Darzynkiewicz Z, Crissman HA (eds) (1990) *Methods in cell biology, Vol. 33: Flow cytometry.* Academic Press, New York
33 Horan PK, Muirhead KA, Slezak SE (1990) Standards and controls in flow cytometry. In: MR Melamed, T Lindmo, ML Mendelsohn (eds): *Flow cytometry*

and sorting. John Wiley & Sons, Inc, New York, 397–414
34 Kachel V, Fellner-Feldegg H, Menke E (1990) Dynamic properties of flow cytometry instruments. In: MR Melamed, T Lindmo, ML Mendelsohn (eds): *Flow cytometry and sorting*. John Wiley & Sons, Inc., New York, 27–44
35 Sharpless T, Traganos F, Darzynkiewicz Z, Melamed MR (1975) Flow cytometry: discrimination between single cells and cell aggregates by direct size measurement. *Acta Cytologica* 19: 577–581
36 Shapiro HM (1991) Quantitative immunofluorescence measurements and standards: practical approach. *Cli. Immunol Newsletter* 11: 49–64
37 Weir DM , Herzenberg LA, Blackwell C (1986) *Handbook of experimental immunology, 4th edition,* Blackwell Sci. Publ., Oxford
38 Loken MR, Parks DR, Herzenberg LA (1977) Two color immunofluorescence using a fluorescence-activated cell sorter. *J Histochem Cytochem* 25: 899–907
39 Vindelov LL, Christensen IJ, Nissen NI (1983) A detergent-trypsin method for the preparation of nuclei for cytometric DNA analysis. *Cytometry* 3: 232–237
40 Landay AL, Muirhead KA (1989) Procedural guidelines for performing immunophenotyping by flow cytometry. *Clin Immunol Immunopathol* 52: 48–60
41 Landay AL (1989) Clinical applications of flow cytometry: quality assurance and immunophenotyping of peripheral blood lymphocytes. *Natl Comm Clin Lab Stand* 9(13)
42 Matsumoto B (1993) *Cell biological applications of confocal microscopy*. Academic Press, San Diego, CA
43 Pawley JB (ed.) (1995) *Handbook of biological confocal microscopy*. Second Edition. Plenum, New York
44 Brelje TC, Wessendorf MW, Sorenson RL (1993) Multicolor laser scanning confocal immunofluorescence microscopy: Practical application and limitations. In: B Matsumoto (ed): *Cell biological applications of confocal microscopy*. Academic Press, San Diego, CA, 98–181
45 Lemasters JJ, Chacon E, Zahrebelski G, et al. (1993) Laser scanning confocal microscopy of living cells. In: B Herman, JJ Lemasters (eds): *Optical microscopy:emerging methods and applications*. Academic Press, San Diego, CA 339–354
46 Sheppard CJR (1993) Confocal microscopy – principles, practice and options. In: WT Mason (ed): *Biological techniques: Fluorescent and luminescent probes for biological activity. A practical guide to technology for quantitative real-time analysis*. Academic Press, San Diego, CA, 229–236
47 Sheppard CJR, Cogswell CJ, Gu M (1991) Signal strength and noise in confocal microscopy: Factors influencing selection of an optimum detecture aperture. *Scanning* 13: 233–240
48 Tsien RY, Waggoner A (1995) Fluorophores for confocal microscopy: photophysics and photochemistry. In: J Pawley (ed.): *Handbook of biological confocal microscopy*. Second Edition. Plenum, New York, 267–279
49 Mason WT (1993) *Biological techniques: Fluorescent and luminescent probes for biological activity. A practical guide to technology for quantitative real-time analysis*. Academic Press, San Diego, CA
50 Longin A, Souchier C, French M, Bryon PA (1993) Comparison of anti-fading agents used in fluorescence microscopy: image analysis and laser confocal microscopy study. *J Histochem Cytochem* 41: 1833–1840
51 Krenik KD, Kephart GM, Offord KP, et al. (1989) Comparison of antifading reagents used in immunofluorescence. *J Immunol Methods* 117: 91–97
52 Haughland RP (1996) *Molecular probes: handbook of fluorescent probes and research biochemicals. Sixth Edition*. Molecular Probes, Eugene, OR

4 Immunological Strip Tests

Ron Verheijen

Contents

Methods and Tools in Biosciences and Medicine
Analytical Biotechnology, ed. by Thomas G.M. Schalkhammer
© 2002 Birkhäuser Verlag Basel/Switzerland

1 Introduction

One of the main goals of investigators in the fields of (clinical) diagnostics and contamination-control programs consists of providing new analytical strategies to obtain accurate data as quickly as possible.

When a new screening test is requested, the method has to be sensitive, specific, fast, cheap, and easy to perform. In general, immunological techniques (immunoassays) can fulfill these high requirements to a great extent. Over the past decade, single-use lateral flow immunoassays have been extremely successful in the laboratory and in outpatient clinic and primary care environments. In this type of assay, all reaction components are impregnated or immobilised on a porous solid phase, usually a nitrocellulose membrane, and are brought into contact with the sample in sequence after addition of a diluent [1–3]. This immunoassay format is also known as strip test, one-step strip test, immunochromatographic test, rapid flow diagnostic, rapid immunoassay (test), lateral flow immunoassay (LFI), on-site test (assay) or near-patient test (NPT). In this chapter, we will use the term strip test.

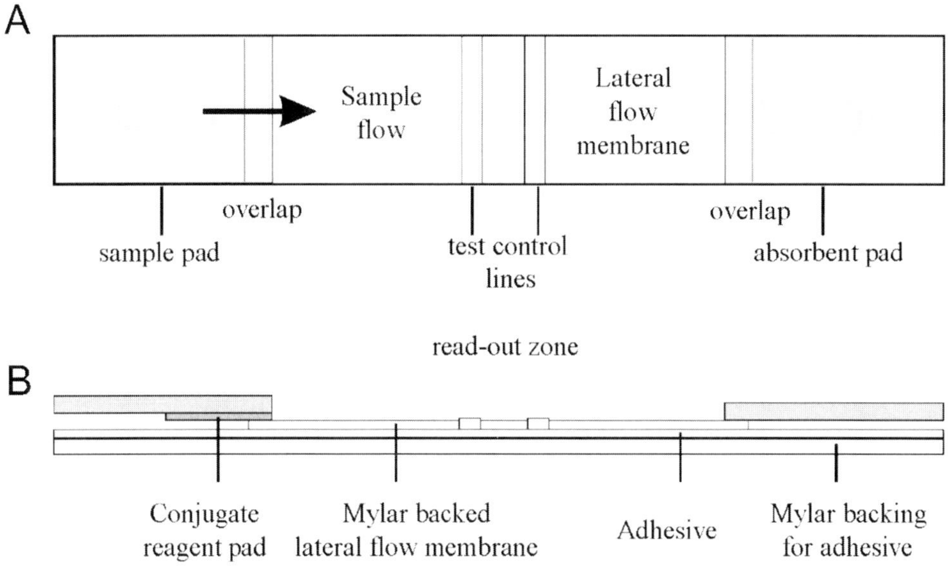

Figure 1 Schematic of a typical test strip, including a Mylar backing plate with adhesive, a sample pad, a conjugate pad, an absorbent pad, and a membrane that incorporates the capture reagents. A. Top view, B. Side view. A complete strip test device consists of a test strip as shown here, packed in a plastic cassette (housing) as shown in Figure 2.

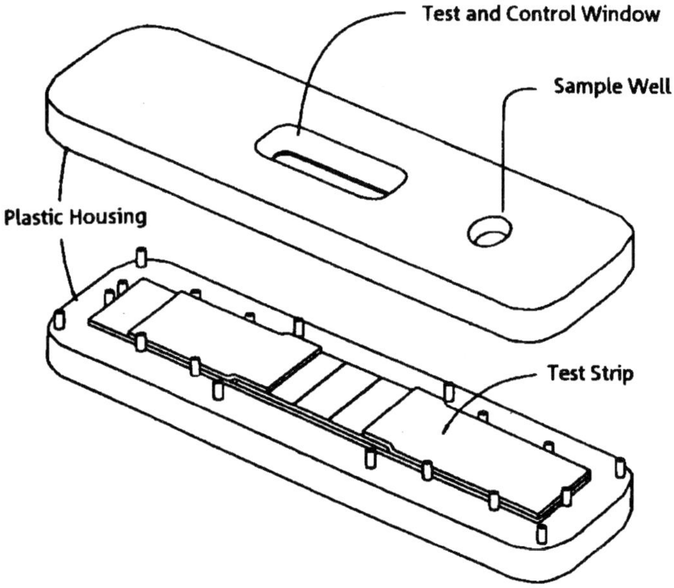

Figure 2 Example of a plastic cassette (housing) used in a strip-test device.

Immunoassays are analytical measurement systems that use antibodies as test reagents. The antibodies are attached to some kind of label and then used as reagents to detect the substance of interest. This label can be either an enzyme for colorimetric detection or a colored collodial particle such as gold (red) [4], carbon (black), silica (several colours), or latex (several colours) for direct visualization of the immunoreaction. In practise, collodial gold particles or gold nanoclusters having a diameter of 25 to 40 nm are probably the most commonly applied labels in strip tests. For this reason, we will focus on the use of such gold nanoclusters.

A typical strip test device consists of a test strip as shown in Figure 1 that is mounted in a plastic cassette (housing) as shown in Figure 2. The test strip is made up of a number of components, including a sample pad, a conjugate pad, an absorbent pad, and a membrane that contains the capture reagents. The main purpose of the housing is to fixate the several components of the test strip and to keep them in close contact with each other. Moreover, the housing determines the dimensions of the sample well and contains the viewing window with readout indications.

When describing the principle of the strip test, one has to distinguise between the direct assay to detect high-molecular-mass components, usually proteins (Fig. 3), and the indirect or competitive assay to detect low-molecular-mass analytes such as drug residues, antibiotics, hormones, etc. (Fig. 4). In both type of tests, the user dispenses a liquid sample (buffer extract, milk, urine, serum, plasma, whole blood, etc.) onto the sample pad. The sample then flows through the sample pad into the conjugate pad, where it releases and mixes with the detector reagent.

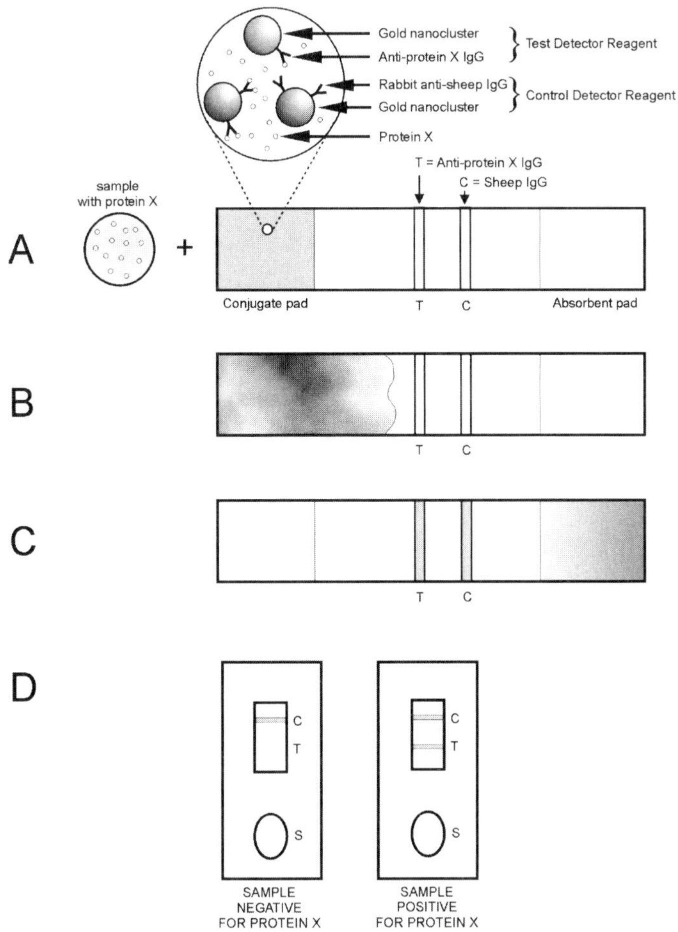

Figure 3 Principle of a strip test for the detection of high-molecular mass-analytes (proteins).
A. Fluid sample is brought onto the sample pad (not shown) and migrates to the conjugate pad underneath. This conjugate pad contains two distinct detector reagents: gold nanoclusters coated with specific anti-protein X IgG (test detector reagent) and gold nanoclusters coated with rabbit anti-sheep IgG (control detector reagent). Protein X present in the sample is bound by the anti-protein X IgG (test detector reagent).
B. The control detector reagent, the protein-X-bound test detector reagent, the unbound protein X, and/or the unbound test detector reagent migrate up the strip with the sample.
C. As the sample passes over the capture zones, the protein X bound to the test detector reagent is captured by the immobilized anti-protein X IgG at line T, whereas the rabbit anti-sheep IgG (control detector reagent) binds to the sheep IgG (control capture reagent) at line C. Note that the immobilized capture antibodies to protein X have to be directed to epitopes other than the detector antibodies to protein X on the gold clusters.
D. Reading the results: No colour is seen at the capture line T if the sample does not contain protein X (negative sample), whereas colour develops if the sample contains protein X (positive sample). Colour develops at the control line (C) if the test has been used properly.

In the direct assay format for detecting high-molecular-mass analytes such as proteins (Fig. 3), the test detector reagent consists of gold nanoclusters coated with specific antibodies to the protein of interest. When that particular protein is present in the sample it will react with the test detector reagent (Fig. 3 A). The formed protein-antibody-gold complexes are mobile and are able to move freely from the reagent pad into the membrane with the flow of the fluid (Fig. 3B). At the test line, the complexes will be captured by immobilized anti-protein antibodies (Fig. 3C). Thus, the presence of the protein of interest in the sample will result in a colored test line (positive sample in Fig. 3D). The color intensity of the test line is proportional to the concentration of the analyte in the sample. When the concentration of the protein of interest is lower than the lowest detection concentration or when the protein of interest is completely absent, no test line will be visible (negative sample in Fig. 3D). Excess sample that flows beyond the test and control lines is taken up in the absorbent pad (Fig. 3C).

In the competitive assay format for detecting low-molecular-mass analytes (Fig. 4), the test detector reagent consists of gold nanoclusters coated with antibodies to the analyte of interest. The test line here consists of a protein-analyte conjugate. The more analyte present in the sample, the more effectively it will compete with the immobilised analyte on the membrane for binding to the limited amount of antibodies of the detector reagent. Thus, the absence of the analyte in the sample will result in a colored test line (negative sample in Fig. 4D), whereas an increase in the amount of analyte will result in a decrease of signal in the readout zone (positive sample in Fig. 4D). At a certain concentration of analyte in the sample, the test line will be no longer visible. The lower detection limit (LDL) is defined as the amount of analyte in the sample that just causes total invisiblity of the test capture line.

Strip-test devices are commercially available for an increasing number of antigens (high and low molecular mass). The first major target analyte for this test format was human chorionic gonadotropin (HCG) for the detection of pregnancy. At present, several test strip assays are available [5–8], e.g., for the detection of hormones (pregnancy, fertility, ovulation, menopause, sexual disorder, thyroid functions), tumour markers (prostate, colorectal, etc.), viruses (HIV, hepatitis B and C), bacteria (*Streptococcus* A and B, *Chlamydia trachomatis, Treponema pallidum, Heliobacter pylori*, etc.), IgE (allergy), and troponin T in cardiac monitoring [9]. All these analytes are measured on the basis of their presence or absence. An extensive review of near-patient testing in primary care has been published by Hobbs et al [8].

So far, these tests are intended mainly for human diagnostics where, especially for the low-molecular-mass analytes, relatively high concentrations of analytes are measured. In food diagnostics, however, much lower detection levels are required. For the detection of veterinary drug residues, for example, detection levels at ppb level (ng analyte/g of sample) are necessary [10–12], as for many of these drugs a maximum residue level (MRL) has been established.

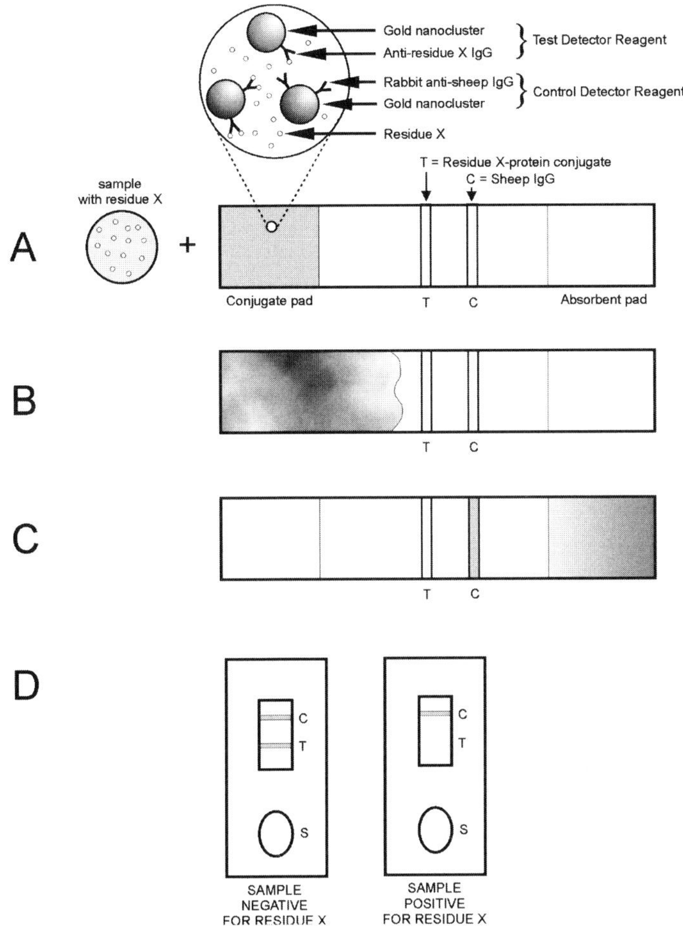

Figure 4 Principle of a strip test for the detection of low-molecular mass-analytes.
A. Fluid sample is brought on to the sample pad (not shown) and migrates to the conjugate pad underneath. This conjugate pad contains two distinct detector reagents: gold nanoclusters coated with specific anti-residue X IgG (test detector reagent) and gold nanoclusters coated with rabbit anti-sheep IgG (control detector reagent). Residue X present in the sample is bound by the anti-residue X IgG (test detector reagent).
B. The control detector reagent, the residue-X-bound test detector reagent, the unbound residue X, and/or the unbound test detector reagent migrate up the strip with the sample.
C. As the sample passes over the capture zones, the anti-residue X IgG of the test detector reagent that is free of residue X is captured by the immobilized residue-X-protein conjugate (test capture reagent) at line T, whereas the rabbit anti-sheep IgG (control detector reagent) binds to the sheep IgG (control capture reagent) at line C.
D. Reading the results: Colour develops at the capture line T if the sample contains no residue X (negative sample), whereas no colour is seen if the sample contains residue X (positive sample). Colour develops at the control line (C) if the test has been used properly.

Although strip tests appear simple, complex interactions among their various components lead to a number of challenges in both the development and manufacturing environments. These challenges are even greater for quantitative tests where the color intensity of the test line must be repeatable between production lots. From a developmental perspective, creating a successful test system means optimizing the interactions among its raw materials, component design, and manufacuring techniques.

It is beyond the scope of this manual to exhaustively describe all possible methods and materials that can be used for the development and manufacturing of strip tests. We will merely describe our experience in making good and reliable strip tests for the detection of both low- and high-molecular-mass analytes. Because of this limitation, some additional reading on new and improved methods is recommended (see Further reading section). Moreover, on a regular basis, BioDot in cooperation with Schleicher & Schuell organize introductory rapid test workshops in Europe, the U.S., and Asia (for information see http://www.biodot.com).

2 Materials

2.1 Materials

Amersham Pharmacia Biotech AB (Uppsala, Sweden):
- CNBr-activated Sepharose 4B (17–0430–01)
- HiTrap Protein G (1 × 5 ml) (17–0405–01)

Amicon (Beverly, MA, USA):
- Centriprep-30 concentrator (4306)

BioDot Inc. (Irvine, Ca, USA):
- BioDot Surfactant Starter Kit
- BioDot Diagnostic Materials Kit

Bio-Rad Laboratories B.V. (Veenendaal, The Netherlands):
- Two-way stopcock (732–8102)

G&L (San Jose, Ca, USA):
- 0.01″ white matte vinyl GL-187 (6.3 cm × 30 cm) (991093)

Gelman Sciences (Ann Arbor, MI, USA):
- 5 µm Acrodisc filter (4489)

Greiner Bio-One B.V. (Alphen a/d Rijn, The Netherlands):
- 50 ml test tubes (210261)

Kenosha C.V. (Amstelveen, The Netherlands):
- Plastic cassettes (housings): 16 × 71 mm for test strips of 5 × 63 mm
- Foil pouches: OPA15/PE15/ALU12/ LDPE75 MU (80 × 144 mm); 403389

Multisorb Technologies Inc (Buffalo, NY, USA):
- MiniPax (sachet containing 1 g of silica gel dessicant)

Millipore Corporation (Bedford, MA, USA):
- MilliQ 185 Plus water purification system

Nalge Nunc International (Rochester, NY, USA):
- Drop-Dispenser bottle (15 ml; 2411–0015)

Pall Gelman Sciences (Champs-sur-Marne, France):
- Cellulose type 133 (8″ × 10″) (AC1071)
- Cytosep 1660 (8″ × 10″) (AC1081)

Pierce (Rockford, IL, USA):
- BCA protein assay reagent (23225)
- PharmaLink Immobilization Kit (44930)

Schleicher & Schuell (Dassel, Germany)
- Nitrocellulose membrane strips (25 × 300 mm) AE 100 12 μm on Mylar 5 backing
- Cellulosic paper GB002 (20 × 20 cm) (10 426 681)
- Cellulose acetate filter FP 030/3 (0.2 μm) (462 200)

Serva Feinbiochemica (Heidelberg, Germany):
- Visking dialyzing tubing; 20/32, ∅ 16 mm (44110)

Varian (Harbor City, CA, USA):
- Varian Bond Elut Reservoir (8 ml) (1213–1015)

Whatman International Ltd (Maidstone, UK):
- Coarse PVA Bound Glass Fibre, grade F 075–17 (A4-size; 676475)
- Cellulosic paper 3MM Chr (20 × 25 cm; 3030–866)
- Cellulosic paper 3MM D28 (80048)

2.2 Chemicals

Caltag Laboratories (Burlingame, Ca, USA):
- Affinity purified goat anti-rabbit IgG (H + L) antibodies (1.25 mg/ml)

Marvel, Premier Brands UK (Moreton, UK):
- Skimmed milk powder

Merck (Darmstadt, Germany):
- Acetic acid (100%) (CH_3COOH; 100063)
- Disodium hydrogen phosphate dihydrate ($Na_2HPO_4.2H_2O$; 106580)
- Glutaraldehyde (25% w/w solution in water; 820603)
- Glycerol (extra pure $C_3H_8O_3$; 104093)
- Hydrochloric acid (32%) (HCl; 100319)
- Methanol (CH_3OH; 106009)
- Potassium carbonate (extra pure K_2CO_3; 104924)
- Potassium dihydrogen phosphate (KH_2PO_4; 4873)
- Polyvinylpyrrolidone (PVP; 107370)
- Sodium acetate (CH_3COONa; 106268)
- Sodium chloride (NaCl; 106404)
- Sodium dihydrogen phosphate monohydrate ($NaH_2PO_4.H_2O$; 106346)

- Sodium ethylene diamine tetra acetate (Na-EDTA, $C_{10}H_{14}N_2Na_2O_8.2H_2O$; 108421)
- Sodium hydrogencarbonate ($NaHCO_3$; 106329)
- Sodium hydroxide (NaOH; 106495)
- Tri sodium citrate dihydrate ($Na_3C_6H_5O_7.2H_2O$; 112005)
- Tris(hydroxymethyl)aminomethane ($C_4H_{11}NO_3$; 108382)
- Tween 20 (Polyoxyethylene sorbitan monolaurate; 822184)
- Sucrose ($C_{12}H_{22}O_{11}$; 107651)

Pragmatics Inc (Elkhart, In, USA):

- Surfactant 10G (p-Isononylphenoxypolyglycidol; 50% aqueous solution)

Serva Feinbiochemica (Heidelberg, Germany):

- Sodium dodecylsulfate (SDS, $C_{12}H_{25}O_4S.Na$; 20760)

Sigma Chemical Company (St Louis, MO, USA):

- 1,4-Butanediol diglycidyl ether ($C_{10}H_{18}O_4$; B 1029)
- Bovine Gamma Globulins (BGG; G 5009)
- Bovine Serum Albumin, fraction V (BSA; A 7888)
- Sodium azide (NaN_3; S 2002)
- Ovalbumin grade V (A 5503)
- Sulfadimidine (SDD, 4-amino-N-[4, 6-dimethyl-2-pyrimidinyl] benzene-sulfonamide = sulfamethazine; S 6256)
- Streptomycin sulfate (S-6501)
- Tetra-chloroauric[III] acid trihydrate ($HAuCl_4.3H_2O$; G 4022)

2.3 Equipment

The BioDot system (BioDot Inc., Irvine, Ca, USA) consisted of two BioJet Quanti3000 dispensers attached to a BioDot XYZ3000 dispensing Platform. An AZCON Sur-Size™ automatic guillotine cutter (Model SS-4) was supplied by AZCON (Elmwood Park, NJ, USA). Sealing equipment (Magneta, Model 421) was purchased from Kenosha (Amstelveen, The Netherlands).

2.4 Solutions, reagents and buffers

Solution 1	Phosphate Buffered Saline (PBS): 5.39 mM Na_2HPO_4; 1.29 mM KH_2PO_4; 153 mM NaCl; pH 7.4
Solution 2	Saturated ammonium sulfate solution: 80 g $(NH_4)_2SO_4$ in 100 ml of water of 20 °C
Solution 3	IAC coupling buffer: 0.1 M $NaHCO_3$ pH 8.3 containing 0.5 M NaCl
Solution 4	IAC blocking buffer: 0.1 M Tris-HCl buffer pH 8.0
Solution 5	IAC washing buffer I: 0.1 M Tris-HCl buffer pH 8.0 containing 0.5 M NaCl
Solution 6	IAC washing buffer II: 0.1 M sodium acetate–acetic acid buffer pH 4.0 containing 0.5 M NaCl

Solution 7 IAC elution solution: 0.1 M acetic acid
Solution 8 Retentate mixing buffer in synthesis BSA-streptomycin conjugate (0.1 M potassium phosphate buffer pH 8.0): 26.5 ml 0.2 M KH_2PO_4 and 473.5 ml 0.2 M K_2HPO_4 is adjusted to 1 l with water
Solution 9 Strip-test membrane-blocking buffer: 0.9 g/l $NaH_2PO_4.H_2O$ pH 7.5 containing 2% (w/v) skimmed milk powder and 0.02% (w/v) sodium dodecylsulfate (SDS)
Solution 10 Strip test membrane washing buffer: 0.9 g/l $NaH_2PO_4.H_2O$ pH 7.5 containing 0.01% (v/v) Surfactant 10G (S24 from the BioDot Surfactant Starter Kit)
Solution 11 Extraction buffer for food products and plant leaves: PBS pH 7.4 containing 0.5% (v/v) Tween 20 and 0.5% (w/v) polyvinylpyrrolidone (PVP)

3 Methods

3.1 Antibodies

Antibodies form the heart of the immunological strip-test device. In practise, they will be obtained either from an animal serum (polyclonal antibodies) or from a culture supernatant (monoclonal antibodies). Both types of antibodies should be purified at least to some extent before being applied in a strip test. For the control capture antibody, an ammonium sulfate precipitation to isolate the total immuunglobulin fraction (Ig fraction) often will be sufficient. For the test and control detector reagents, as well as for the test capture reagent in the direct assay format, however, merely antibodies that have been purified by immunoaffinity chromatography (IAC), should be used. For an even better performance of the test, one may decide to use antigen-specific purified antibodies as test reagents. These can be prepared by additional IAC of the IgG fraction using a column in which the antigen has been immobilised onto a chromatographic support. Figure 5 shows the various steps in the production and purification of both polyclonal and (mouse) monoclonal antibodies.

It is important to determine the cross-reactivities of the antibodies with the sample matrices to be used in the assay. Moreover, knowledge about the behaviour and properties of the antibodies in a specific sample matrix is essential in order to obtain an optimal performing strip test. When testing urine samples, for example, one has to know the binding properties of the antibodies at the wide range of possible pH values in such samples.

It goes without saying that in a direct assay where monoclonal antibodies are used for the detector and capture reagent, each of the two antibodies should be directed against a different epitope on the antigenic protein. When using polyclonal antibodies, however, the same pool of antibodies may be applied for both the detector and the capture reagent.

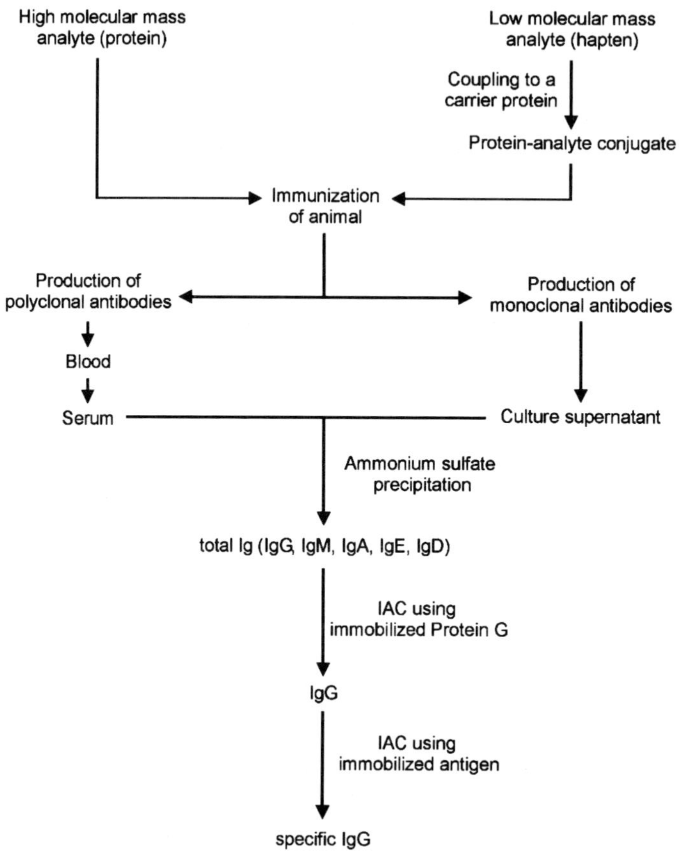

Figure 5 Flowchart showing the various steps in the production and purification of both polyclonal and monoclonal antibodies.

3.2 Isolation of total Ig by ammonium sulfate precipitation

Protocol 1 Isolation of total Ig by ammonium sulfate precipitation

1. When starting from a crude rabbit or sheep serum, 10 ml of serum is first diluted with 20 ml of PBS (Solution 1), after which 30 ml of a saturated ammonium sulfate solution (Solution 2) is added drop-wise with constant stirring at ambient temperature.
2. When starting from a culture supernatant, an equal volume of a saturated ammonium sulfate solution is added directly with constant stirring at ambient temperature.
3. After standing for 30 min., the antibodies are collected by centrifugation for 10 min. at 10,000 × g at ambient temperature. The use of a swinging-bucket rotor is recommended.
4. After discarding the supernatant as completely as possible, the pellet is dissolved in as small a volume of PBS as possible (a few ml).

5. The solution is then exhaustively dialysed against PBS at 4 °C for at least 48 h
6. The dialysed fraction can then be concentrated by using a centriprep-30 concentrator according to the manufacturer's user guide.
7. Finally, the protein content is determined, e.g., by performing a BCA protein assay according to the manufacturer's user guide and using bovine gamma globulins (BGG) as standard.

Notes:

- When isolating sheep IgG for the control capture line, the dialysed fraction is now purified sufficiently and can be stored in small aliquots at –20 °C until used. For other applications the antibodies should be further purified by IAC.
- Before use, some dialysis tubing has to be boiled for 5–10 min. in the presence of sodium ethylene diamine tetra acetate (Na-EDTA). The obtained sterile tubing can be stored for several months at 4 °C in 50% (v/v) glycerol in water.

3.3 Isolation of IgG by immunoaffinity chromatography (IAC)

HiTrap Protein G columns (Amersham Pharmacia Biotech) contain Protein G immobilised to Sepharose High Performance as chromatographic support. Like Protein A, Protein G binds specifically to the Fc region of IgG. However, the range of polyclonal IgGs that bind strongly to protein G is much wider, and adsorption takes place over a wider pH range. This makes IAC using Protein G ideal for fast purificaton of specifically the IgG fraction from a serum or culture supernatant. For the use of HiTrap Protein G columns, the manufacturer's user guide should be followed.

Note: For the production of (mouse) monoclonal antibodies, the hybridoma cells are often cultured in the presence of 1% to 10% foetal calf serum. This implies that the antibody fraction obtained after IAC purification on a HiTrap Protein G column of such samples will contain a relatively large amount of bovine IgG as well.

3.4 Isolation of specific antibodies by IAC

IAC using immobilised antigenic proteins

CNBr-activated Sepharose 4B is a preactivated gel for immobilization of ligands containing primary amines such as proteins. Such a gel with immobilized proteins can be used to isolate specific antibodies to the proteins from a serum or culture supernatant.

Protocol 2 Preparation of the column

1. 2 g freeze-dried powder of CNBr-activated Sepharose 4B is swollen in 15 ml 1 mM hydrochloric acid for about 10 min.
2. The swollen gel is transferred into an 8 ml Varian column with a porous bed support (frit) at the bottom of the colomn and sealed with a two-way stopcock.
3. The gel is subsequently washed with 400 ml 1 mM hydrochloric acid and 100 ml coupling buffer (Solution 3).
4. The gel is transferred into a 50 ml Greiner tube and mixed with 7.5 ml coupling buffer containing 20 mg of the proteins of interest (5–10 mg protein/ml gel).
5. The coupling solution is then gently rotated end over end for 2 h at ambient temperature or overnight at 4 °C. Do not use a magnetic stirrer!
6. After coupling, the gel is either centrifuged for 2 min. at 200 × g or allowed to settle out for 5–10 min.
7. The supernatant is removed and the gel is washed three times with 30 ml coupling buffer, each wash followed by either centrifugation or standing at the bench for 5–10 min.
8. After removing the last washing solution, any remaining active groups are blocked by adding 15 ml blocking solution (Solution 4) to the gel and gently rotating the mixture end over end for 2 h at ambient temperature.
9. The gel is then transferred to the column, after which the upper-frit is placed.
10. The gel is washed with three cycles of alternating pH. Each cycle consists of a wash with 20 ml washing buffer I (Solution 5) followed by a wash of 20 ml washing buffer II (Solution 6).
11. Finally, the column is washed with 30 ml PBS, pH 7.4.

Protocol 3 Binding of antibodies

1. The column is loaded with ammonium-sulfate-precipitated and HiTrap Protein G immunoaffinity-purified antibodies in PBS, pH 7.4. The capacity of the column is high enough to cope with an amount of total IgG originating from 25 ml crude animal serum. The sample is loaded three times onto the column.
2. The column is washed with 20 ml PBS, pH 7.4

Protocol 4 Elution of antibodies

1. Bound antibodies are eluted with 15 ml elution solution (Solution 7).
2. The eluate is immediately brought at neutral pH by addition of 1 M sodium hydroxide.
3. The eluate is concentrated to about 5 ml by using a centriprep-30 concentrator, according to the manufacturer's user guide.
4. Finally, the protein content is determined, e. g., by using the BCA protein assay according to the manufacturer's user guide.

Storage:

- The column is stored at 4 °C in PBS, pH 7.4, containing 0.05% (w/w) sodium azide.

Notes:

- Be sure that none of the reagents in the coupling reaction contains sodium azide, as this will seriously disturb the reaction.

IAC using immobilised small ligands
For the purification of antibodies to low-molecular-mass ligands, the ligand of interest has to be coupled to a chromatographic support. Because the chemical nature of small ligands usually varies widely, there is no such thing as a general immobilization protocol. Often, coupling of small ligands to a support appears to be far more difficult than coupling of a protein as described in the section above. An overview of small ligand immobilization techniques is described in [20].

The PharmaLink Immobilization Kit enables the coupling of a large variety of difficult-to-immobilize compounds. It utilizes the Mannich reaction in which certain active hydrogens of the ligand are condensed with formaldehyde and an amine-containing support (the PharmaLink gel). Particular hydrogens in ketones, esters, phenols, acetylenes, α-picolines, quinaldines, and a host of other compounds can be aminoalkylated using the Mannich reaction. For an overview of the possibilities of the kit and a step-by-step guide to the procedure, see the manufacturer's instruction manual.

When the ligand of interest contains an alifatic amine group, coupling can be performed easily by using CNBr-activated Sepharose 4B as described in section 3.4 (protocol 2), now adding 1 to 10 μmoles of the ligand/ml gel. When the ligand contains an aromatic amine group, however, coupling is more successful when using a modified PharmaLink protocol as described by Verheijen et al. [11]. The example describes the immobilization of sulfadimidine (SDD; sulfamethazine), a sulfonamide drug that is coupled by its aromatic amine group.

Note:

- Be sure that none of the reagents in the coupling reaction contains sodium azide, as this will seriously disturb the reaction.

3.5 Preparation of 40 nm gold nanoclusters (G40)

Preparation of gold nanoclusters is based upon reduction of $HAuCl_4$ and has been described for a wide variety of reagents, including formaldehyde, white phosphorus, citric acid, ascorbic acid, tannic acid, and hydrogen peroxide [13, 14]. The size of the individual nanoclusters can be manipulated easily and depends on several factors such as the chemical nature of the reducing reagents, temperature, pH, and concentration of the reagents. Protocols have been published for the preparation of colloidal gold particles with a diameter ranging from 3 to 150 nm. The most commonly used method is the reduction of gold chloride with sodium citrate as described by Frens [15].

The procedure for preparing a gold nanoclusters with an average particle diameter of 40 nm (G40) is as follows:

Protocol 5 Preparation of gold nanoclusters (40 nm)

1. 100 ml 0.01% (w/v) tetra-chloroauric[III]acid trihydrate ($HAuCl_4.3H_2O$) is heated to boiling under reflux conditions.
2. 1 ml of 1% (w/v) trisodium citrate dihydrate is added to the boiling solution under constant stirring.
3. In about 25 s the slightly yellow solution will turn faintly blue (nucleation). After approximately 70 s the blue colour then suddenly changed into dark red, indicating the formation of monodisperse sperical particles.
4. The solution is boiled for another 5 min. to complete reduction of the gold chloride. The optical density, measured at 540 nm (OD540), of such G40 suspensions will be about 0.9.
5. Supplemented with 0.05% (w/v) sodium azide, the obtained G40 suspensions can be stored at 4 °C for several months.

Notes:

- All reagents should be prepared in highly purified water, e. g., by using a Milli-Q 185 Plus water purification system (Millipore) yielding water with a resistivity of > 18.2 MΩ.cm.
- Another frequently used particle size of gold nanoclusters is that of 25 nm. For the preparation of such particles, the amount of added trisodium citrate solution should be increased to 1.5 ml. A G25 suspension will be somewhat lighter red in colour than a G40 solution.
- Suppliers of gold nanoclusters are:

British BioCell International, Golden Gate, Ty Glas Avenue,Cardiff CF4 5DX, UK, http://www.british-biocell.co.uk, e-mail: info@britishbiocell.co.uk; and Aurion, Costerweg 5, 6702 AA Wageningen, The Netherlands, http://www.aurion.nl, e-mail: info@aurion.nl

3.6 Preparation of the detector reagents

The test and control detector reagents consist of G40 nanoclusters coated with the appropriate antibodies. The amount of antibody attached to the individual gold nanocluster depends upon many factors, including the size of the gold nanocluster, the pH at which the coupling is performed, the concentration of the antibody added, and the electrical charge of the antibody [16, 17]. Coupling will take place only if the negative charge of the unstabilized gold nanocluster can be compensated by a net positive charge of the antibody. Because of the lability of unstabilized gold nanoclusters and its sensitivity to salts, the salt concentration in the coupling solution should be kept as low as possible. The amounts of electrolytes necessary to induce flocculation depend upon the valences of the particular ions. In order to eliminate the same amount of negative surface charge of colloidal gold and to induce flocculation, cations such as K^+, Ba^{2+}, and Al^{3+} have to be added in equivalents amounts of 1000:10:1 to cause the same effect.

The minimal protecting amount (MPA), i. e., the minimal amount of protein needed to protect the sol from salt-induced precipitation, can be determined by the procedure of Horisberger and Rosset [18]. In this procedure, small amounts of unstabilized gold nanoclusters (0.5 to 1.0 ml) are added to a series of increasing concentrations of the protein of interest in a constant volume (100 μl). After 5 min., 100 μl of a 10% (w/v) sodium chloride solution is added to the mixture. If the colour changes from red to violet and finally to blue, the protection was incomplete and aggregation, followed by flocculation of unstabilized gold nanoclusters, was induced by the added salt. Spectrophotometric analysis of a suspension of gold nanoclusters at 520 nm results in a more precise estimate of the degree of stabilization of the sol. In general, a final concentration of 10% to 100% above the MPA is used for coating.

When using the G40 nanoclusters made in our lab, the MPA was always found to vary around 5 μg antibody/ml G40 suspension. Coating was therefore performed with 6.0 to 7.5 μg antibody per ml of G40 suspension.

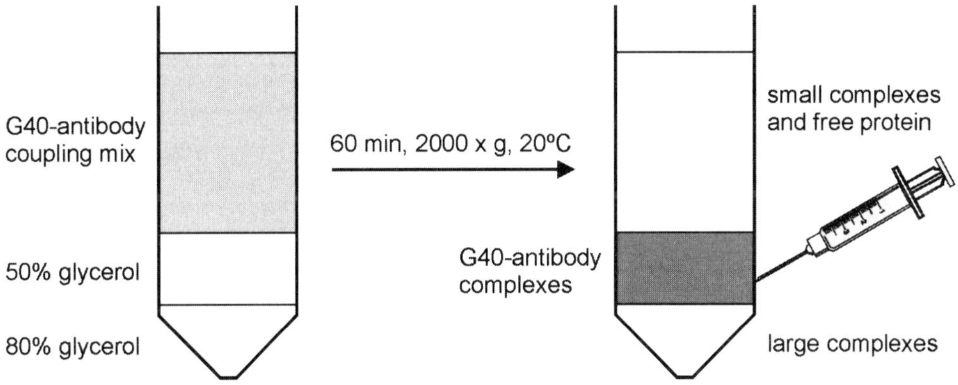

Figure 6 Isolation and purification of the detector reagents. After coating of the G40 gold nanoclusters with antibodies, followed by blocking with milk proteins, the obtained mixture is centrifuged through a 50% (v/v) glycerol layer that is mounted onto a 80% (v/v) glycerol cushion. The centrifugal speed and/or running time is adjusted such that the right-sized coated antibody clusters forming the detector reagents are found almost entirely in the 50% glycerol layer. After centrifugation, antibodies and/or blocking proteins that are not bound to the nanoclusters (free proteins), as well as relatively small protein nanocluster complexes, will be found in the upper water layer, whereas the relatively large, purple-colored protein nanocluster complexes will be present in the 80% glycerol layer and on the bottom of the centrifuge tube. In order to avoid contamination of the detector reagent with free protein and/or large antibody nanocluster complexes, the detector reagent is removed sideways out of the centrifuge tube with a syringe.

The general procedure for coating G40 nanoclusters with antibodies is as follows:

Protocol 6 Coating of gold nanoclusters

1. 50 ml G40 suspension (OD540 ~ 0.9) is adjusted to pH 8.5 with a 0.2 M potassium carbonate solution (K_2CO_3).
2. Coating of the G40 clusters is performed by incubating X µg purified antibody (X > MPA) per ml G40 for 20 min. at ambient temperature while being gently swirled on a shaking platform. While adding the antibodies, the sol should be stirred vigorously in order to obtain as uniform a coating as possible.
3. After coating, the sol is further stabilised by adding 5 ml 1% (w/v) skimmed milk powder that has been adjusted to pH 8.5 with 0.2 M potassium carbonate.
4. The mixture is gently swirled for another 60 min. at ambient temperature on a shaking platform.
5. The suspension is split in half, and each half (about 27.5 ml) is centrifuged for 60 min. at 2500 × g and 20 °C in a 50 ml Greiner tube over a discontinuous glycerol gradient consisting of 5 ml 80% (v/v) glycerol, of pH 8.5, and 7.5 ml 50% (v/v) glycerol, pH 8.5, using a swinging-bucket rotor. Under these conditions the antibody-coated G40 nanoclusters will be found almost entirely in the 50% (v/v) glycerol layer.
6. Particles are harvested by piercing the centrifuge tube with a syringe needle and removing the dark-red layer sideways out of the tube. In this way, the antibody-coated G40 nanoclusters will not be contaminated with free protein (antibodies and blocking proteins) and very small complexes present in the upper water layer and/or with large protein cluster complexes present in the lower 80% glycerol layer (Fig. 6).
7. The obtained 4–5 ml of coated G40 nanoclusters are diluted to 15 ml with water that has been adjusted to pH 8.5 with 0.2 M potassium carbonate.
8. The solution is then transferred into a centriprep-30 concentrator.
9. The glycerol is removed in about five centrifugational runs in which the retentate after each run is filled up to 15 ml with water of pH 8.5.
10. In an additional run, the suspension is then further concentrated to 2 ml.
11. After addition of 400 µl of a 6% (w/v) skimmed milk powder solution in water of pH 8.5 (1% (w/v) end concentration) and 12 µl of a 10% (w/v) sodium azide (0.05% (w/v) end concentration), the suspension is stored at 4 °C until used.

Notes:

- When using other volumes and/or different sized gold nanoclusters, one has to change the centrifugal speed and/or the running time such that the purified detector reagent is found almost entirely into the 50% glycerol layer.

- The washes to remove (most of) the glycerol are necessary to prevent problems with drying of the conjugate pads. Moreover, glycerol reduces the flow of the detector reagent over the membrane.

3.7 Preparation of the capture reagents

Control capture reagent
The control capture reagent consists of a total Ig fraction which can be prepared by ammonium sulfate precipitation of sheep serum as described in section 3.2.

Test capture reagent in the direct assay
In the direct assay, the test capture reagent consists of polyclonal or monoclonal antibodies to the antigenic protein of interest that have been purified by immunoaffinity chromatography (HiTrap Protein G) as described in section 3.3.

Test capture reagent in the indirect (competitive) assay
In the competitive assay, the test capture reagent consists of a ligand-protein conjugate in which the ligand of interest has been coupled to a protein, usually ovalbumin or bovine serum albumin (BSA). Similar to the immobilization of low-molecular-mass ligands to a chromatographic support (section 3.4), a successful coupling of small ligands to a protein largely depends on the available reactive groups on the ligand. Moreover, the low solubility of a ligand in an aqueous solution may be an additional problem. A large variety of coupling protocols have been described [19, 20]. One of these procedures is the use of glutaraldehyde as linker molecule between an amine group of the ligand and the ε-aminogroup of the lysin residues of the protein. As an example, the coupling of the sulfonamide drug SDD by its aromatic amine group to ovalbumin will be described.

Protocol 7 Coupling of a sulfonamide drug to ovalbumin

1. 0.1 mg SDD/ml aqueous solution is prepared by dissolving 10 mg of SDD in 1 ml of methanol which is then added to 100 ml of PBS (Solution I).
2. Solution II is prepared by dissolving 300 mg of ovalbumin in 30 ml of PBS, after which the solution is filtrated over a 5 µm Acrodisc.
3. The reaction mixture consists of 30 ml of Solution I, 30 ml of Solution II and 30 ml of PBS, to which 900 µl 25% (w/w) glutaraldehyde solution is added slowly under vigorous constant stirring.
4. Approximately 10 min. after adding the glutaraldehyde, the clear solution will take on a yellowish appearance. The mixture is stirred for 4–5 h at ambient temperature.
5. The conjugate mixture is exhaustively dialysed against PBS for 4–5 days at 4 °C.
6. The dialysate is then concentrated to 50 mg/ml using a centriprep-30 concentrator and stored in 1 ml aliquots at –20 °C.

Another useful method is the coupling of the analyte to an epoxy-activated protein. As an example, the synthesis of a BSA-streptomycin conjugate is described in which the antibiotic streptomycin is coupled to BSA by using the two epoxy groups of 1,4-butanediol diglycidyl ether as reactive groups. First, one epoxy group is coupled to the ε-amino group of lysine residues as well as to the hydroxyl group of tyrosine residues [21]. Thereafter, the remaining epoxy group of the obtained epoxy-activated BSA is coupled to the streptomycin molecule. In this way, 1,4-butanediol diglycidyl ether provides a 12-atom spacer between BSA and the streptomycin molecule. The procedure is as follows:

Protocol 8 Coupling of epoxy-activated BSA to streptomycin

1. A solution of 630 mg BSA in 10 ml water is brought on pH 11.2 with 0.5 M sodium hydroxide.
2. After adding 152 mg 1,4-butanediol diglycidyl ether, the solution is gently stirred for 20 h at 27 °C.
3. During the first few hours of the incubation, the pH has to be checked every half hour and kept between 11.1 and 11.2 by adding small amounts of 0.5 M sodium hydroxide.
4. After the incubation, the brownish solution is transferred into a centriprep-30 concentrator and centrifuged for 15 min. at 1500 × g and ambient temperature.
5. The retentate is mixed with retentate mixing buffer (Solution 8) to an end volume of 15 ml and centrifuged again in a centriprep-30 concentrator.
6. After five such washing cycles, the remaining 10 ml of epoxy-activated BSA solution is allowed to react with 376 mg streptomycin sulfate at pH 10.8 during an incubation of 48 h at 27 °C under constant stirring. During this incubation, a yellow precipitate is formed.
7. The conjugate mixture is exhaustively dialysed against PBS for at least 48 h at 4 °C.
8. The yellow precipitate is spun down by centrifugation for 10 min. at 3500 × g and 4 °C.
9. The yellow supernatant is concentrated to approximately 3.5 ml, divided into small aliquots and stored at -20 °C.

3.8 Choice of membrane

A proper choice of membrane is essential for the good performance of a strip-test. Several different strip test membranes are available, varying in capillary flow rate (sec/4 cm), tensile strength (the presence or absence of a plastic membrane support), thickness, and surface quality. The parameters that affect capillary flow rate are the membrane's pore size, pore size distribution, and porosity. Millipore gives an overview of how these parameters are related [22].

The thickness of the membrane is of influence on the bed volume, the dispensed line-width and the tensile strength.

The advantage of a membrane that has been cast directly on a plastic support is the large increase of the membrane's tensile strength properties, which makes its handling much easier. Moreover, the presence of a membrane support prevents the migration of adhesives from the backing plate into the membrane layer. On a direct-cast membrane, however, it is not possible to apply reagents to the belt side, i. e., the side of the membrane that is the smoothest and is relatively defect free.

Nitrocellulose is probably the most commonly used polymer for strip test membranes. The pore size of such membranes varies between 5 and 20 µm, which is large compared to a pore size of 0.2 to 1.2 µm of a nitrocellulose membrane used for protein blotting. A large pore size implies a small membrane surface area and, consequently, a low-protein binding capacity, i. e., 20 to 30 µg IgG/cm^2 instead of the 110 µg IgG/cm^2 for a blotting membrane with a pore size of 0.45 µm.

Nitrocellulose membranes bind proteins electrostatically; the strong dipole of the nitro-ester interacts with the strong dipole of the peptide bonds of the protein. Because nitrocellulose membranes are completely neutral, their binding properties are independent of the pH of the immobilization solution. However, the pH might have an effect on both the solubility and immobilization efficiency of a particular protein.

When using nitrocellulose membranes that bind proteins electrostatically, detergents such as Tween-20 and Triton-X 100 should not be used or should be used only in very low concentrations (< 0.01% v/v). Ionic detergents, such as sodium dodecylsulfate (SDS), are compatable with this type of membrane at concentrations up to 0.5% (w/v).

Depending on the choice for a batch or a continuous manufacturing process, membranes are available in different formats, i. e., as small strips (30 × 2.5 cm), sheets (20 × 20 cm), or rolls (30 cm × 100 m).

A partial list of suppliers of nitrocellulose membranes is shown below:

- Schleicher & Schuell GmbH, Postfach 4, D-37582 Dassel, Germany http://www.s-und-s.de, e-mail: filtration @s-und-s.de
- Millipore Corporation, Bedford, MA, USA http://www.millipore.com/healthcare, e-mail: tech_service@millipore.com

Schleicher & Schuell is provider of the AE- and FF-membrane series, Millipore of the Hi-Flow and Hi-Flow Plus membranes.

3.9 Immobilization of the capture reagents onto the membrane

A crucial step in manufacturing strip tests is the proper application of reagents to the membrane. A partial list of suppliers of reagent application equipment is shown below:

- BioDot, 17781 Sky Park Circle, Irvine, California, 92714 USA http://www.biodot.com, e-mail: webmaster@biodot.com

- IVEK, Fairbanks Road, North Springfield, Vermont, 05150 USA http://www.ivek.com, e-mail: ivek@ivek.com
- Camag, Sonnenmattstrasse 11, CH-4132 Muttenz, Switzerland http://www.camag.com, e-mail:info@camag.ch

For the preparation of our strip-test devices, a BioJet Quanti3000 attached to a BioDot XYZ3000 Dispensing Platform (Fig. 7) is used for dispensing the test and control capture antibodies onto nitrocellulose membrane AE 100 (12 μm) strips (300 × 25 mm) on a Mylar 5 support. Prior to use, the antibody stock solutions (5 mg IgG/ml) are filtered over a 0.2 μm FP 030/3 disposible cellulose acetate filter. A typical amount of antibody used for both the control and test capture reagent is 0.5 μg/cm membrane. The distance between control and test line largely depends on the readout indications on the viewing window of the housing, but is usually about 5 mm.

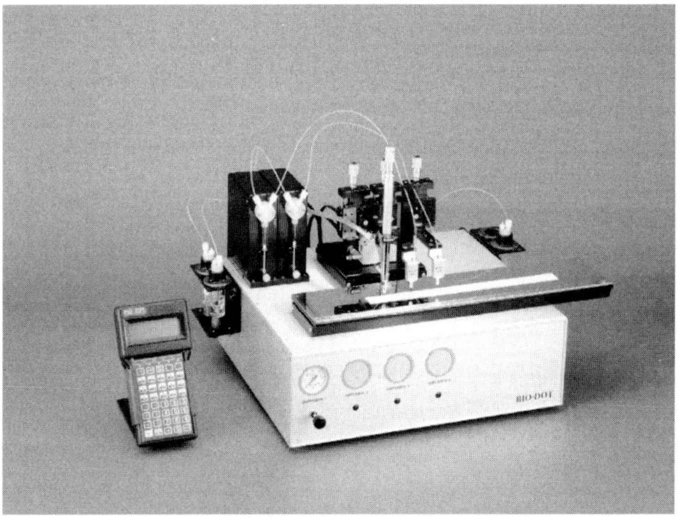

Figure 7 The BioDot XYZ3000 Dispension Platform with two BioJet Quanti3000 devices for dispending the test and control detector reagent onto the nitrocellulose membrane strips (Courtisey of Chris Flack, BioDot).

3.10 Drying of the membranes

After dispensing the sample and/or control lines, the membrane is dried for 1 h at 50 °C. Drying a membrane completely is essential in order to obtain a proper fixation of the proteins that have been deposited. Drying efficacy is a function of temperature, humidity, and time. If the membrane is backed, then the drying times must be extended.

The conditions that are typically required to achieve complete drying of a membrane are 0.5 h, 1–2 h, and > 16 h for drying temperatures of 15–25 °C, 35–40 °C, and 45–50 °C, respectively [22].

3.11 Blocking and washing of the membranes

Blocking is often necessary in order to prevent non-specific binding of the analyte and/or detector reagent to the membrane.

After drying the membrane strips, blocking is performed by incubating the strips for 30 min. in blocking buffer (Solution 9).

Because an excess of blocking agent can interfere dramatically with the membrane's capillary flow properties, unadsorbed blocking agent has to be removed by washing the membrane in a weak buffer solution. To promote a uniform re-wetting of the blocked membrane, the presence of a low concentration of surfactant in the washing solution is recommended. In case one would like to compare the effects of various surfactants, the BioDot Surfactant Starter Kit may be very useful. Using the nitrocellulose membrane AE100, washing can be performed three times for 5 min. each with washing buffer (Solution 10).

After drying for another 2 h at 50 °C, the membranes can be either further processed or stored under dry conditions, i. e., sealed in plastic in the presence of desiccator, until use. Storage may be performed at ambient temperature (for a few weeks) or at –20 °C (for a longer period).

Note:

- The term surfactant is an acronym for surface active agent and generally refers to wetters, solubilizers, emulsifiers, dispersers, or, detergents. Surfactants almost always act to reduce attractive interactions between "like" particles and bring them to "unlike" surfaces.

- An alternative for the above-described blocking of the test strip directly after dispensing of the readout lines is the "blocking on the fly" technique, by which the blocking agent is included in with the conjugate pad impregnation mixture. When the sample then enters the conjugate pad, the solubilized blocking agent will run with the front along the membrane, binding to non-specific sites.

3.12 Choice of the conjugate pad

The conjugate pad performs multiple tasks, the most important being the delivery of the assay detector reagent, which is deposited into the pad material typically in conjugation with some combination of blocking agents and/or surfactants. When sample flows from the sample pad into the conjugate pad, the detector reagent should be released consistently, quickly, and quantitatively and then float with the sample front. Such an ideal-acting conjugate pad thus consists of a well-balanced combination of the proper conjugate pad material and the proper surfactants.

Materials that are commonly used to make conjugate pads include glass fiber filters, paper (cellulosic) filters, and surface-treated (hydrofilic) propylene filters. A number of these materials can be found in the BioDot Diagnostic

Materials Kit, whereas the BioDot Surfactant Starter Kit contains a variety of surfactants that one may try for his own specific application.

In our strip tests, Whatman Glass Fiber Paper with binder F 075–17 was chosen to be the most appropriate material for a conjugate pad because of the ease with which the material is wetted, the high capacity to absorb fluid, the uniform distribution of the detector reagent on the pad, and the low amount of detector reagent that remains on the pad after running the assay.

3.13 Composition of the detector reagent on the conjugate pad

For a membrane strip of 5 mm in width, a conjugate pad of 5 × 5 mm may be applied. Using Whatman Glass Fiber Paper with binder F 075–17, such a pad can be loaded with 12 to 13 µl of conjugate pad impregnation mixture. Besides the two detector reagents, the choice of other key ingredients appears to be a matter of trial and error. Moreover, the required concentration of the test and control detector reagent in the mixture largely depends on the quality of these reagents.

As an example, the composition of a conjugate pad impregnation mixture is given for a test that has been developed in our laboratory for the detection of Pepino Mosaic virus (PepMV) in tomato plants:

- 215 µl rabbit anti-PepMV IgG coated G40 nanoclusters (test detector reagent)
- 45 µl rabbit anti-sheep IgG coated G40 nanoclusters (control detector reagent)
- 130 µl 40% (w/v) sucrose in water
- 10 µl 1% (v/v) Surfactant 10G in water

Each conjugate pad (5 × 5 mm) was loaded with 13 µl of this mixture. After drying for 2 to 3 h at 50 °C, the conjugate pads were stored at -20 °C until used.

Another example is the composition of a conjugate pad impregnation mixture for a test that has been developed in our laboratory for the detection of (dihydro)streptomycin in milk:

- 70 µl of anti-(dihydro)streptomycin IgG coated G40 nanoclusters (test detector reagent)
- 150 µl of rabbit anti-sheep IgG coated G40 nanoclusters (control detector reagent)
- 70 µl of 12% (w/v) sucrose in water
- 10 µl of 1% (v/v) Surfactant 10G in water
- 100 µl of water

Each conjugate pad (5 × 5 mm) was loaded with 10 µl of this mixture. After drying for 2 to 3 h at 50 °C, the conjugate pads were stored at -20 °C until used.

3.14 Choice of the absorbant pad

Absorbent pads are placed at the end of the test strip. The major advantage of using an absorbent pad is that assay sensitivity can be increased by virtue of the increase in the total volume of sample that is analyzed. Without an absorbant pad, the total volume of sample analyzed is determined by the bed volume of the membrane. Most absorbent pads are made from paper filters.

Cellulose type 133 (Gelman), cellulosic paper 3MM (Whatman), or cellulosic paper GB002 (Schleicher & Schuell) are all excellent materials for use as absorbent pad.

3.15 Choice of the sample pad

Sample pads are placed at the beginning of the test strip. Their major task is to absorb the sample and provide a uniform flow of the sample fluid from the sample pad via the conjugate pad onto the membrane. Moreover, the sample pad acts as a filtration device by removing particles, e. g., fat particles in milk, from the sample solution. The pad may be pre-treated with a special buffer solution in order to adjust the pH or viscosity of the sample solution. Most sample pads are made from paper filters. When using aqueous sample solutions like urine, buffer, etc., cellulose paper type 133 (Gelman), 3MM Chr (Whatman), 3MM D28(Whatman), or GB002 (Schleicher & Schuell) can be used as sample pad. However, these materials are pretty thick and and therefore the sample pads will be easily clogged when using samples with high viscosity and/or that contain a high concentration of particles or sample debris. For milk samples as well as for food and/or plant extracts, cytosep 1660 (Gelman) is a good choice.

When using whole-blood samples, special blood separation filters will be needed as sample pad. A partial list of blood separation media manufacturers is provided below:

- Ahlstrom Filtration, Inc, 122 W. Butler Street, P.O. Box A, Mount Holly Springs, Pen 17065, USA http://devicelink.com/company98/a/a00012.html, e-mail: feedback@devicelink.com
- Spectral Diagnostics Inc., 135–2 The West Mall, Toronto, Ontario M9C 1C2, Canada http://www.spectraldiagnostics.com, e-mail: info@spectraldiagnostics.-com

Furthermore, there are two interesting patents in this field that should be mentioned (see also http://www.birmingham.ac.uk/WATL/rapid-assay.html):

- US5916521 Lateral flow filter devices for separation of body fluids from particulate materials. Describes the use of an asymmetric filter wherein red cells and plasma flow laterally along the filter in order to improve the volume of plasma obtained without clogging the filter.
- PCT WO97/34148A1 Immunoassay Device. Describes various devices where the analytical membrane also acts to separate plasma from whole blood.

3.16 Test production

As shown in Figure 1, the test strip consists of a backing plate (laminated card) on which the individual components are pasted. The backing plate in our tests consists of 0.01″ white matte vinyl GL-187 (6.3 cm × 30 cm). Test production can be performed in two ways. One way is to first assemble complete sheets by pasting the components (membrane and the various pads) in lengths of 30 cm onto the backing plate. The cards are then cut into appropriate-sized strips, either by hand or by using an automated cutting device that will cut the completed cards into pieces of uniform width, e. g., 5 mm.

Another way that is especially suited for the production of a relatively small number of test strips is to first cut the individual components to their appropriate length and then followed assemble them into complete test strips.

The test strips are then ready to be placed into a plastic cassette (housing) as shown in Figure 2. For storage, the strip tests are packed into a foil pouch in the presence of a sachet with some kind of dessicant, e. g., 1 g silica gel MiniPax (Fig. 8). In this way, the strip tests can be stored for several months at ambient temperature without serious loss of quality.

There are many companies all over the world that can supply you with plastic cassettes (housings), foil pouches, backing cards, and drying agents. However, the minimal purchase of these products is usually very high, i. e., in the range of 10,000 or more. Especially when situated in the developing phase and/or when having no commercial interests, starting with smaller numbers of such strip test components will often be more desirable. Relatively small numbers of the above components, as well as sealing and cutting equipment, can be provided by Kenosha C.V. (Amstelveen, The Netherlands).

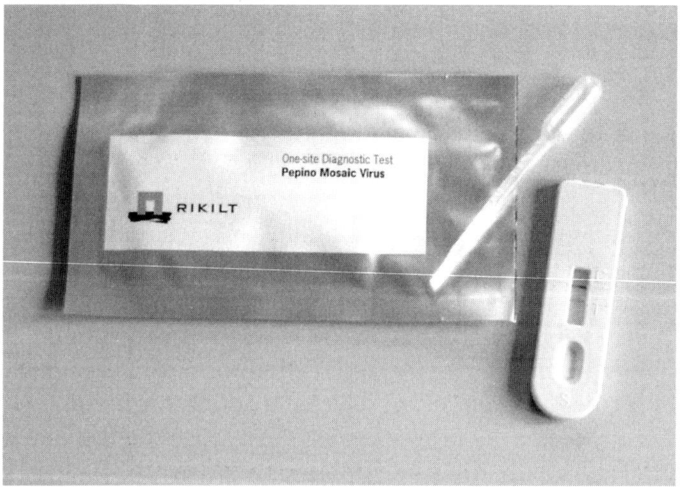

Figure 8 For storage, the strip test is packed into a foil pouch in the presence of a sachet with some kind of dessicant, e. g., 1 g silica gel MiniPax. In this way, the strip tests can be stored for several months at ambient temperature without serious loss of quality. When the assay will be used to test liquid samples, a small pipet may be included as well.

3.17 Performing the test

When using liquid samples (milk, urine, serum, etc.), three or four droplets are brought directly into the sample well of the horizontally placed strip-test device. Usually, this is performed with a small plastic pipet that is provided with the test. Such a pipet will assure a pretty constant volume of sample per added drop. Of course, one is free to add a contant volume of sample by using a microliter pipet (recommended for quantitative/semi-quantitative tests).

When starting from solid sample material (plant leaves, solid food products, etc.), one has to make an extract of the sample first. This can be performed by placing a certain amount of sample material into a small bottle with dropper cap that contains a few milliliters of extraction buffer and five or six stainless steel ball bearings (Fig. 9). After having shaken the bottle firmly for 20 to 30 s, a few drops of extract are brought into the sample well. It should be noted here that the type of sample matrix and/or the type of extraction buffer will greatly influence the performance of the test. A good extraction buffer to start with is Solution 11.

After having the sample brought into the sample well, the sample will be absorbed by the sample pad, and the test begins. Usually, the colored sample front will appear in the viewing window within 30 s.

The results should be read after the control line is clearly visible in the viewing window. Usually, this will take 5 to 10 min., depending on the type of membrane used, the dimensions of the test strip, and the fluidity of the sample. Moreover, the dimensions of the sample well will greatly influence the test speed.

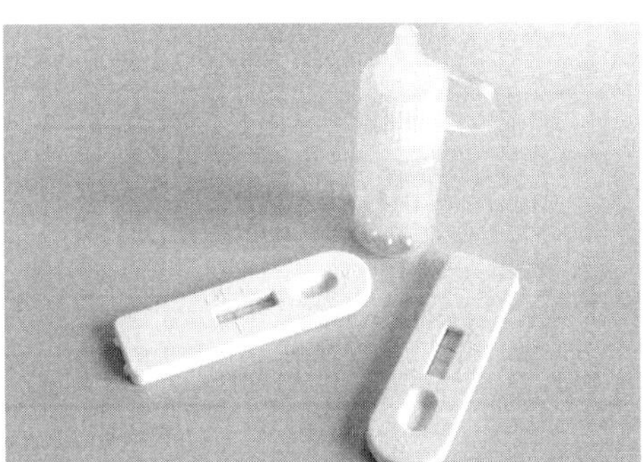

Figure 9 When the assay will be used to test solid samples (plant leaves, solid food, or feed samples, etc.), a simple extraction procedure has to be included. This can be achieved by placing a certain amount of sample material into a small bottle with dropper cap that contains a few milliliters of extraction buffer and five or six stainless steel ball bearings. After having shaken the bottle firmly for 20 to 30 s, a few drops of extract are brought into the sample well.

3.18 Interpretation of the results

The possible readouts are shown in Figure 3 (direct test) and Figure 4 (competitive or indirect test). When the test is performed properly, the control test line will always be visible. In the competitive assay, a certain amount of analyte in the sample will prevent the formation of a colored test line. In the direct assay, however, a certain amount of analyte in the sample will result in the appearance of a test line.

Orginially, strip tests were designed to be judged by the naked eye. However, one tends to apply optical reading equipment for interpretation of the test results. An example of such equipment is the BioDot Test Strip Reader (Model TSR3000), which is a reader designed for the qualitative and/or quantitative measurement of a test strip itself or within a housing. The system has a CDD array camera to take a high-resolution image of the test strip. The user can then define a window and measure the intensity of light reflected from the test strip. The great advantage of such a system is that the results can be stored, analyzed, and compared in a spreadsheet format. See also the article by Tisone et al. [23] on image analysis for rapid-flow diagnostics.

3.19 Patents related to strip tests

Finally, it should be mentioned that there are many patents that cover a number of technologies and materials that are commonly used in strip-test devices. Patents are a vital aspect of research, particularly if commercial interests are involved.

When preparing this manuscript it became clear that writing a meaningful overview of patents in this field would be difficult and complex because of technical and legal implications. However, in order to comply with international trade and licensing agreements and to prevent possible legal problems after product launch, it would be prudent to review the patent literature prior to commercialization. A useful source of patent information can be the Diagnostic Club patent abstract service (contact Tony Towen, Diagnostics Club Administrator; e-mail: towen@waitrose.com).

4 Troubleshooting

In developing a strip test, multiple reagents, materials, and techniques are brought together in order to create a sensitive, specific, fast, cheap, and easy to perform assay for the detection of either low-molecular-mass analytes (hormones, antibiotics, drugs of abuse etc) or high-molecular-mass components (antibodies, bacteria, viruses, cancer markers, food allergens, etc.). Moreover, the test should be applicable for a wide variety of sample matrices such as whole blood, plasma, serum, urine, saliva, milk, food extracts, plant leaves, and even sweat. All these aspects in developing a strip test rely upon one another for their efficacy; often changing just one of these items can lead to completely unexpected and/or undesired results. The possible causes of a poorly performing strip test are numereous. The most common difficulties encountered in creating a strip test are listed in Table 1, which consists of the troubleshooting list by Schleicher & Schuell [24], extended with our own experience on the subject.

Table 1 The most common difficulties encountered in creating a strip test and the possible solutions. Note that various problems mentioned in the table are related to each other, e.g., the item "No sample front visible" is a more profound definition of the item "Migration problems".

Problem	Possible solutions
Nonspecific binding	Evaluate absorption
(Heterophile reactions)	Add IgG
	Add surfactants
	Antibody fragmentation
	Change pH
	Add buffers
General nonspecific reactions	Add a blocking agent
	Change the blocking agent
	Add or change surfactant
	Smaller particles production
	Change pH
	Change buffer make up
	Decrease reagent/strength
	Change conjugation procedure
Lack of sensitivity	Increase quantity capture material
	Increase quantity of the conjugate
	Increase strength of conjugate
	Increase particle size
	Decrease or change blocking agent
	Decrease flow rate
	Decrease rate of conjugate release
	Use higher affinity antibodies/stronger antigens
	Repurify reagents to up activity
	Add activity enhancer
	Check for analyte absorption
	Increase capacity of the absorbant pad
Lack of specificity	Repurify the reagents
	Try new reagents
	Check pH of the sample
Migration problems	Try smaller particles
	Try other type of particles
	Change production protocols
	Change or increase blocking agents
	Add a surfactant
	Block membrane directly
	Look for leeching of reagents from one component to another
	Check the viscosity of the sample

Problem	Possible solutions
Stability issues	Change the conjugate pad material Change preparation of conjugate pad – increase or add surfactants – increase or add polymers – change pH of buffers – add or increase sugars to conjugate mixture Look for moisture problems Try new membrane materials Add preservatives to membrane block Add preservatives to capture protein prior to dispensing Look for leeching effects Look for degradation of materials on the sample pad
No control line visible	Flooded test. Repeat with new device and add fewer drops Check the control capture reagent Check the control detector reagent Check pH of the sample
Faint lines	Check dispensing equipment and/or dispensing procedure Check the capture reagents Check the detector reagents Check membrane blocking, drying and washing procedure Look for membrane anomalies Check pH of the sample
No sample front visible	Sample volume is too small, add more sample Make sure the sample pad is not clogged by sample material (fat, debris)
Detector reagents are not completely released from the conjugate pad	Change conjugate pad material Increase sample volume Add or increase sugar concentration on conjugate pad

5 Applications

In principle, there are no limitations for the type of analyte that can be detected by using a strip test. This is demonstrated by the large number of commercial strip tests for the detection of a wide variety of analytes. Most strip tests, with special emphasis on the pregnancy tests, find their application in human clinical diagnostics. A systematic review of near-patient tests in primary care has been published by Hobbs et al. [8]. This lists several hundred commercial devices. Furthermore, strip tests have been developed in the field of veterinary diagnostics [11, 12], food testing, and plant pest and disease diagnosis (see CSL, York, UK: http://www.csl.gov.uk).

In most human diagnostic applications, relatively high concentrations of analyte are measured. It has been shown, however, that strip tests are also applicable in the low ppb (ng analyte/g sample) detection region, which is often required for the detection of veterinary drug residues in food diagnostics [11]. As for pesticides and other environmental contaminants, LDLs are required in the ppt (1 ng/kg) region rather than in the ppb region; to our knowledge, no strip tests have been described for the detection of analytes at such low concentrations.

Multianalyte devices also have been developed for the simultaneous detection of several analytes, e. g., drugs of abuse [25].

6 Remarks and conclusions

The proper performance of a strip test for the detection of either low-molecular-mass analytes (indirect or competitive assay) or high-molecular-mass analytes (direct assay) can be summarized as being a delicate balance between all of the applied components. Changing just one of these components can lead to completely unexpected and/or undesired results. The sample matrix also will greatly influence the performance of the test.

The strip test has some important advantages, i. e., the test is cheap, fast (test results can be obtained in 5 to 10 min.), and easy to perform. Moreover, all reagents are included in the test device and no expensive equipment is required for using the test. A disadvantage of the strip test is the fact that large number of samples can not be tested simultaneously, as in an ELISA. Also, the high specificity of a strip test may be considered a disadvantage as well. Especially for general screening purposes in food diagnostics, there is a demand for (immuno)assays with which one is able to detect a whole group of analytes, e. g., several aminoglycosides or sulfonamides, rather than one specific member of such a group. A possible solution for this problem is the application of generic antibodies that are directed against the generic part of a specific group of analytes [26].

In principle, there are no limitations for the type of analyte that can be detected by using a strip test. However, when detection limits in the ppt region are required, the strip test in its present form is simply not sensitive enough. Another drawback of the strip test is its nonquantitative character. In practise, interpretation of the test results will allow merely two possible conclusions: "the analyte is detectable" or "the analyte is not detectable". For analytes where an MRL has been established, the LDL of the strip test should be at least as high as or lower than the MRL value.

Despite the above-mentioned limitations, the strip test has its own specific place in the field of immunological screening assays. The development of strip tests and related assays is increasing and constantly improving. A most interesting development in the field of rapid assay devices is provided by the Wolfson Applied Technology Laboratory (WATL) of the university of Birmingham, UK (http://www.birmingham.ac.uk/WATL/rapid-assay.html). Their work involves devices based on flow in porous media such as nitrocellulose membranes. These devices are classified in first- and second-generation devices. The first-generation devices have channels formed by printing so that liquid flows in two dimensions, distinct from conventional strip devices in which the flow is essentially unidirectional (patent US 5354538; Liquid Transfer Devices). Their second-generation devices additionally include a switchable barrier to control the flow in the channels (patent WO 01/25789A1; Fluid-flow Control Device). This is made by printing down a resin, typically polycoumarone-co-indene resin, and switching with a surfactant such as octyl-β-D-glucopyranoside. Undoubtly, applications of this type of rapid assay device will be heard of in the near future.

7 Acknowledgments

The author gratefully acknowledges Schleicher & Schuell (Dassel, Germany) for permission to use their data on troubleshooting as included in Table 1, Chris Flack from BioDot Inc. (Irvine, Ca, USA) for providing Figure 7, and Dr. Christian Mayer from TU-Delft (Delft, The Netherlands) for preparing Figures 8 and 9.

8 Further reading

Anonymous. Millipore's short guide for developing immunochromatographic test strips, Millipore (Lit. no. TB500)

Harvey M, Kremer R, Vickers L (1999) Guide to diagnostic rapid test device components. In: L Vickers (ed): *Schleicher & Schuell diagnostic components*. Schleicher & Schuell GmbH, Dassel, Germany, no. 706

Hermanson GT, Krishna Mallia A, Smith PK (eds) (1992) *Immobilized affinity ligand techniques*. Academic Press, San Diego, CA

Hobbs FDR, Delaney BC, Fitzmaurice DA et al. (1997) A review of near patient testing in primary care. *Health Technology Assessment* 1(5) (http://www.ncchta.org/ see under publications)

Price CP, Thorpe GHG, Hall J, Bunce RA (1997) Disposable integrated immunoassay devices. In: CP Price, DJ Newman (eds): *Principles and practices of immunoassays (2nd edition)*. MacMillan Reference Ltd, London, 579–603

References

1 Zuk RF, Ginsberg VK, Houts T et al. (1985) Enzyme immunochromatography: A quantitative immunoassay requiring no instrumentation. *Clin Chem* 31: 1144–1150

2 Bunce RA, Thorpe GH, Keen L (1991) Disposable analytical devices permitting automatic, timed sequential delivery of multiple reagents. *Anal Chim Acta* 249: 263–269

3 May K (1994) Unipath ClearBlue One Step™, Clearplan One Step™ and Clearview™. In: D Wild (ed): *The immunoassay handbook*. Macmillan Press, London, 233–235

4 Leuvering JHW, Thal PJHM, Van der Waart M, Schuurs AHWM (1980) Sol particle immunoassay (SPIA). *J Immunoassay* 1(1): 77–91

5 Anonymous (1997), Syllabus of a two-day seminar on solid phase membrane-based immunoassays, Paris, September 25–26, Millipore Corporation, Bedford, MA

6 Anonymous (1997), Syllabus of the latex course, london, October 1–3th, , Organised by Bangs Laboratories, Inc., Fishers, IN

7 Price CP, Thorpe GHG, Hall J, Bunce RA (1997) Disposable integrated immunoassay devices. In: CP Price, DJ Newman (eds): *Principles and practices of immunoassays (2nd edition)*. MacMillan Reference Ltd, London, 579–603

8 Hobbs FDR, Delaney BC, Fitzmaurice DA et al. (1997) A review of near patient testing in primary care. *Health Technology Assessment* 1(5) (http://www.ncchta.org/ under publications)

9 Mueller-Bardorff M, Freitag H, Scheffold T et al. (1995) Development and characterization of a rapid assay for bedside determinations of cardiac troponin T. *Circulation* 92(10): 2869–2875

10 Haasnoot W, Kim KA, Cazemier G et al. (1996) Evaluation of a sol particle immunoassay (SPIA) based single-step strip test for the detection of sulfadimidine residues. In: N Haagsma, A Ruiter (eds): *Proceedings of the EuroResidue III Conference*. University of Utrecht, Faculty of Veterinary Medicine, Utrecht, 461–465

11 Verheijen R, Stouten P, Cazemier G, Haasnoot W (1998) Development of a one step strip test for the detection of sulfadimidine residues. *The Analyst* 123: 2437–2441

12 Verheijen R, Osswald IK, Dietrich R, Haasnoot W (2000) Development of a one step strip test for the detection of (dihydro)streptomycin residues in raw milk. *Food Agric Immunol* 12: 31–40

13 Jennes L, Conn PM, Stumpf WE (1986) Synthesis and use of colloidal gold-coupled receptor ligands. *Meth Enzymol* 124: 36–47

14 Leunissen JLM, De Mey JR (1989) Preparation of gold probes. In: AJ Verkley, JLM Leunissen (eds): *Immuno-gold labeling in cell biology*. CRC Press, Boca Raton, FL, 1–16

15 Frens G (1973) Controlled nucleation for the regulation of the particle size in monodisperse gold suspensions. *Nature (London) Phys Sci* 241: 20–22

16 Goodman SL, Hodges GM, Trejdosiewicz LK, Livingston DC (1981) Colloidal gold markers and probes for routine application in microscopy. *J Microsc*op 123: 201–213

17 Horisberger M, Vauthey M (1984) Labelling of colloidal gold with protein. A quantitative study using β-lactoglobulin. *Histochem* 80(1): 13–18

18 Horisberger M, Rosset J (1977) Colloidal gold, a useful marker for transmission and scanning electron microscopy. *J Histochem Cytochem* 25: 295–305

19 Erlangen BF (1980) The preparation of antigenic hapten-carrier conjugated: a survey. *Methods in Enzymology* 70: 85–104

20 Hermanson GT, Krishna Mallia A, Smith PK (eds) (1992) *Immobilized affinity ligand techniques*. Academic Press, San Diego, CA

21 Lommen A, Haasnoot W, Weseman JM (1995) Nuclear magnetic resonance controlled method for coupling of fenoterol to a carier and enzyme. *Food & Agricultural Immunol* 7: 123–129
22 Anonymous, Millipore's short guide for developing immunochromatographic test strips, Millipore (Lit. no. TB500)
23 Tisone TC, Rodriquez M, Queeney P (1999) Image analysis for rapid-flow diagnostics. IVD Technology Magazine, (http://www.devicelink.com/ivdt/archive/99/09/010.html)
24 Harvey M, Kremer R, Vickers L (1999) Guide to diagnostic rapid test device components. In: L Vickers (ed): *Schleicher & Schuell diagnostic components.* Schleicher & Schuell GmbH, Dassel, Germany, no. 706
25 Buechler KF, Moi S, Noar B et al. (1992) Simultaneous detection of seven drugs of abuse by the Triage™ panel for drugs of abuse. *Clin Chem* 38: 1678–1684
26 Haasnoot W, Kohen F, du Pré J, et al. (2000) Sulphonamide antibodies: from specific polyclonals to generic monoclonals. *Food & Agricultural Immunol* 12: 15–30

5 Biosensors

Thomas G. M. Schalkhammer

Contents

Methods and Tools in Biosciences and Medicine
Analytical Biotechnology, ed. by Thomas G.M. Schalkhammer
© 2002 Birkhäuser Verlag Basel/Switzerland

1 Introduction

1.1 Basis

A biosensor is an analytical device which uses biologically material to detect chemical species directly without the need for complex sample processing. The major impetus driving the significant increase of interest in biosensors over the past decade has been the attraction of utilizing the high specificity and sensitivity offered by biomolecules and biological systems.

A biosensor consists of a biological sensing element, such as an antibody, enzyme, or cell, that is in contact with a physical or chemical transducer, such as an electrode or optical detector. Measurement of the desired analyte is achieved by selective transduction of a parameter of the biomolecule-analyte reaction into a quantifiable electrical signal.

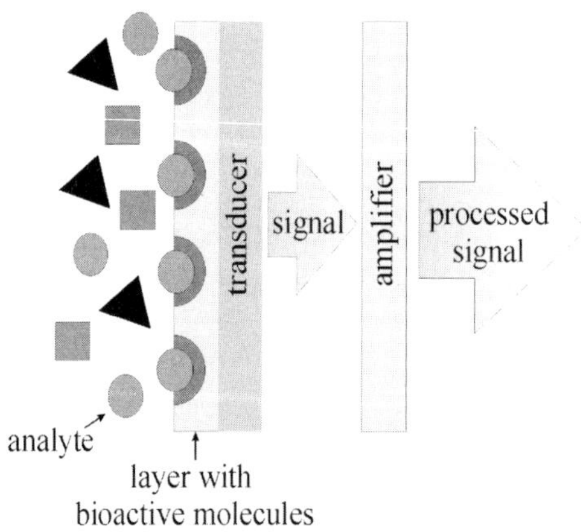

Figure 1

The most important charcteristics of any biosensor are specificity and sensitivity toward the target analyte. The specificity of a biosensor is based primarily on the properties of the biological component because this is the component responsible for analyte interaction at any biosensor. The sensitivity of the integrated device, however, is dependent on both the biological component and the transducer. Additional to the biomolecule-analyte interaction a transducer needs to convert the chemical event with high efficiency for subsequent detection.

When comparing chemical sensors and bio-devices, the inherent advantage of biosensor technology is the significantly higher specificity that can be achieved as a direct result of optimized macro-molecular recognition. This phenomenon is best illustrated by an antibody-antigen interaction. Minor chemical modifications of the molecular structure of the antibody can dramatically modify its affinity for the antigen. Similarly, enzymes such as lipases, nucleases or proteases will recognize their natural substrate with a far higher affinity than other components in the biological microenvironment. Molecules structurally that are related, e. g., amides instead of esters in the case of lipases, can elicit a small amount of cross-reactivity if present in high concentrations. Biosensors outperform all chemical sensors regarding specificity and to some extent sensitivity of detection.

1.2 Biological components of biosensors

Biomolecules used for biosensors need to have specific recognitive interactions concerning the substance of interest. Table 1 gives a survey of bio-recognitions employed as analyte-recognizing sensor layer.

Table 1

Analyte	**Bioreceptor**
Nucleic acid	Nuclei acid (DNA/RNA), pseudo-DNA e. g. PNA, DNA/RNA-binding proteins, intercalating dyes
Proteins	Antibodies, phage display proteins, adaptamers
Glyco-proteins	Lectines
Enzymes	Inhibitors, effectors, substrates
Coenzymes	Enzyme
Receptors	Ligands, Inhibitors
Metal ions	Complexing agents, emzymes with metals as co-factors
Membranes	Liposomes, fusion proteins
Cells	Metabolic trigger
Lymphoid cells	Mitogenes

1.3 Transducers

Besides the concept of the first amperometric biosensors dating back to 1962, various measurement modes and transducers have been established within the past few decades. Table 2 gives a short summary of some transducer systems used in biosensor technology nowadays.

Table 2

Transducer system	**Measurement mode**	**Typical applications**
1. Electrodes		
Enzyme electrode	Amperometric (current)	Enzyme catalyzed reactions, enzyme immunoelectrodes
Field effect transistors	Potentiometric (voltage)	Enzyme substrates, ions, gases, enzyme substrates, immunological analytes
Ion selective electrode	Potentiometric (voltage)	Ions in biological media, enzyme electrodes
Gas sensing electrode	Potentiometric (voltage)	Gases, enzymes, cell or tissue electrodes
Impedimetric	Impedance	Enzyme immunoelectrodes
Conductimetric	(Nano-) Conductance	Artificial cells
2. Piezo-electric crystals, surface acoustic devices (SAW)	Mass change	Volatile gases, vapors, oil, immunological analytes
3. Opto-electronic and fiber optics	Absorbance, fluorescence	pH, enzyme substrates, enzyme-transduced immunological analytes
4. Surface plasmon and waveguide devices	Resonant signal (e. g. angle)	DNA, RNA, proteins (primarily macromolecules)
5. Thermistors, diodes	Thermometry, calorimetric (heat)	Cells, tissue, exothermic enzymatic reactions

2 Sensors

2.1 Electrode-based biosensors

Amperometric enzyme electrodes

Amperometric biosensors, commonly referred to as “enzyme electrodes”, combine the specificity and selectivity of enzymes with the sensitivity of electrochemical detection methods. The reaction of the analyte with the bioactive component is chosen to produce an electro-active species. The use of enzymes, which catalyze biospecific reactions, enables the production of reactive molecules, which either can be oxidized or reduced and thus quantified electro-chemically. The enzyme is immobilized at or near the electrode (often

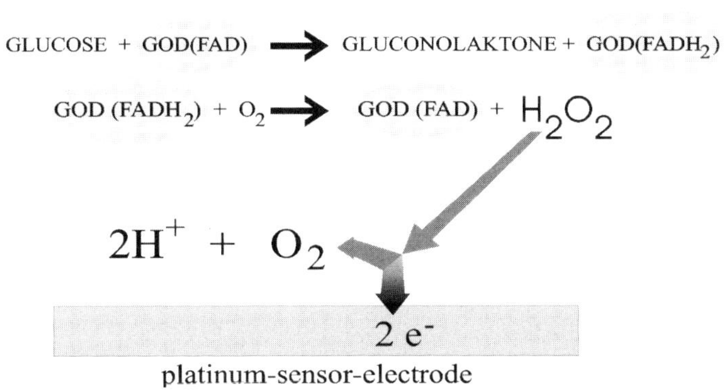

Figure 2 Glucose sensor based on enzymatic conversion of glucose to hydrogen peroxide and amperometric detection

referred to as working electrode). A potential is applied between the working electrode and a counter electrode in the same solution [1–6].

The oxidation or reduction of the analyte (or an enzymatically generated intermediate molecule) proceeds at the working electrode. The current generated is thus proportional to the analyte concentration. A well-defined and reproducible potential is necessary to avoid unwanted redox-reactions at the electrode. In order to minimize the drift of potential caused by the current, a third electrode, quite often a silver/silver chloride electrode, is necessary. Most electro-analytical devices perform the measurements using a three-electrode arrangement.

The desired potential of the working electrode is adjusted via the reference electrode and thus is stabilized throughout the experiment, while the current of interest is measured between working electrode and counter electrode. The advantages of this method are that no shift in potential can be induced by the current produced during the reaction of the analyte. Moreover, neither potential changes at the counter electrode nor the resistance of the electrolytes in the surrounding solution can influence the signal.

Electrodes

Working electrode (Biosensor = enzyme membrane electrode) In corrosion testing, the working electrode is the sample to be studied, whereas, generally, the working electrode is not converted while being studied, but the electrochemical reactions of interest occurs at the working electrode.

Reference electrode The reference electrode is used to measure the working electrode potential. A reference electrode should have a constant electrochemical potential as long as no current flows through it. The most common lab references are the saturated calomel electrode (SCE) and the silver/silver chloride (Ag/AgCl) electrodes. For nano-electrodes a pseudo-reference (a piece of silver) is often used.

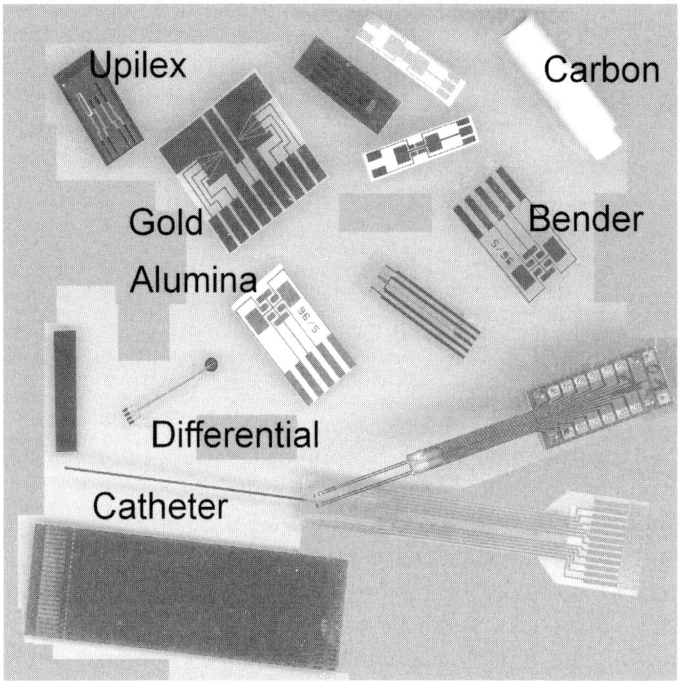

Figure 3 Various electrochemical thin-film sensors

Auxiliary electrode The auxiliary electrode is a conductor that completes the cell circuit. The auxiliary (counter) electrode in lab cells is generally an inert conductor like platinum or graphite. The current that flows into the solution via the working electrode leaves the solution via the auxiliary electrode.

The electrodes are immersed in an electrolyte (an electrically conductive solution). The collection of the electrodes, the solution and the container holding the solution are referred to as an electrochemical cell.

In constructing electrochemical biosensors at least the following topics have to be taken into consideration:

* Surface to which the bio-component should be coupled.
* Coupling technique to be applied
* Stability of the immobilized bio-system under measurement and storage conditions
* Change of redox-behavior of the electrode due to coupling procedures
* Redox-active substances in the analyte solution

Sensor surfaces

Thin-film and thick-film techniques offer a wide variety of possibilities in designing miniaturized electrochemical sensors. Thin-film technology especially is able to provide the high purity and reproducibility required for the electrode surface. The striking difference of monolayer thin-film biosensors compared to conventional types of biosensors is the eminent importance of surface chemistry in the former. Micro-structuring of sensor substrates by thin-film technologies has become state of the art, as a result of great advances in the field of microelectronics. Silicon, ceramics, or polymeric supports (polyimide, polycarbonate, etc.) can be used as sensor substrates. After standard cleaning procedures, metals are evaporated by an electron gun in high vacuum and used to coat the substrates with the metal (some tens of nm thick). Structuring of these thin films is performed by a lift off technique with photo resists. Figure 4 gives a typical thin-film platinum sensor layout.

Various metals can be used as electrochemical electrode surfaces for thin-film biosensors. The most frequently used metals are platinum, gold, rhodium, palladium and iridium. Because of a high degree of surface oxidation palladium, and iridium electrochemistry are rather complex. Gold electrodes are sensitive to destruction by complexing agents such as e. g. chloride ions. Conducting glass (e. g. tin-/indium oxide) also can be used as electrode for thin-film biosensors but the electron transfer at the interface and the ion mobility are lower than the transfer on metal electrodes. Reference electrodes are often silver films deposited electrochemically, via sputter coating, or as a thick-film paste.

Setup a thin-film electrode

By using this procedure, electrodes down to an outer diameter of 0.1 mm can be constructed using standard lithography equipment [1].

Protocol 1 Setup of a thin-film electrode

1. Sodium silicate glass sheets of 0.3 mm thickness or Upilex 0.1 mm foils are used as electrode carriers.
2. Standard cleaning procedures are performed with detergents, ultra-sonication, and organic solvents.
3. Titanium is evaporated by an electron gun in a high vacuum (5.10^{-7} mbar) instrument (e. g. Balzers) and used to coat the substrates with a thin film up to a thickness of 80 nm as an adhesion layer. Alternatively, titanium is coated via sputter deposition at 1.10^{-3} mbar. Care is necessary to avoid any oxygen ad-layer on titanium. Otherwise, the next metal layer will not adhere.
4. A platinum layer up to a thickness of 60 nm being evaporated on top of the titanium film acts as an electrochemical electrode.

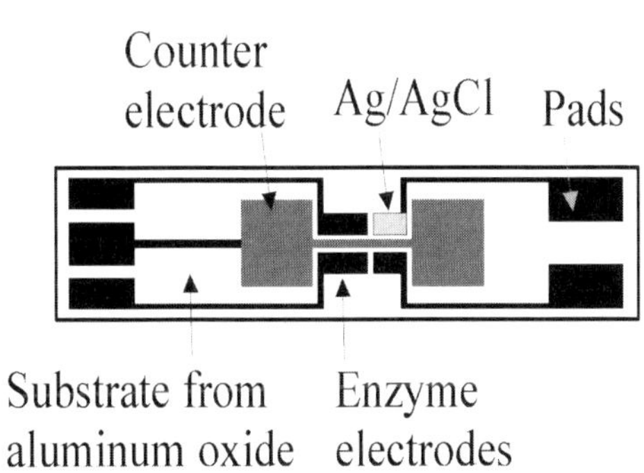

Figure 4 A multi-electrode thin-film sensor

5. Structuring of the metal thin films is performed by a lift off technique with AZ photoresist (e. g., 5218E).
6. Electrodes are isolated by a 1000 nm siliconnitride layer.
7. The SiNx film is structurized by plasma etching.
8. The platinum surface is cleaned by etching with oxygen plasma (30 watt) for 3 min.
9. If necessary the surface can now be hydrophobized by dipping into a 10% solution of hexamethyldisilazane in xylene for 15 min.
10. A Ag/AgCl reference electrode can be deposited by evaporating and structuring a 1000 nm silver film.
11. The AgCl layer is formed via chlorination with 10 mM $FeCl_3$ (15 min).

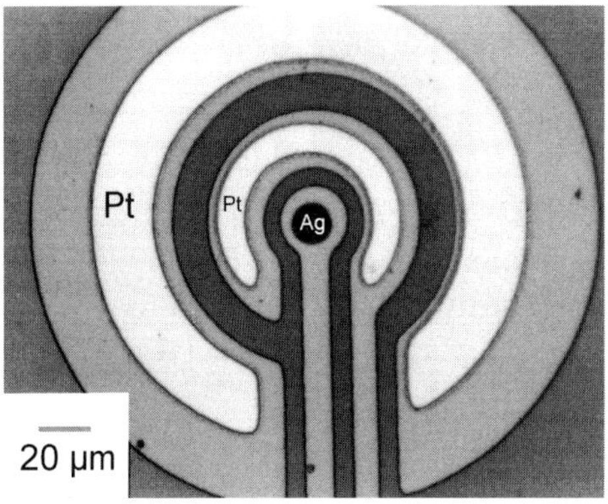

Figure 5 A miniaturized multi-electrode thin-film sensor

Metal-surface interactions
When using metal electrode surfaces, it has to be kept in mind that reactive compounds may bind at the surface thus causing electrochemical interferences, which had often been neglected in the past [7].

Inference by low-molecular-weight substances: Because most enzyme electrodes use hydrogen peroxide generated via an oxidase immobilized at the chip surface (e. g., gluocose oxidase, glutamate oxidase, lactate oxidase, etc.) the hydrogen peroxide response of a clean metal electrode (e. g., platinum) is the desired electrochemical process. Nevertheless, in additional to an *in situ* electrochemical activity of a variety of molecules such as ascorbic acid or uric acid, many thiols or peptides found in biological matrices interact irreversibly with the electrode surface.

By using platinum as electrode material, the electrochemical surface can be modified by anionic and cationic inorganic compounds. Chemisorption of thio- and amino compounds is the main process. A significant decrease in electrochemical response to hydrogen peroxide is observed when sensor electrodes come into contact with solutions of iodine, iodide, thiocyanate, sulphide, and thiols. Minor or no effects are found with bromide, chloride, nitrate, and chlorate.

Inference by protein adsorption on sensor surfaces: In some cases, proteins adsorb very strongly onto sensor surfaces. To test protein binding enzymatically, active molecules can be used. Glucose oxidase (GOD) is adsorbed irreversibly on most metal surfaces. This is a chemisorption because the protein could not be desorbed in a vital state, e.g. by treating the electrode with saturated NaCl, 3M KCl, or buffers of high or low pH. Moreover, GOD could not be replaced by highly concentrated BSA-solutions. The high enthalpy of chemisorption, caused by the interaction of various functional groups on the protein surfaces with the top metal layer atoms, results in a nearly irreversible coating of the metal electrodes.

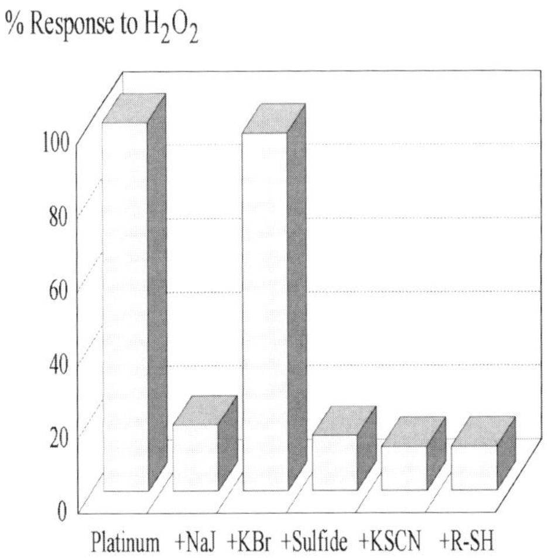

Figure 6 Modification of the electrochemistry of thin-film Pt-electrodes after incubation in solutions containing NaJ, KBr, sulphide, KSCN, and thiols in trace amounts.

Chemisorption is a limiting factor in most *in vivo* application but, on the other hand, is used to setup enzyme electrodes. Metal electrodes can be covered via adsorbing e.g. GOD. When comparing various enzyme preparations (Roche, Sigma both raw and pre-purified), it turned out that neither GOD with the lowest catalase content nor with the highest purification grade or the highest glucose oxidase activity in solution gave sensors with the highest response. Three preparations of GOD having varying degrees of purity were obtained for adsorptive and covalent immobilization by using size exclusion chromatography to purify GOD, thereby splitting the active GOD peak into fractions and concentrating them by ultrafiltration (using a membrane with a cutoff of 30,000 Dalton) up to a concentration of 10 mg/ml glucose oxidase. Using these preparations, it could be shown that adsorptive immobilisation on a native platinum surface (having a negative net charge) is highly influenced by impurities being present in low but effective concentrations spread over the molecular weight fractions of 180,000 down to about 80,000 Dalton. When using an amino-silanized metal surface (having a positive net charge), only impurities with a molecular weight lower than 140,000 Dalton influenced GOD adsorption.

When using adsorption from aqueous buffer solutions, functionally active monolayers of avidin or streptavidin form spontaneously, and irreversibly , on freshly evaporated gold and silver surfaces. The observation that both proteins form monolayers indicates that metal-sulphur interactions via cysteine residues are not a dominant factor, since cysteine is present in avidin but absent in streptavidin. This indicates that multiple ligand interactions with functional groups (NH_2, COO^-) are vital for the immobilization process. Chemisorbed layers of avidin or streptavidin are a straightforward means to bind a wide variety of biotinylated enzymes as a basis for electro-catalytic conversion.

For some proteins temperature is another important factor influencing adsorption. The adsorption of Glucose Oxidase (which is irreversibly adsorbed at platinum surfaces) shows a high temperature dependence resulting in significantly higher sensor response with increasing temperature during surface binding (0 to 37 °C: 2.3-fold increase). This may be related to conformation changes of the protein in the bulk solution at higher temperatures or at the sensor surface.

Inference by protein adsorption in complex systems: Among the most complex systems in which protein adsorption can be observed are biological fluids such as plasma or whole blood. The surface of a protein is complex in nature, having differences in characteristics such as hydrophobicity and charge. This complicates the prediction of how a protein will interact with the surface. The major factor influencing protein adsorption is surface energy. The total adsorption process is dynamic and can be divided into five steps:

(1) Diffusion or transport to the surface
(2) Binding via adsorption
(3) Structural re-arrangements at the surface
(4) Desorption or exchange with other proteins
(5) Diffusion or transport from the surface

The exchange processes between blood proteins at solid surfaces is thought to be a major factor in determining the behaviour of surfaces in contact with blood. Because of the exchange processes, the composition of the adsorbate changes with time. High-molecular-weight proteins present at low concentrations replace low-molecular-weight proteins. The kinetics and sequence of exchange varies with the type of surface and with the degree of dilution of the sample.

Since adsorption behaviour of proteins from blood is one of the major interferences in blood glucose determination using amperometric biosensors, several groups tried to investigate which of the blood-protein components are responsible for this interference. Serum fractionated over a molecular sieve column to yield fractions of different molecular weight is a means to search for the major interfering components or fractions. The interference of the blood

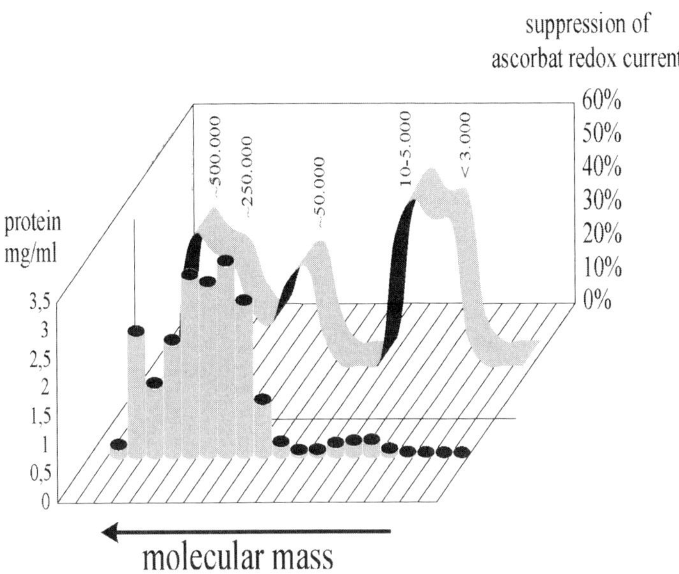

Figure 7 Inhibition of electrochemical response by serum fractions monitored via anodic oxidation of ascorbic acid at a platinum electrode.

protein fractions is estimated by measuring the suppression of the current caused by anodic ascorbate oxidation at the electrode.

Varying and strong suppression of redox reactions at the electrode surface in complex media (contrary to buffers or simple fermentation media) suggest the use of a semi-permeable membrane chemically altered to suppress unwanted protein adsorption at the electrode surface and to repel interfering substances that poison the electrode.

Inference by adsorption of oligo-saccharides to electrode surfaces: In addition to proteins and low-molecular-weight substances oligosaccharides (e. g., dextrans) also form adsorptive layers on platinum surfaces. The adsorption can be estimated by cyclovoltammetry because it significantly decreases redox currents of the electrode (e. g., PtO-reduction peak) This adsorption is dependent on the types of ions present (e. g. PO_4^{3-} compared to Cl^- shows no significant effect), on the ionic strength of the solution, and on the chain length of the oligosaccharide. Thus, a 10-fold decrease of adsorption is observed using solutions containing 100 mM NaCl. It also can be seen that high-molecular-weight dextrans (100,000 to 500,000 D) are nearly irreversibly adsorbed, while smaller ones (100 to 100,000 D) are adsorbed only weakly. Determination of adsorption can be done by cyclovoltammetry, i. e. comparing the PtO-reduction peak before and after cycling the electrode in dextran-containing solutions (500 to 700 mV, 53 mV/s).

Therefore, it can be concluded that the absorbability of a molecule at a metal electrode is determined not only by the functional groups and characteristic structural entities but also by size and diffusion properties.

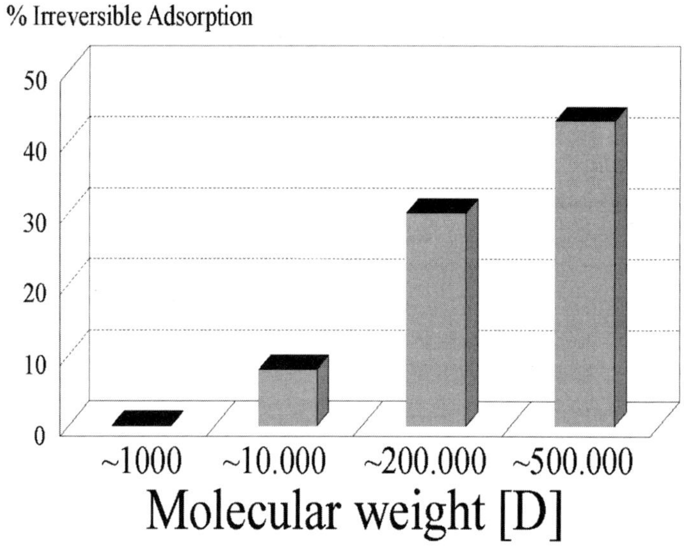

Figure 8 Irreversible adsorption to electrochemical electrodes

Covalent coupling techniques

There are various principles that can be used to couple proteins at or near metal surfaces [8–12]:

* *Entrapment* (include enzyme beyond a semi-permeable membrane covering the electrode surface)
* *Adsorption of proteins* (cited above)
* *Covalent coupling* of proteins to the sensor surface (see chapter 1)
* *Co-cross-linking* of proteins in gellayers

For most biosensors, covalent coupling or cross-linking is preferred because they offer several fundamental advantages. Thermal and long-time stability of chemically bound biomolecules are often much higher than the stability of adsorbed or entrapped species. Moreover, there is also nearly no physical loss of enzyme during long-time measurement because of the chemical coupling. Beyond that, covalent and biorecognitive binding allows the construction of protein monolayers as a basis for immune electrodes.

Before coupling a molecule, it is necessary to activate the metal electrode surfaces. Various agents and techniques have been developed to allow a covalent coupling of proteins.

Protocol 2 Activation of electrode surfaces

1. Detailed protocols are given in chapter 1.

Protocol 3 Coupling of proteins

Phosphate buffer is applicable for most enzymes and coupling techniques. Nevertheless, some immobilization techniques require another buffer to obtain a higher yield (see chapter on immobilization).

1. To couple protein to the activated surface, the electrodes are immersed in a solution of 5 mg/ml protein in 0.1 M phosphate buffer, pH 7.0, for 2 h.
2. The electrodes are rinsed with 4-molar saline to eliminate adsorbed protein.

Covalent coupling proceeds within a few minutes, often being limited by the diffusion process of the enzyme to the electrode surface. Typically, a yield of 90% surface coverage is obtained at the conditions given above within 10 to 15 min. Further incubation is often required to form a dense enzyme film.

Membranes
The glucose analyser from Yellow Springs Instruments was one of the first analysers employing cellulose acetate multi-membrane technology for glucose monitoring. A high oxidase loading and low interfering currents are among the main advantages. However, the thick membrane used in this sensor assembly leads to an increased response time and limited use in biomedical applications. Moreover, only automated techniques such as thick and thin film technology are able to produce the high numbers required in a cost-effective way. Whereas thick-film technology such as screen-printing of enzyme pastes is able to produce simple and low-cost sensors, highly integrated sensors require litho-

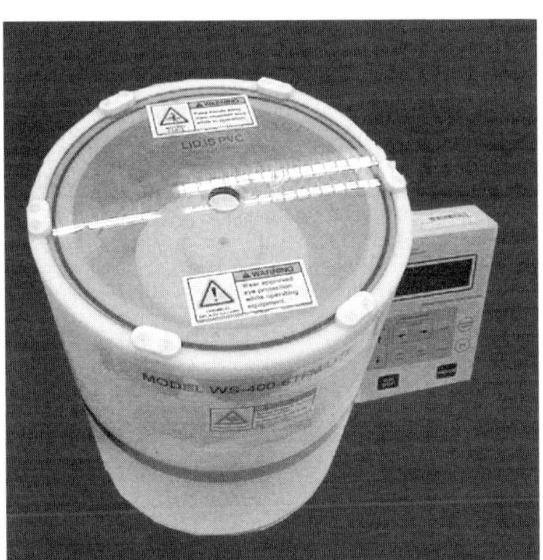

Figure 9 Photo-resist spin coater

graphically structured membrane technology using thin-film technology. Dip-, spray- and spin coating (see Fig. 9) are employed to form the enzyme layers on the biochips substrate [13].

When using oxidase-based biosensors (e. g., GOD), three types of membranes are recommended:

1. a hydrogen-peroxide-permeable ultra-filtration layer with a low cutoff directly on top of the metal electrode to suppress of interfering substances,
2. an enzyme containing membrane
3. a top membrane for diffusion limitation (linearization of response) and biocompatibility

The use of glutaraldehyde for the preparation of enzyme membranes was described 20 years ago. Most of these membranes are composed of the desired oxidase mixed with carrier proteins like BSA and cross-linked with glutaraldehyde. The gels are not well defined and sometimes are subjected to irreversible shrinking or swelling.

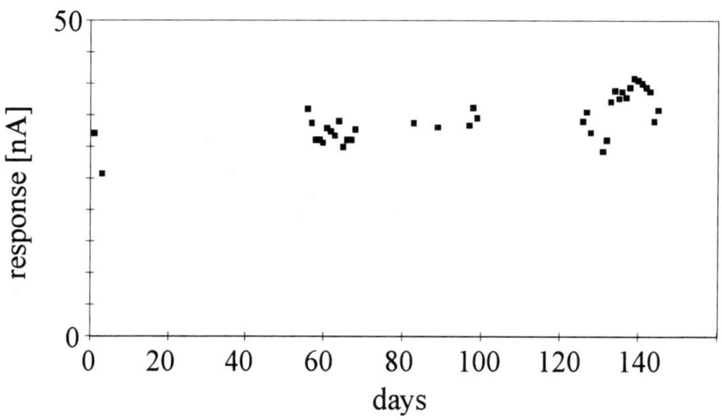

Figure 10 Stability of an electrochemical glucose sensor

An advanced technique is based on a new type of glutaraldehyde-diamine (amine) polymers. At neutral and slightly alkaline pH, amines reacting with glutaraldehyde do not form single enamine bonds but lead to linear high-molecular-weight polymers. Employing the reaction procedure described above and using mixtures of amines and diamine compounds results in stable cross-linked polymers of the desired properties. By using 1,6-diaminohexane as reacting and cross-linking amine (1 M) and glutaraldehyde (2.5 to 3 M) at an pH of 7 to 9, a hydrophilic gel will result. Two fundamental procedures are given below.

Spin-coating techniques are the most practical for forming thin and ultra-thin membranes (0.01 to 10 μm). Many preformed polymers can be dissolved to obtain highly viscous fluids, which can be used directly to spin membranes onto the surface of sensor substrates. Water-soluble polymers can be further cross-linked by a variety of multifunctional reagents (2,6-bis-(4-azidobenzylidene)-4-

methylcyclohexanone, epichlorhydrine, isocyanates, etc.) to obtain hydrophilic but water-insoluble networks. Being embedded in a hydrophilic network with hydrophilic amides or polyols increases the stability of a variety of enzymes.

Protocol 4 Coating with diamine / glutaraldehyde membrane

1. An aqueous solution of 1,6-diaminohexane (15%, adjusted to pH 7.5 with acetic acid) 5 µl , 0.1 M phosphate buffer (0.1 M, pH 7.0) 5 µl and glutaraldehyde (25%) 10 µl is mixed.
2. The solution is immediately used for coating the sensor by casting or spin coating.
3. To prevent the polymer from peeling off, the sensors should by dipped into Triton X-100 0.1% before coating with the polymer layer.
4. After cross-linking for 60 min. at room temperature, an oxidase can be coupled to the membrane by simply dipping the coated sensor into a buffered solution of the enzyme 0.1–10 mg/ml for 60 min.
5. Before use, the electrode needs CV-cycling (up to oxygen evolution potential) to remove inferring amines and aldehydes blocking the electrochemical response of the metal surface.
6. To achieve a high stability of the immobilised enzyme under dry storage conditions, a further layer of protein, e. g., hemoglobin, should be immobilised on top of the enzyme layer by dipping the sensor into buffered glutaraldehyde (5%, pH 7.0) for 30 min. at 25 °C followed by washing and incubating with hemoglobin (1%, pH 7.0, 30 min., 25 °C).

Sensor stability using the gel of protocol 4 and glucose oxidase was tested by comparing results obtained with 500 mg/l glucose standard over 150 days. The test sensor was stored at room temperature in FIA buffer between test runs. During the this period, significant decline of the sensor response could be observed. This excellent stability is due to the optimized design of the sensor as well as to the fact that the FIA system provides optimal conditions for repetitive analysis. Sample influence on sensor stability is minimized by short contact time (minimal sensor surface contamination and fouling) and relatively high flow rates during buffer rinsing cycles.

Protocol 5 Coating with diamine / glutaraldehyde / enzyme (GOD) membrane

1. An aqueous solution of 1,6-diaminohexane (15%, adjusted to pH 7.5 with acetic acid) 5 µl, a buffered solution of an oxidase (0.1–10 mg/ml, pH 7.0) 5 µl and glutaraldehyde (25%) 10 µl is mixed. If required, a redox-mediator such as ferrocene carboxylic acid is added to the mixture.
2. The solution is used immediately for coating the sensor by casting or spin coating.

3. To reduce protein interference or to prevent the loss of the redox mediator from the gel, a membrane with a cutoff smaller than the size of the mediator may be glued to the chip by putting the membrane into the wet (unpolymerized) gel.
4. Before use, the electrode needs CV-cycling (up to oxygen evolution potential) to remove inferring amines and aldehydes blocking the electrochemical response.

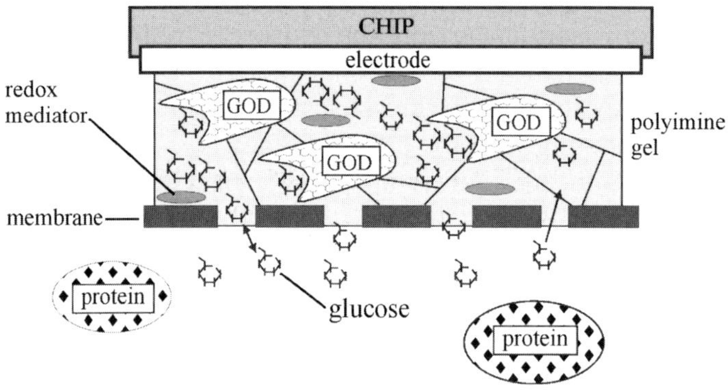

Figure 11 Redox-mediated gel-based glucose sensor

Protocol 6 Coating with poly-HEMA biocompatibility membrane

1. Glycerol 500 µl, water 500 µl, 2-OH-ethylmethacrylate (HEMA) 100 µl, aqueous solution of ammoniumperoxodisulfate (10%) 10 µl and tetraethylmethylenediamine 2 µl are mixed.
2. The mixture is immediately used for spin or dip coating of the electrode.
3. The polymerization takes about 45 min.
4. Before using for measurement, the electrodes are washed and equilibrated with buffer for 30 min. to swell the polymer layer.
5. The coated electrodes should be stored in a vapor-saturated atmosphere at 4 °C.

Photoprocessing of polymethacrylate membranes can be performed using a related method. A solution of a methacrylate polymer is mixed with additional monomer, multifunctional cross-linking monomer (e. g., a bis- or trisacrylate), and photo-inititator (e. g., Irgacure 651 a benzilketal) and is applied to the wafer by spin coating. Polymerisation and cross-linking are induced by irradiation with light and unexposed polymer developed by the solvent used for polymer dissolution.

Immobilization via photo-cross-linked gel

To apply mass production using thin film techniques it is necessary to structure enzyme layers. Since prepolymerized gels cannot be photo-structured easily, a photosensitive-gel layer being compatible with aqueous solvents is necessary [14–16].

Spinning techniques are the most practical for forming thin as well as thick membranes (0.01 to 50 μm). Polymers can be dissolved in an appropriate solvent to obtain highly viscous fluids, which can be used directly to spin membranes onto the surface of sensor substrates. Water-soluble polymer films can be further cross-linked by a variety of multifunctional reagents to obtain hydrophilic but water-insoluble networks.

Whereas standard photo-resist technology does not use swellable polymers, in a biological application water-based polymer systems are essential for creating chemically permeable structures with good adhesion.

Photosensitive bisazido compounds offer the possibility to combine membrane and micro structuring technologies. Sulfonated bisazidostilbenes are suitable for the cross-linking of preformed polymers containing amino groups, double bonds or $HC\text{-}R_3$ structures. Azidostilbenes can be photoactivated in the near UV range (< 385 nm), resulting in reactive nitrenes. A yellow coloring of the polymer film due to water-soluble by products can monitor the proceeding of the photoreaction.

Glucose oxidase, lactate oxidase, glutamate oxidase, and glutaminase (*Bacillus subtilis*, pig kidney, *E. coli*) have been incorporated and co-cross-linked with the polyvinylpyrrolidone polymer in spincoated films.

Protocol 7 Coating with a polyvinylpyrrolidone / photo crosslinker membrane

1. Water 900 μl, PVP (Mw 360,000) 90 mg, and 10 mg of 2,5-bis-(4-azido-2́-sulfobenzal)-cyclopentanone are mixed in the dark (or low light).
2. After addition of the desired protein (up to 5%) in water or buffer, the solution is spin coated onto the wafer. If required by enzyme stability, use a low concentration of a buffer but avoid crystals in the film (turbid); do not use amines!
3. The layer is dried and exposed to near UV radiation. Any UV source is acceptable, but exact timing needs to be optimized (typically between 1 s and 1 min.). Avoid short wave UV to minimize degradation of the biocomponent.
4. The wafer is developed in water or buffer for 2 min.
5. Additional layers using other enzymes and proteins can be coated layer by layer.
6. If the film 'crackly' cross-linking needs to be reduced. If the film peels from the electrode the electrode, might be amino-silanized prior to film deposition.

Immobilization via electro-deposited polymers

Given the well-defined structure of the microelectrodes, electrochemically prepared polymer coatings are efficient carriers for various enzymes. Because of the simple preparation at moderate pH and ion strength (e.g., pH 6.0 , 1 M KCl) polypyrrole is suitable for standard enzymes as e.g. glucose oxidase. A number of glucose sensors have already been constructed that deal with the immobilization of glucose oxidase by copolymerisation with pyrrole and adsorption on or inclusion in pyrrole polymers [17–19].

Sensors using polypyrrole as an electrochemical active electrode and a mediated electron transfer (by, e. g., ferrocene, quinone) are independent of oxygen, but sometimes suffer from low operational stability as a result of mediator leakage and destruction of the mediator and the polypyrrole matrix by oxygen and hydrogen peroxide.

Considering the limits of enzyme monolayer electrodes, three fundamental advantages of electrodes covered by microporous polymers are

* significant increase of response per unit area that is due to the porous surface
* permeation control by the polymeric layer for interfering electroactive substances (e. g., ascorbate, thiols, etc.) resulting in a distinct increase in selectivity.
* oxygen demand of the sensor is significantly decreased because of the high permeability of the bulk polypyrrole for oxygen and a significantly lower permeability for glucose (extend linear range).

A tuning of electrode behavior can be done by the use of modified polypyrrole layers with or without significant conductivity and redox activity under working conditions. For optimal performance, the electrocatalytic reaction should take place at the metal surface; otherwise, high background currents and low selectivity will be obtained.

For the coating of platinum electrodes with polymeric layers of 1- and 3-substituted pyrroles, e. g., 1-(carboxyalkyl)-pyrroles, 2-(1-pyrrolo)-acetylglycine, 1-alkylpyrroles, 1-(4-carboxybenzyl)- pyrrole, 1-(4-nitrophenyl) pyrrole, 4-(3-pyrrolo)-4-ketobutyric acid, 3-((keto 4-nitrophenyl) methyl) pyrrole, are useful monomers. By electrochemical oxidation and polymerisation of these monomers in organic solvents, various types of polymers can be deposited at the electrode surface. Homopolymers, heteropolymers, and sandwiched types of these polymer layers may be used. The various nano-structures obtained by polymerisation of the modified pyrroles allow designable properties of the resulting enzyme sensors. Satisfactory homopolymer layers can be formed only with pyrrole derivatives having no bulky side groups directly attached to the pyrrole ring. Also, it is vital to increase hydrophobicity of carboxy-substituted pyrroles by using long chain derivatives, otherwise, the polymer films will be soluble in the neutral and slightly alkaline buffered solutions the sensor should be used in.

To overcome the problems due to increased solubility of, e. g., carboxylated polypyrroles, copolymerisation with pyrrole is an efficient way to get stable and nevertheless reactive polymer films. When synthesizing these heteropolymers, it is important to pay attention to the fact that the compound with the more positive polymerisation potential will be incorporated into the polymer to a much lower extent than its concentration in the polymerising solution would normally allow.

When comparing copolymers of all the above-mentioned modified pyrrole monomers with pyrrole it turned out that the most hydrophilic compounds gave the most promising copolymers for sensor preparation. 2-(1-Pyrrolo)-acetylglycine forming only thin water-soluble polymer films can be copolymerised with pyrrole to obtain porous films having optimal enzyme load.

COOH and nitro groups that are stable against oxidation under polymerizing conditions have proved to be optimal to obtain a modified polypyrrole layer that can be used for covalent coupling of enzymes using water-soluble carbodiimides and chloranil as activating reagents.

For the coupling of the enzyme, the pore size of the polymer layer must be at least 10 to 20 nm. Because of the fractal growth of the polymer on the electrode during polymerization slight changes in cycling speed, water content, purity of reagents, etc. do have a great influence on the pore size and inner surface of the polymer film. It is therefore vital to optimize the polymerization of each sensor lot and polymerization mixture.

The activated electrodes are reacted immediately with glucose oxidase and the glucose sensors thus obtained are stored at 4 °C.

A substituted polypyrrole layer acts as an effective barrier for interfering redox active compounds such as ascorbic acid, bioactive amines, sulfhydryl-containing peptides and proteins, but has a high permeability for, e. g., hydrogenperoxide.

A characterization of a typical electrode is given below:

* Porous polypyrrole immobilized glucose oxidase has a 10 to 50-fold greater response/sensor area than monolayer electrodes because of a significant increase in inner surface.
* The temperature dependence of the signal is 5% / °C and the base current is 1.5 to 3 nA/mm^2
* The response time is in the range of a few seconds, increasing if a further membrane covers the electrode.
* The response of polypyrrole-coated electrodes is nearly independent of fluctuations in the test solution.
* A high reproducibility requires a good lab standard when immobilizing on the internal surface of a polymer layer.

The unspecific response of interfering redoxactive substances (e. g., ascorbic acid, phenacetine, etc.) can be suppressed by using either differential measurement with a double-electrode structure or the polypyrrole type of electrodes exhibiting permeation-selective properties. Both techniques can be combined for optimal performance (see Fig. 12).

A typical property of an 2-(1-Pyrrolo)-acetylglycine polymer-coated glucose sensor (GOD) is the electrochemical response as a function of voltage has the same shape as on a native platinum electrode. Under optimum conditions, the electrodes working at a potential of 500 mV/(Ag/AgCl) have a linear response up to 25 mM without any further diffusion limiting membrane. By covering the electrode with a polymer membrane, the linear range for glucose can be extended to more than 60 mM.

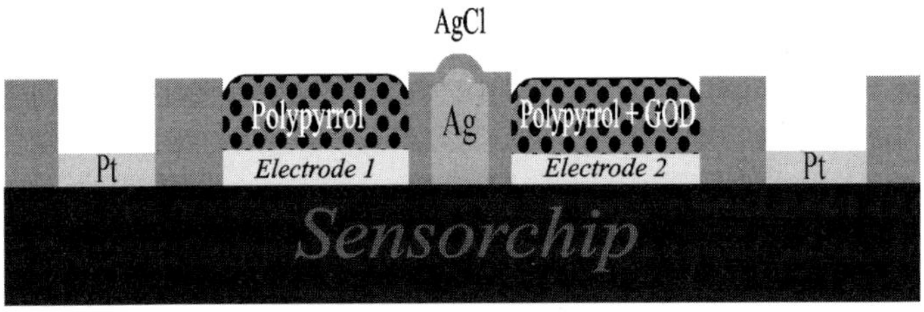

Figure 12 2+2-electrode polypyrrole glucose sensor

Protocol 8 Preparation of a polypyrrol amperometric glucose sensor

1. Cleaning of the electrode: Platinum thin-film electrodes are cleaned by ultrasonication in distilled water, rinsed with acetonitrile, and dried carefully under dust-free conditions.
2. The electrodes are cycled five times in acetonitrile/2.5% $LiClO_4$ between -500 and 1800 mV (100 mV/s) *versus* Ag/AgCl using a three electrode configuration in an electrochemical cell. A potentiostate is employed for generating the required voltage sweep.
3. Coating with polymerised substituted homopolypyrroles: A 0.5% solution of the monomer (see Tab. 3) in acetonitrile containing 2.5% of $LiClO_4$, NR_4BF_4 or NR_4PF_6 (R = Me, Et, Bu) is dried over Na_2SO_4 or $CaCl_2$ for several hours.
4. The clear solution (filter if necessary) is transferred to a glass cell and bubbled with argon for 10 min.
5. The electrode immersed in the de-aerated solution is cycled 10–30 times between – 300 mV and the potential listed given in Table 3 (100 mV/s).
6. If the substituted pyrrole is polymerisable, the formation of a thin brown or black polymeric layer is observed. Most of these polymers showing redox activity in acetonitrile/$LiClO_4$ were not redoxactive in aqueous solutions at pH 7.0.
7. Polypyrrole coated electrodes having COOH groups are incubated with a saturated solution of N-cyclohexyl-N′-[2-(N-metylmorpholino)-ethyl]-carbodiimide-4-toluenesulfonate for 30 min. at 25 °C without shaking. The electrodes are rinsed several times with water and immediately reacted with a 5 mg/ml enzyme solution in buffer pH 5–7; 0.1 M for 2 h.
8. Substituted polypyrrole films having nitro groups are reduced using a solution of 1% $SnCl_2$ in 10% HCl for 30 min. forming a film with pending amino- groups. After excessive rinsing with diluted HCl, distilled water and ethanol, these electrodes are dried. A variety of amino-reactive reagents such as 1,4 – arenequinones can be employed to couple the enzyme (see chapter 1).
9. The sensors thus obtained are rinsed with buffer and water and stored at 4 °C.

Table 3 Monomers useful for electro-polymerization

Compound	U [Ag/AgCl]*	redox peaks+
1. 1-(5-Carboxypentyl) – pyrrole	1200 mV	350–830 mV
2. 1-(10-Carboxydecyl) – pyrrole	1200 mV	350–830 mV
3. 2-(1 – Pyrroleo) – acetylglycine	1200 mV	550–800 mV
4. 1-Dodecylpyrrole	1200 mV	550–780 mv
5. 1-(4-Carboxybenzyl)- pyrrole	1200 mV	450–750 mV
6. 1-(4-Nitrophenyl) pyrrole	1400 mV	730–770 mV
7. 1-(4-Carboxyphenyl) pyrrole	only copolymer	
8. 4-(3-Pyrroleo)-4-ketobutyric acid	1300–2000 mV	
9. 3-((Keto 4-Nitrophenyl) methyl) pyrrole	1800 mV	850–950 mV

** to obtain stable polymer films*   *+ in Acetonitrile / LiClO4*

Stability, solubility and permeability of polymer films

Compound ->	1	2	3	4	5	6	7	8	9
stability of the polymer film in aqueous buffer pH 7	+	+	–	+	+	+	+	–	+
redoxactivity of polymer film in aqueous buffer pH 7	+	+	–	+	–	–	–	–	–
permeability for hydrogenperoxide	+–	+-	0	–	+–	+–	–	–	0

For most sensor applications, it is necessary to make a copolymer having a high content of unsubstituted polypyrrole in order to obtain thick and stable substituted polypyrrole films. Since the polymerisation speed of the substituted pyrroles is significantly lower than that of unsubstituted pyrrole, it is, nevertheless, necessary to have a 10:1 excess of the substituted monomer.

Protocol 9 Preparation of a polypyrrole copolymer amperometric glucose sensor

11. Follow protocol 8
12. To obtain a stable copolymer film, the electrolyte of the above-mentioned composition is complemented with additional 0.05% pyrrole.
13. The electrode is cycled up to 10 times between –300 and 1400 mV (100 mV/s). A black polymer layer of varying thickness is obtained.

For coupling of the enzymes, a variety of methods might be employed to activate, e. g., the terminal carboxylic acid groups. The use of water-soluble carbodiimides may be recommended for carboxy-modified pyrroles.

Redox-mediated electrodes

Because oxidases use molecular oxygen as a co-substrate, it is sometimes necessary to add chemical-replacing oxygen in the enzymatic reaction. These molecules are “redox mediators” because they mediate the electron transfer from the biomolecule to the electrode surface [20]. A beneficial effect of these mediators is the independence of the sensor signal on oxygen saturation of the solvent, e. g., the oxygen level in a fermentor. Moreover, the linear range is often extended. Nevertheless, it is difficult to keep redox mediators around the enzyme. Some tricks are given in Figure 13. Whereas free mediators such as ferrocene carboxyl acids or quinones are soluble and thus lost by diffusion processes, encapsulated or immobilized forms are more stable. Nevertheless, even immobilized redox-transfer systems suffer from the instability of most mediators being destroyed by oxygen or redox-reactions at the electrode surface.

A simple technique is to add the mediator, to the gel mixture prior to polymerization. To test the efficiency of the mediator the oxidase sensor and all solutions are flushed with argon for 10 min. A remaining analyte response is due to a mediated and no longer oxygen–hydrogen peroxide redox transfer.

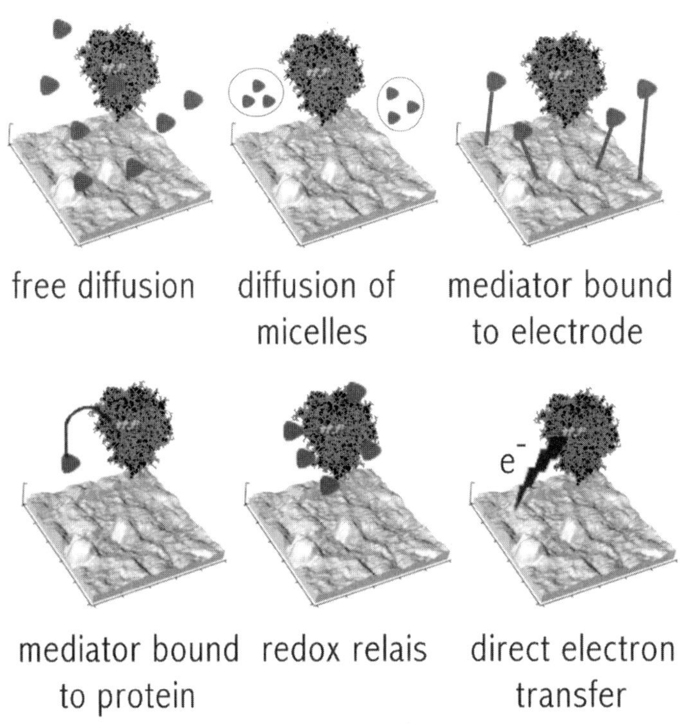

Figure 13 How to keep the redox-mediator near the enzyme

Nano-column and nano-well biosensors

When departing from the classical biosensor geometry nano columns or nano wells, place the immobilized reagent adjacent to rather than on top of the transducer. The operation of the system is reminiscent of the "dipstick" technology (see chapter 4). For some applications, the basic channel geometry had been optimized in analytical flow channels to achieve a signal enhancement by including parallel electrodes capable of electrochemical recycling. The integrity of kinetic performance is ensured by laminar flow, with a Poiseuille velocity profile. The nano cells are high performance types of the flow-through cells that are not stand-alone but are connected with a pumped flow system in flow injection analysis (FIA) and liquid chromatography (LC).

With standard FIA, the necessity for continuous flow lines, solution reservoirs, and a pumping system generally renders the technique too complex for small disposable devices that can operate at the site of analysis, although some systems have been presented. Thus, portable channel cells for biosensor applications all require pumping or do not include a continuous flow period in their operation. Often the cell is filled by capillary action, but the flow stopped once the cell is completely full.

Figure 14 Nano-devices *versus* standard microtiter plates

Amperometric immunosensor

Enzyme-immunosensor systems based on the use of enzymes as labels follow the same basic principles as enzyme immunoassyas combined with an enzyme biosensor. The sensor systems differ from standard immunoassays in the degree of automation; in the detection principle, often a flow through solid phase to which the antibody is bound, and the reusability of the biolayer. Among others,

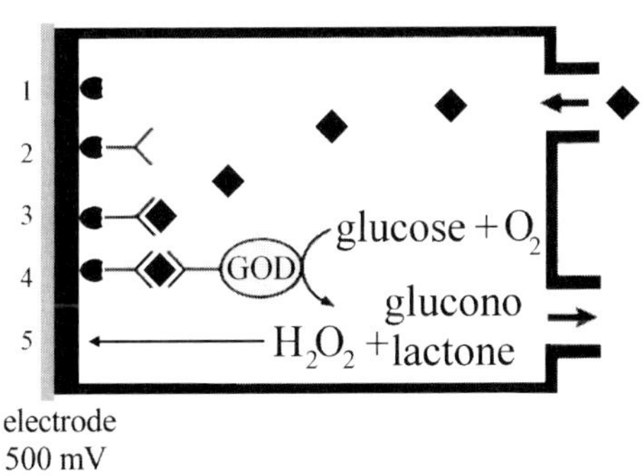

Figure 15 An amperometric flow-through immunosensor

glucose oxidase, alkaline phosphatase, lipase, and urease have been used as labels with electrochemical detection. The sensor-system protocol is similar to the corresponding microtiter plate assay–antibody immobilization, the addition of sample, the addition of enzyme tracer, the washing of the sensor, the addition of enzyme substrates, incubation, and detection. Biosensor systems add a final regeneration step, which allows reuse of the biolayer by removing the analyte from the sensor. This can be achieved by varying pH and adding solutions such as ethylene glycol or chaotropes such as urea or thiocyanate. Nevertheless, these regeneration protocols stress the immobilized protein, thus leading to a limited reusability of the sensor device. To compensate for this loss, antibodies are often bound to proteins, such as protein A or protein G, allowing removal by the regeneration protocol and a fresh loading. Automation is achieved with standard (ml) and increasingly with micro or nano (µl or nl) flow systems. Systems have been described that often are suitable for direct coupling to a sample line for online monitoring (see below).

The most important element in these automated systems is quite often the affinity column, which contains the biorecognitive component. All solutions are pumped through or incubated in this column.

Several setups have been published using, e. g., flow-through column filled with beads, glass capillaries, silicon-chips, or microtiter wells with attached electrochemical devices (see Fig. 16). Even a thin-layer structure of a few µm in diameter and an electrode can serve as affinity reactor.

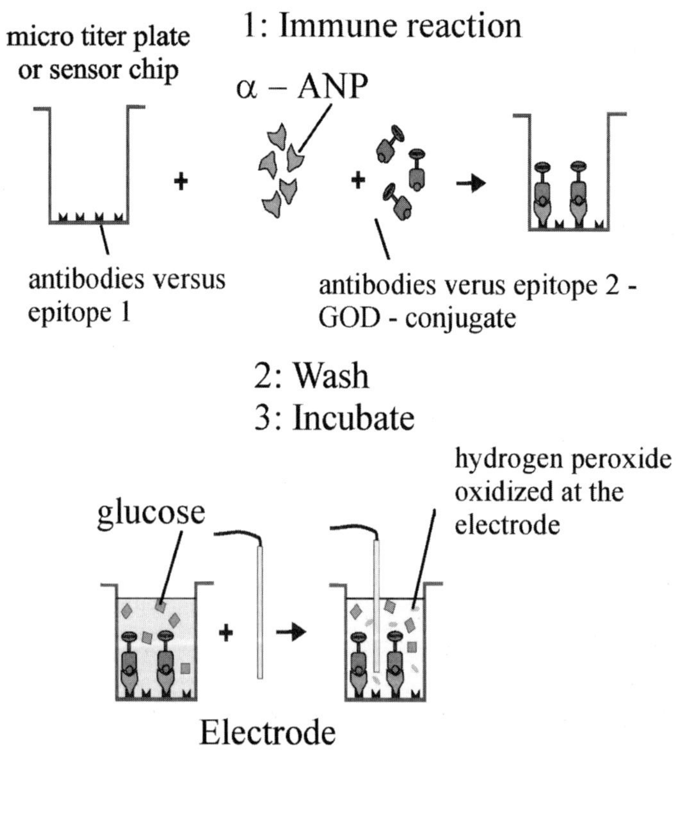

Figure 16 An automated electrochemical amperometric immune assay

Amperometric DNA-sensor

Enzyme-transduced DNA sensor systems follow the same basic principles as amperometric immunosensors. The sensor systems differ from standard immunosensors in the degree of thermostability required, sometimes up to 90 °C, to hybridise and re-elute DNA from the chips surface to achieve a reusability of the bio-layer. One way to circumvent thermostability problems is the use of short oligonucleotides or the use of organic modifiers such as formamide.

A vital element of such an enzyme-based device is to achieve a low unspecific background adsorption, often limiting the performance of the device. Adsorption of DNA on several types of silanized, oxidized, and unmodified sensors have been studied by several groups. It turned out that only a mercapto-silane layer provides a very low adsorption background [21]. Nevertheless, it should be kept in mind that thiol groups may interact with biological molecules other than DNA to form thiol bonds as well as nucleophilic reaction products.

Activation of the thiol groups of mercaptosilane, coating by nucleophilic substitution with iodoacetic acid, and coupling of aspartic acid with carbodiimide modify adsorption on the sensor surface. Amino-silane coated sensors have a very high unspecific adsorption that cannot be reversed by washing. It also should be noted that slight scratches in the thin film sensor surface can irreversibly adsorb significant amounts of DNA.

To achieve a stable sensor matrix, DNA-oligonucleotides are bound similar to DNA biochips. For covalently coupling ss-DNA three strategies can be used:

1. Random coupling using the intrinsic but not very reactive amino groups of the oligo nucleotide bases.
2. Selective immobilization using a natural terminal phosphate.
3. Artificial mostly terminal linkers with NH_2 or SH groups.

For coupling via amino-linker, see chapter 6.

Protocol 10 Coupling of SH-oligonucleotides to electrodes with mercaptopropyl–triethoxysilane coating

1. Metal surfaces are oxidized (as describe in the chapter 1) to introduce functional hydroxy groups.
2. The oxidized electrodes are incubated in a 5% aqueous solution of trimethoxypropyle-mercapto-silane, pH 3.5, at 37 °C for 30 min.
3. The silylated surface is cleaned with water and ethanol.
4. Using an alternative procedure the oxidized electrodes are incubated in a 7% aqueous solution (containing 50% acetone) of 3-mercaptosilane, pH 4, at 37 °C for 30 min.
5. To enhance silane layer stability, the electrodes are further cross-linked by drying them at 110 °C for 15 min.
6. P-chloranil or p-halogenanils such as p-fluoranil or p-bromanil are used to activate the electrode surface by incubating the silanized sensors with a 1% solution for 30 min. at 25 °C.
7. After activation with quinones, the sensors can be stored at 4 °C in the dark.

8. To couple SH-terminated oligonucleotides, the sensor is incubated in an aqueous solution for 10 min. up to 2 h (time depends on concentration of oligo-solution). If long coupling times are required, oxygen protection, e. g., by argon flushing is necessary.
9. Note: most SH-linkers require a deprotection step often using incubation with silver salts (see manual of supplier!).

For non-terminal immobilisation, mercaptosilane-coated electrodes are activated by either iodoacetic acid and carbodiimide or chloranil, thioacetic acid, and carbodiimide. Both procedures have a high affinity for all primary amino groups.

Protocol 11 Coupling via activation by iodoacetic acid and carbodiimide

1. Metal surfaces are SH-silanized as descibed in protocol 10.
2. Mercaptosilane-coated electrodes are incubated with iodoacetic acid (1%, pH 8.0) for 30 min. at room temperature.
3. Sensors are rinsed with water and activated with water-soluble carbodiimides, e. g. by incubating the sensor in an aqueous solution of N-cyclohexyl-N′-[2-(N-methylmorpholino)- ethyl]- carbodiimide- 4-toluene sulfonate (CDI) (60 mg/ml) for 90 min. at room temperature.
4. After activation with CDI the sensor is coupled immediately with an oligonucleotide in aqueous solution (60 min.).

Protocol 12 Coupling via activation by chloranil, thioacetic acid, and carbodiimide

1. Metal surfaces are SH-silanized as descibed in protocol 10.
2. To activate the electrode surface the silanised sensors are incubated with a solution of chloranil (1% in toluene) for 30 min. at 25 °C, washed with toluene and acetone.
3. The chips is reacted with thioacetic acid (10% solution buffered to pH 6.0) for 30′ at room temperature and coupled with carbodiimide as listed above.
4. After activation the sensors is coupled immediately with an oligonucleotide in aqueous solution (100 pM/ml, 2 h) .

For immobilization of unmodified oligonucleotides mercaptosilane coated electrodes are activated by iodoacetic acid, carbodiimide, diaminohexane (as a spacer) and carbodiimide.

Protocol 13 Activation by iodoacetic acid, carbodiimide, diamine, and carbodiimide

1. Electrodes are reacted with iodoacetic acid and carbodiimide as described.
2. To activate the electrode the chips are incubated with 0.2 M 1,6-diaminohexane (pH 6.0, 25 °C) for 1 h.
3. Oligos with a reactive terminal phosphate group are coupled by incubating in 0.1 M imidazol, 0.1 M water-soluble carbodiimide, and oligonucleotide (100 pM/ml) at 25 °C for 12 h.
4. A terminal phosphate group is not present in most standard DNA fragments and needs to be introduced by an enzymatic kinase reaction (see DNA handbook).

Coupling via chloranil (see chapter 1) should selectively immobilize SH – modified oligonucleotides since it prefers linkage to thiols that suppress most active amino groups. Thus, chloranil is able to immobilize SH-modified oligonucleotides with very high selectivity.

Random coupling procedure (about 50% less effective) is able to immobilise any DNA via amino groups. Terminal immobilisation by phosphate coupling is not as effective but because a higher mobility of the immobilised strands of DNA, the hybridisation properties are superior to non-terminal immobilization. It should be kept in mind that phosphate-amine bonds do have a limited stability regarding extreme pH or temperature. Thus, regeneration of the chip by heating or organic solvents may result in a loss of the bound oligonucleotide.

The biorecognition process of a DNA biosensors is similar to biochips (see chapter 6), while the detection process of enzyme-labeled biosensors is similar to enzyme assays (see chapter 6) and amperometric biosensors.

Potentiometric biosensors

Potentiometric measurements involve non-faradayic electrode processes with no net current flow and operate on the principle of an accumulation of charges at an electrode surface. The charge redistribution process results in a significant potential at that electrode. The potential is given by the

Nernst equation: $E = E_0 \pm (RT/nF)\ln a_1$,

where E_0 is the standard potential for $a_1 = 1$ mol l^{-1}, R is the gas constant, F is the Faraday constant, T is the temperature in K, and n is the number of charges per ion. This potential is proportional to the logarithm of the analyte activity (concentration) present in the sample and is measured relative to an inert reference electrode, also in contact with the sample. A standard setup includes a reference electrode, (inert) and one working electrode both in contact with the sample. The advantage of potentiometric methods is the broad dynamic range (10^{-6}–10^{-1} mol/l) of measurable concentrations caused by the logarithmic correlation of signal and analyte concentration. The sensor is inexpensive, portable, and is well suited for *in situ* measurements. This type of configuration has several drawbacks, however. Disadvantages are the lower sensitivity compared to amperometric readout and the rather poor selectivity. Furthermore, the analyte response curve is influenced by the buffer capacity of the solution, which must be adjusted in the test sample [22, 23].

A simple potentiometric sensor, the pH electrode, was used to measure the activities of enzymes, such as urease, penicillinase, glucose oxidase and acetylcholinesterase, that either produce or consume protons.

Ion-selective potentiometric sensors are setup from a semipermeable membrane, which is only permeable for either anions or cations and impermeable for the counterion. The membrane divides sample and reference solution. A concentration gradient of the analyte leads to the formation of a potential at the membrane. Gas-detecting sensors can be constructed of an ISE with a gas-permeable membrane. A biosensor can be achieved by immobilization of an enzyme to the membrane. Reaction of the analyte with the immobilized enzymes leads to changes in the concentration of the molecules that can pass through the membrane barrier. This sort of biosensor is used for the detection of NH_3 from urea or creatine and for the detection of CO_2 from urea or amino acids.

A miniaturization and further development of ion selective electrodes are ISFET techniques (ISFET = ion selective field effect transistor). A biosensor of this kind is an insulated gate field-effect transistor (MOSFET), in which the usual gate metal electrode has been replaced by a suitable sensitive membrane and a reference electrode. Enzyme-sensitive field-effect transistors can be fabricated from ISFETs by applying a thin layer of enzyme-loaded gel onto the ion-selective membrane. In practice, a dual-gate pH-ISFET chip is normally employed so that one of the FETs can act as a reference for the ENFET when its gate is coated with an enzyme-free membrane. If the difference between the two drain currents is monitored, the signal is insensitive to changes in the solution pH, temperature, or electrical noise, as the reference FET and ENFET respond almost equivalently to changes in sample potential and only enzyme-generated pH changes are measured.

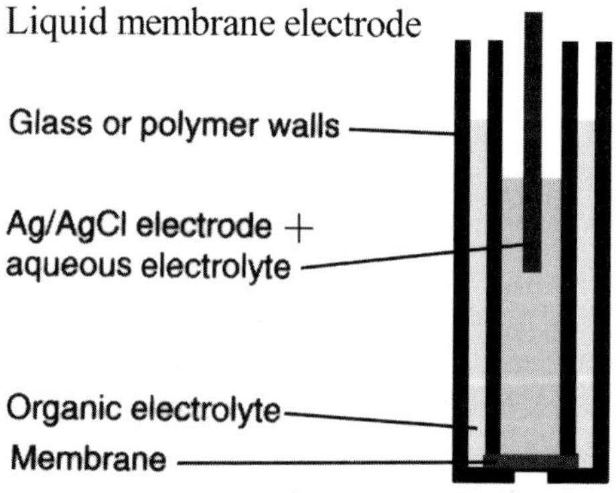

Figure 17 A potentiometric ion sensor

Conductimetric biosensors

The change of conductivity of a solution in correlation with the concentration of the analyte is determined. The derived signal is proportional to the square root of the concentration of interest. Enzyme reactions, which produce or consume ionic species, will change the conductance of the solution to a greater or lesser extent. Planar inter-digitated electrode configurations have been reported as conductometric transducers for biosensors. Enzymes such as urease, which catalyzes the production of ammonia and carbon dioxide, have been used in these devices. Nevertheless, because the conductance is sensitive to temperature, redox processes and double layer charging differential methods must be used. Urease-coated electrodes in combination with a control electrode coated with an inactive protein have been used to measure urea.

The main disadvantage is that ionic species produced must significantly change the total ionic strength. This increases the detection limit to unacceptable levels and results in potential interferences from variability in the ionic strength of the sample. The dynamic concentration ranges from 1 to 100 µM for these devices.

Whereas conductivity measurement offers little advances in standard biosensor construction, novel sensors related to biological cell membranes get increasing attention.

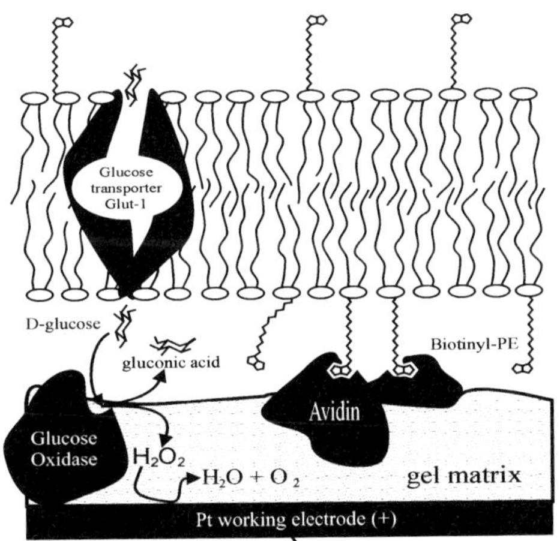

Figure 18 A supported membrane glucose sensor using a glucose transporter and an enzyme

Ion-channel biosensor

A biosensor that mimics biological membrane functions has caught the attention of pharmaceutical companies engaged in drug discovery. At least three groups using a supported membrane or ion-channel switch biosensor have published data [24–27]. The unique feature of this sensor is that it gives a functional test of the interaction between molecules and a receptor, respectively and an ion channel in a lipid membrane. Changes of ion flux across the membrane are directly detected as a change in the membrane's electrical conductance with a sensitivity approaching single molecule events. The sensors are fabricated with chip technology, but novel strategies are required to achieve sensitivity and even more critical long-term stability.

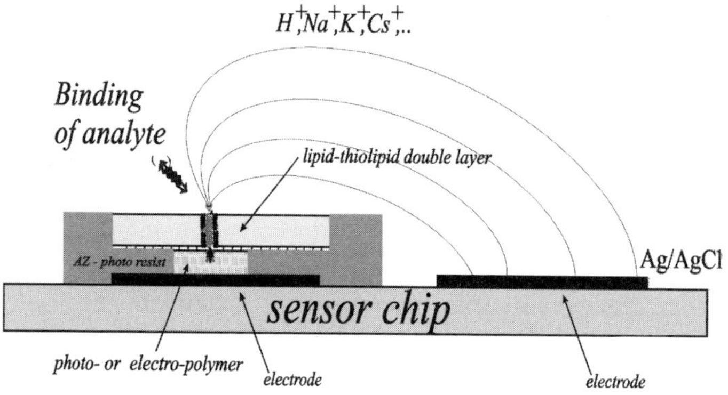

Figure 19 Setup of a supported membrane ion channel sensor using thin-film technology

The basic constructive element of an ion channel biosensor is a highly insulating synthetic lipid bilayer membrane bound to a support with a well-defined ionic reservoir between the membrane and the substrate. A reservoir for ions is necessary and may be assessed by using ionophores such as valinomycin. The absence of a reservoir causes the impedance spectrum to be essentially linear over the 1 Hz to 1 kHz range. The dimensions of the reservoir are important in determining the magnitude of current that flows across the membrane.

The use of a ligand coupled to or near an artificial gated membrane ion channel is the basic strategy. Binding of protein or DNA/RNA analytes at the membrane channels results in an on/off-response of the channel current that is channel closure or distortion.

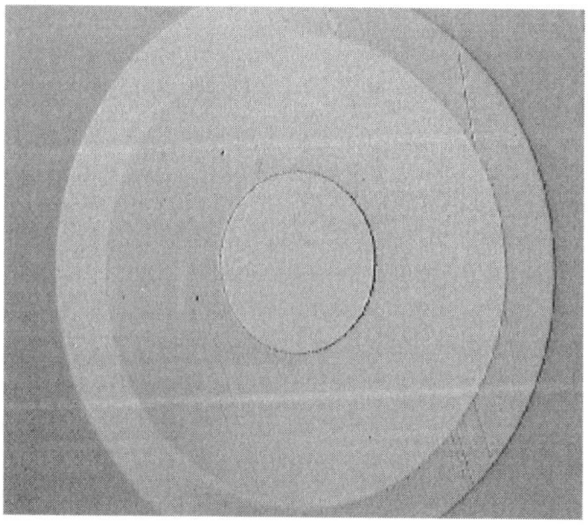

Figure 20 Artificial supported patch membrane on a thin-film chip

The sensor consists of stable, trans-membrane channels with a ligand bound covalently either at the peptide channel entrance or near to it, a sensor chip with a photo-structured hydrophobic polymer frame, a hydrophilic ion-conducting membrane support, a lipid membrane incorporating the engineered ion channels, and a current amplifier (see Fig. 21). Detection of channel opening or closure can be obtained either by directly monitoring membrane conductivity or by using a transient pulse of pH or ion concentration within the membrane compartment. This change can be induced by electrochemical or optical means and its decay is directly correlated to the permeability of the membrane.

To obtain the stable sensor membrane, the lipid-layer has to be attached on a support and the floating of the second lipid membrane on top of the first one has to be prevented [33–36].

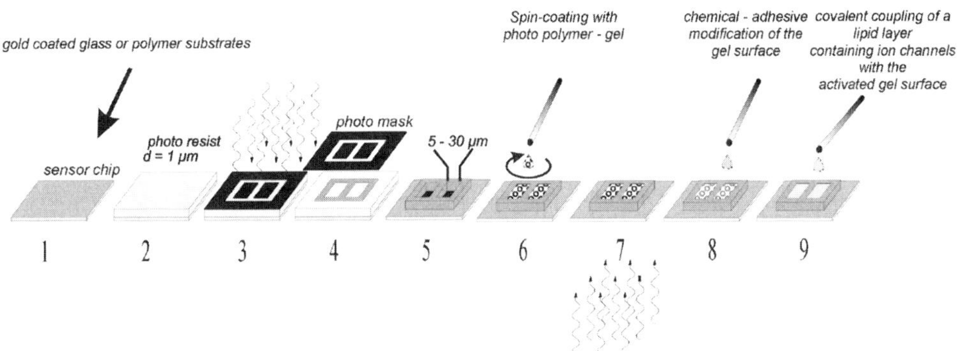

Figure 21 Schematic production line of supported membrane thin-film chips

Protocol 14 Photolithography of membrane frames

1. To setup a supported membrane sensor, lithographic steps are necessary.
2. Spin coat the resist to form a homogeneous film (~ 500 nm to 2 μm thick).
3. Align wafers to mask and expose.
4. Typical exposure times are 30 to 60 s.
5. Cover wafer box with Al foil when carrying exposed wafer into etching room.
6. Pour 100–500 ml of developer solution into a beaker.
7. Turn off the white lights in the etching room; develop and etch only in yellow light.
8. Dip the exposed wafer into the developer for 40–60 s; it takes more time if the developer has been used often. Agitate slowly. Observe the reddish liquid forming near the wafer. It is the dissolved photo-resist.
9. Rinse in DI water for 10 s in each of three dedicated rinse beakers successively, then spray rinse.
10. Spin dry.
11. Inspect under microscope with yellow filter.

12. If not completely developed, repeat.
13. Hard bake the photo-resist in an oven for 20 min. at 120 °C. To cross-link the resist ultra-bake at 160–180 °C to convert the resist into a plastic frame.
14. Hard baking drives off the rest of the solvent, making the photo-resist physically hard, more adhesive, and less permeable to chemicals.
15. Use photo-cross-linked polyvinyl-pyrrolidone gel or electro-polymerized polypyrrol as membrane support. Continue with protocol 7 or 8.
16. Coat lipid membrane on the hydro-gel layer similar to black-lipid-membrane devices with painted or folded lipid membranes. For optimal adhesion, a few percent of reactive lipids need to be added to serve as membrane anchors.
17. Keep in mind that Ag/AgCl electrodes and sub-membrane compartments of sufficient size are necessary if using DC for readout. Using AC-impedance readout gold electrodes and small sub-membrane compartments might be used.
18. To test the performance of the device use Gramicidin channels and a standard membrane mixture of DOPC 0,05 µg/µl + lecithin 0.5 µg/µl, subphase: 100 mM KCl, potential: U = 300 mV. Time resolution = 100 ms, Gramicidin ion channels (~ 5000 molecules), Gramicidin is inhibited by bivalent heavy metal ions, for typical readout, see Figure 22.

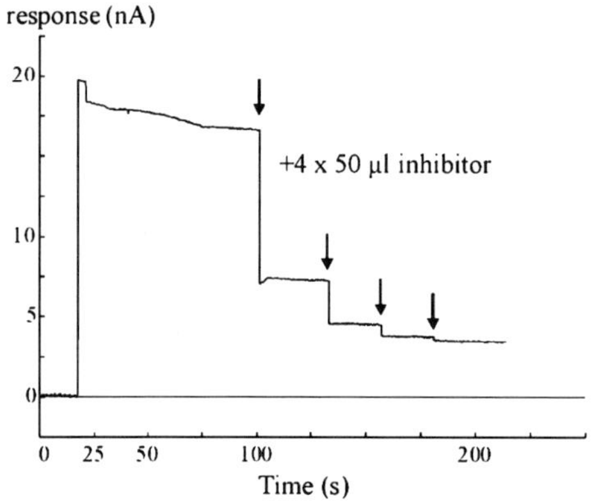

Figure 22 Read out from ~ 3000 ion channels in an artificial supported biomembrane

Capacitive biosensors

Capacitance measurement also has been investigated as a novel direct approach to biorecognitive binding. When an immunosensor based on a capacitive transducer is constructed, the size of the bound molecule *versus* the size of the binder at the chip surface is of vital importance. Moreover, the assembled recognition layer should be electrically insulating to prevent interferences from redox reactions inducing high faradaic background currents. On the other hand, it should be as thin as possible in order to achieve high sensitivity [28].

Different immobilization procedures onto different substrates have been reported in the literature. Antibodies have been grafted covalently via silanization to quartz, coated via a Langmuir-Blodgett film or coupled via silanization to tantalum oxide electrodes.

Self-assembled monolayers of thiols, sulfides, and disulfides on gold electrodes have been widely studied, and alkane thiols are known to form insulating, well-organized structures on metal substrates. The binding formed between the sulfur atom and a heavy metal is very strong, and the formed self-assembled monolayers are stable in air and water. Low–molecular weight antigens bearing a sulfide group had been synthesized and immobilized via chemisorption to form a self-assembled monolayer (SAM) on a gold electrode. A thiolated peptide can be immobilized as a self-assembled layer on gold. Micro-contact printing and photolithography can be used to pattern surfaces with functionalized self-assembled monolayers for biosensor production.

A typical setup uses a flow cell with a volume of 1 to 2 ml. The working electrode is a gold-coated sensor. The carrier solution is pumped with a flow rate of ~ 0.5 ml/min through the flow cell. An injector with a loop of 250 µl is connected to the flow system.

When an antigen binds to the antibody immobilized on the electrode, there will be an additional layer decreasing the total capacitance. The binding between the antigen and antibody is therefore detected directly and no label is necessary for the antigen. The physical basis for the response thus arises from displacement of the polar water by much less polar molecules.

To monitor the antibody-antigen interaction at these electrodes capacitance, impedance measurements and cyclic voltammetry might be employed. The capacitance changes can be evaluated from the transient current response obtained when a potentiostatic step is applied to the electrode. Another technique relies on the evaluation of the currents of sinusoidal waves, usually called impedance spectroscopy. The two methods give almost the same results in terms of equivalent capacitances and resistances. The potentiostatic pulse method is faster and more convenient. The measuring setup consists of a three-electrode system, with a reference electrode connected to a fast potentiostate. The resting potential should be around 0 mV *versus* Ag/AgCl reference electrode. A potential step of around 50 mV is applied, and the current transient that followed is sampled. An identical current transient of opposite direction steps the potential back to the rest value. Taking the logarithm of the current gives an almost linear curve from which R and C can be calculated. The detection limit is around 10^{-14} M or 0.5 pg/ml of analyte.

It should be kept in mind that any part of the surface that allows the aqueous solution to penetrate the isolation layer where the recognitive reaction takes place would act like a short-circuiting element. The capacitance will therefore increase because of the higher dielectric constant of the penetrating aqueous solution. Thus, pinholes in the film as well as instability induced by the applied potential induce an immanent drift in these nano-isolated devices.

2.2 Piezoelectric biosensors and surface acoustic devices

Piezoelectric biosensors

Piezoelectricity was discovered in 1880 by the Curie brothers. Some anisotropic materials (e. g., quartz or tourmaline) exhibited electric dipole formation when under mechanical stress. Nowadays, ceramic piezoelectrics (e. g., barium titanate) and, more recently piezopolymers are applied widely [29].

Sensors are based on the measurement of change in resonant frequency of a piezoelectric crystal as a result of mass changes on its surface. These are caused by the interaction of test species or analyte with a bio-specific agent immobilized on the crystal surface. The frequency of vibration of the oscillating crystal normally decreases as the analyte binds to the receptor coating the surface.

The change in resonance frequency can be calculated from:

$\Delta f = -2.3 \times 10^6 f^2 \Delta m/A$.

The sensitivity is proportional to the square of the frequency; thus, a reduction in the size of the resonator increases sensitivity significantly.

Bulk piezoelectric sensors contain quartz wafers in the form of 5 to 20 mm disks, squares, or rectangles, which are about 0.1 to 0.2 mm thick and are sandwiched between two gold or silver electrodes.

Surface acoustic wave devices

Such sensors generally operate by the propagation of acoustic-electric waves, either along the surface of the crystal or through a combination of bulk and surface. Two sets of metal electrodes are put to the surface of a piezoelectric crystal substrate with very little space between them.

An electrical signal is applied to one set of electrodes, thus generating an acoustic wave. The wave is received by the other set of electrodes situated a few millimeters away on the opposite end of the substrate and converted back to an electrical signal. The bioactive compound is immobilized between the two sets of electrodes on the crystal surface.

Binding of the analyte leads to a change in the delay time of the registration of the acoustic signal by the second set of electrodes.

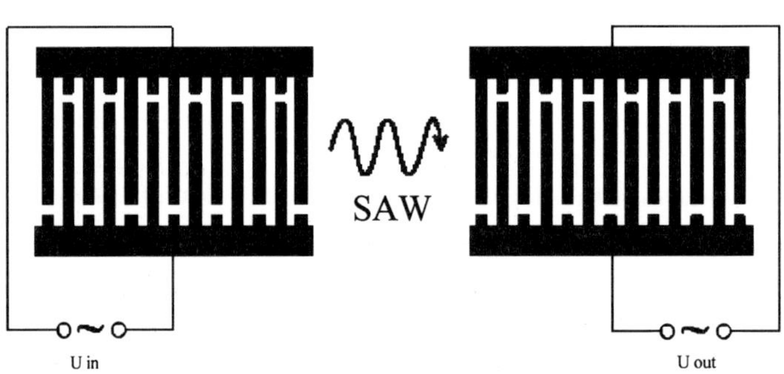

Figure 23 Setup of a surface acoustic wave device

2.3 Optical, fiber optic, and waveguide devices

Standard optical biosensors modify fundamental properties such as light absorption, frequency, or light emission [30]. In a fiber-optic setup, the light is delivered to the immobilized sensing layer at the tip of the fiber and absorbance, luminescence, or fluorescence is monitored either via a second fiber or by the same fiber, optically bifurcated. The bioactive substance is immobilized at the end or at the surface of the fiber core. This method is straightforward and is applied to devices to be used in critical environment such as a sterile fermenter, within the human body, or in drilled holes tens of meters beyond ground level.

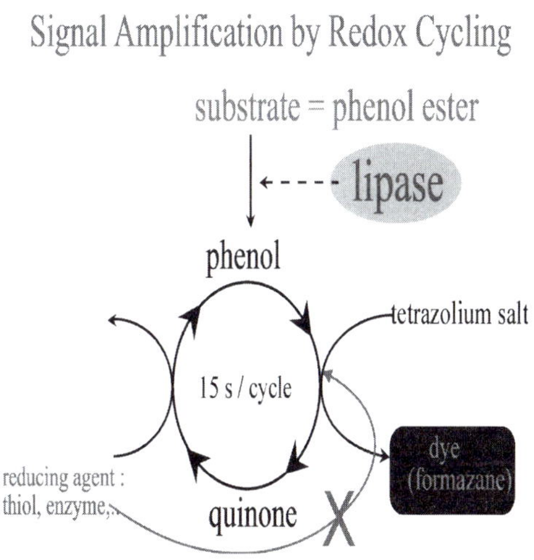

Figure 24 Substrate cycling to increase the sensitivity of an optical biosensor

Optical probes, in contrast to labels, are not to be inert respond to their micro-environment or to a chemical species. Probes responding to a chemical species (such as an ion, to pH or oxygen) may also be referred to as indicators. Numerous probes have become available in the past few years and have substantially contributed to the success of spectroscopy in biosciences.

While absorbance measurement is not sensitive enough for standard applications and suffers from short light-pass-limitation in small devices, some tricks such as redox cycling (see Fig. 24) are able to boost the sensitivity by several orders of magnitude. Given an optimal setup, even immunological tests might be automated using standard color transducers.

Fluorescence sensors

Intrinsic fluorescent molecules such as NADH or fluorophore-labeled biomolecules can be surface-immobilized and thus used for biosensors [31, 32]. A number of bioanalytical assays are based on the fact that NADH is fluorescent, while NAD+ is not. Thus, all enzymatic reactions based on NAD/NADH are transducable by fluorescence analysis, and this is widely exploited in practice. Care is necessary, though NADH has to be excited at 350 nm, which can cause

UNIVERSITÄTSBIBLIOTHEK
BODENKULTUR
Wien
Peter-Jordanstraße 82
A-1190 Wien

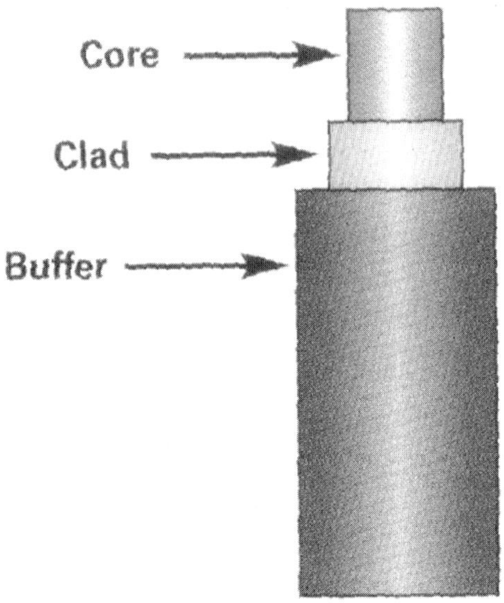

Figure 25 Optical fiber

substantial background fluorescence from other biomaterial. On the other hand, most assays are performed in the kinetic mode. Thus, it is the relative signal change that is measured rather than the total intensity. FAD is another strongly luminescent coenzyme, which found application because both the oxidized (FAD) and the reduced form (FADH2) display fluorescence. Their excitation is at around 450 nm, and fluorescence peaks at 512 nm. Both NADH and FAD have been shown to be useful for purposes of chemical sensing.

To illustrate the setup of such sensors, a flow through detection system using NADH-converting enzymes and the intrinsic fluorescence of NADH is illustrated in Figure 26. The enzymes are trapped together with a NADH-derivative (size-enlarged by coupling to a sugar or PEG-molecule). The low-molecular-weight analyte is able to pass the membrane whereas all macromolecules stay within the sensor compartment. To regenerate the NAD or NADH, a second enzyme is coupled to the first and a substrate pulse to this enzyme is used to regenerate the NAD/NADH pool.

A nitric-oxide-selective biosensor can be setup by incorporating cytochrome *c'*, a hemoprotein that is highly selective for nitric oxide. The cytochrome *c'* is attached to the tip of an optical fiber. In this sensor, cytochrome *c'* is labeled with a fluorescent dye called Oregon green. The dye alone is unaffected by nitric oxide, but it reports changes in the cytochrome *c'* as it binds nitric oxide. The sensor exhibits a fast, linear and reversible response to nitric oxide.

Fluoroimmunoassays

Fluoro-immunoassays have been studied for almost two decades as alternatives to radioimmunoassays. Fluorescence detection is rapid, nearly as sensitive and safer than radioisotopic methods. Moreover, fluorescence is well established in other fields as e.g. DNA analysis. Fluorescence immuno-sensors use optical

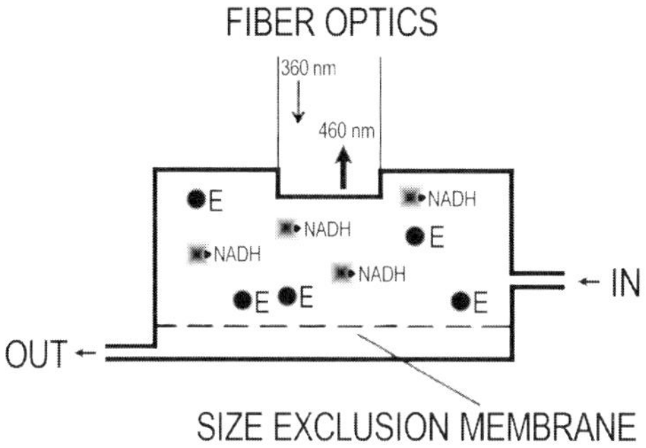

Figure 26 Optical fiber sensor using the fluorescence of NAD/NADH

fibers or integrated optics. Light propagating through an optical fiber or a planar layer (waveguide) generates an evanescent field. If a fluorescent dye is bound within this field, fluorophore is excited with an evanescent field of light of a gas or semi-conductor laser or a Xenon lamp. If the optical properties are resonant for the structure used, emitted fluorescent light is collected by the waveguide. Thus, the use of optical fibers or waveguides as transducers in fluorescence affinity sensors allows discrimination of the bound *versus* the unbound tracers through the use of an evanescent field. Most of the fluorescent immunosensor systems are based on either an inhibition assay format or a sandwich format, depending on the analyte. Nevertheless, labeled analyte analogs have been used for specific applications.

Fiber-optic biosensor based on molecular beacons

A novel type of optical-fiber, evanescent-wave DNA biosensor is based on the use of molecular beacon DNA probe. The molecular beacons (see chapter 2) are oligonucleotide probes, which are auto-quenched and become fluorescent upon hybridization with target DNA/RNA molecules. Labeled beacons can be designed and immobilized on an optical fiber core surface via, e. g., biotin-avidin or biotin-streptavidin interactions. Contrary to most other types of DNA sensors, this type of DNA sensor based on beacons does not need labeled-analyte or intercalation reagents. It can be used to directly detect, in real-time, target DNA/RNA molecules without using competitive assays.

The sensor response is rapid and can be used to study the hybridization kinetics of the DNA by changing the ionic strength of the hybridization solution and target DNA concentration.

The detection limit of the molecular beacon evanescent wave biosensor is around 1 nM. To enhance sensitivity, the beacon DNA biosensor can be combined with a polymerase chain reaction.

Nano-optical sensors

Using near-field optics, a novel nanofabrication technology has been developed for preparing structures. Photonanofabrication, based on an optical near field, controls the size of material grown at the end of a light transmitter, such as a fiber tip by photochemical reactions. These reactions are initiated and driven by light. By using the near field of an optical fiber, the diffraction limit of resolution is surpassed and structures down to a few nanometers are feasible. A straightforward technique synthesizes materials only in the presence of light and "bonds" it only to the area where light is emitted. The key to photonanofabrication is a near-field photochemical reaction. Thus, the size of the luminescent probe is defined by the light-emitting aperture and is independent of the wavelength of the light. The photochemical reaction only occurs in the near-field region, where the photon flux and the absorption cross-section are the highest.

Protocol 15 A submicrometer optical-fiber pH

1. To setup a submicrometer optical-fiber pH sensor a, near-field photopolymerization process is employed.
2. The fiber tip is first pretreated by silanizing with a 3-(trimethoxysilyl)propyl methacrylate solution.
3. The fiber then dried under nitrogen for 1 h.
4. The silanized tip end is activated with a solution of benzophenone in cyclohexane, 1 w/v.
5. The pH sensor is prepared by incorporating a fluoresceinamine derivative, acryloylfluorescein, into an acrylamide-methylene bis(acrylamide) copolymer, which is attached covalently to a silanized tip surface by photopolymerization.
6. The sensitized end of the optical fiber is placed in the monomer solution, which ensures size control of the formed polymer.
7. The optical beam from the tip initiates the polymerization process.
8. A microscope is used to control the rate and size of polymer formation.

This process enables the incorporation of pH-sensitive dye molecules covalently bonded to the fiber tip surface, yielding the nanometer-sized sensors. To test the applicability of the pH sensor for measurements in small volumes, the sensors can be positioned in porous polycarbonate membrane holes (~ 10 μm) filled with buffer.

Sensitivity is high yielding a measurable value as a result of about 3000 hydrogen ions. Submicrometer optical pH sensors can be created using pyranine-doped sol-gel in 1-μm-tip-diameter micropipettes. To cope with limitations of the chromophore stability, an internal calibration method, based on ratiometric measurements, should be employed.

The covalent bonding of the sensing elements means that the analytes have immediate access to the dye on the sensor tip, which gives the sensors among the fastest response times reported optical-fiber sensors (~ 50 ms). The intracellular sensors require only attoliters of sample and zeptomoles of analyte. The sensing occurs in the near-field regime of the optical excitation, thereby greatly increasing the sensitivity.

A standard application of fiber sensors is measuring intracellular calcium ions. Many physiological processes are triggered, regulated, or influenced by this ion. Excellent intracellular dyes have been developed for sensitive Ca^{2+} determination by using photopolymerized gels or sol-gel glass doped with calcium green. This sensor is capable of measuring calcium concentrations as low as 10 nM. The sensor is highly calcium-selective and uses a long-wavelength fluorescence. Calcium green can be derivatized with an amine group for covalent bonding, resulting in faster response times for monitoring concentration fluctuations. The sensor can follow a change in free Ca^{2+} concentration, up to ~ 40 nM against a background of Mg^{2+} with a selectivity for calcium over magnesium of around $10^{-4.5}$.

Numerous ion-selective optical sensors have been prepared by incorporating sensing components into a hydrophobic liquid polymer film, such as PVC. Such sensors contain three components:

- an ionophore,
- a chromo-ionophore, which is a pH indicator dye,
- and ionic additives to maintain constant ionic strength.

Such sensors rely on the coupled response of the ionophore and chromoionophore and therefore are referred to as ion-correlated.

Ultra-small optical sensors have advantages over measurements made with a conventional fluorescence microscope. Sensors provide localized measurements, and no dye is loaded into the biological sample. The sensors are not affected by sample thickness because the dye molecules are only on the sensor tip. A detectable signal change is easier to obtain with these sensors because there are only a few dye molecules (~ 100) on a nano probe. A 10% change in the 100 molecules on the tip needs only 10 analyte molecules. Thus, the detection limit is lower and auto-fluorescence is avoided.

Optical sensors: leaching and bleaching

Features that may worsen optical sensor response (except resonant techniques such as SPR) include chromophore or fluorophore leaching and photodamage. Methods for overcoming these disadvantages make use of the shorter response times of small sensors and forward optical signal collection with small samples. Nevertheless, photobleaching will always be a major consideration for any optical measurement.

The problem is more severe for smaller optical probes because the number of sensing molecules is much smaller. However, the size reduction result in higher sensitivity and lower detection limits. Moreover, smaller sample sizes and faster response times can be obtained. While leaching can be avoided by covalent immobilization of the dye, a variety of strategies have been employed to reduce photobleaching: elimination of oxygen and other chemical species, limited light intensity, use of stable dye molecules, use of ratiometric detection, sensing molecules that can be regenerated, inclusion of a second reference dye, lifetime instead of intensity measurements, and many more. In general, the photostability of the dye limits the lifetime of optical devices.

Contrary to standard optical devices, surface plasmon-based sensing became popular within the last year as the method of choice to monitor biorecognitive binding in label-free systems.

Surface plasmon resonance (SPR)

A surface plasmon is a collective phenomenon exciting electrons at the interface between a conductor and an insulator. Surface plasmon-based effects are found in a variety of phenomena: the energy loss of electrons within a thin metal film, the color of suspensions of small metallic nano-clusters, and the dips in the intensity of light reflected from metal coated chip. Applications are based on the interaction of light with surface plasmons in thin films, to make optical modulators, switches, and sensors.

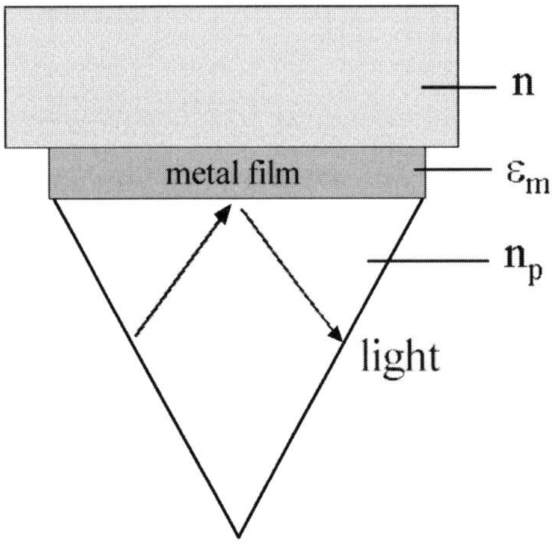

Figure 27 SPR 3-layer system with prism

Biosensors use these means to transduce the properties of a dielectric biolayer bound to the metal surface.

Surface plasmons in a regular thin-film are non-radiative electromagnetic modes and thus cannot be generated directly by light. Conversely, surface plasmons cannot decay spontaneously into photons. The reason for the stability and non-radiative nature of surface plasmons is that the conversion of light into plasmons cannot simultaneously satisfy energy and momentum conservation.

To enable energy input and output, roughening or corrugating the metal surface, or using metal island films is effective. Another method is to increase the effective wave vector and hence momentum of the light by using a prism or grating coupling technique (see Fig. 28). In case of resonance the field strength is maximal at the metal interface and about 8 times enhanced to the non-resonant case. In the non-resonant setup the field is maximal at or near to surface adjacent to the prism.

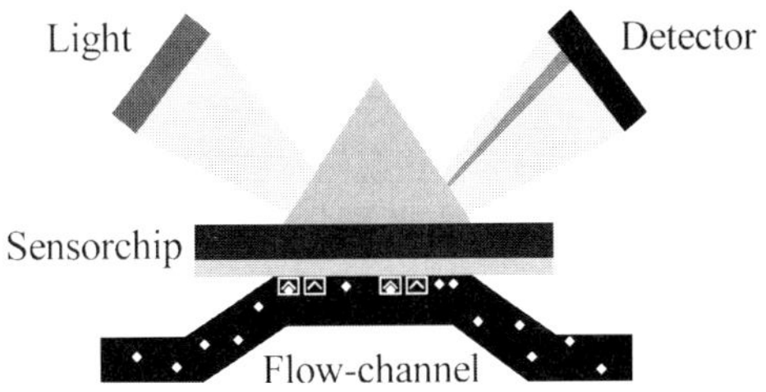

Figure 28 SPR flow cell

For sensor application the metal surface at which the SP is generated is covered with a dielectric thin film. The presence of even very thin films of molecules at the chip measurably alters the behaviour of the surface plasmon dip in reflectivity resonance shifting the incident angle at which resonance occurs. This effect can be used to make devices.

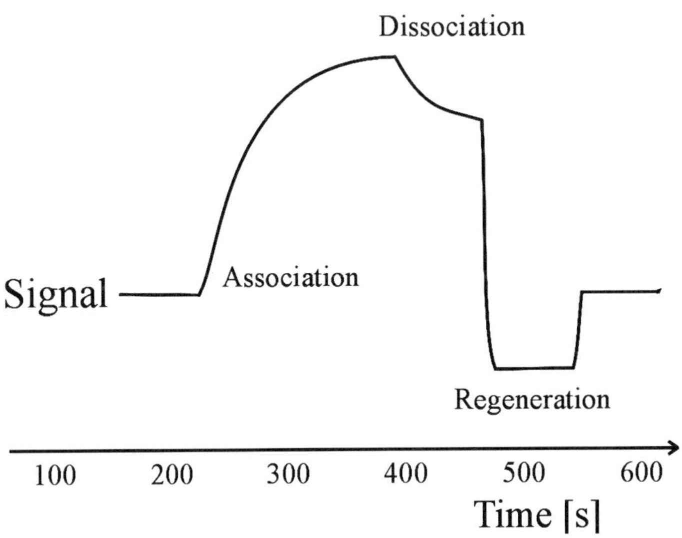

Figure 29 Kinetic binding and dissociation data obtained from a SPR instrument

The angle position of surface plasmon coupling (dip in reflectivity) can be used to characterize the dielectric over-layers of molecules such as biomolecules or polymers.

SPR is now standard to study protein ligand interaction as well as DNA hybridisation and even is useful to screen for single-nucleotide polymorphisms [37–42].

Within the past few years surface plasmon resonance seems intent on being a player in the array arena. Novel techniques are under way for use it microarray-based measurements of e. g., RNA–DNA hybridization.

New chips are available using microarrays that were fabricated on gold-coated glass microscope slides. Thiol-modified DNA probes of 15 to 25 nucleotides long with a poly-T spacer are arranged in a grid of squares, with a surface

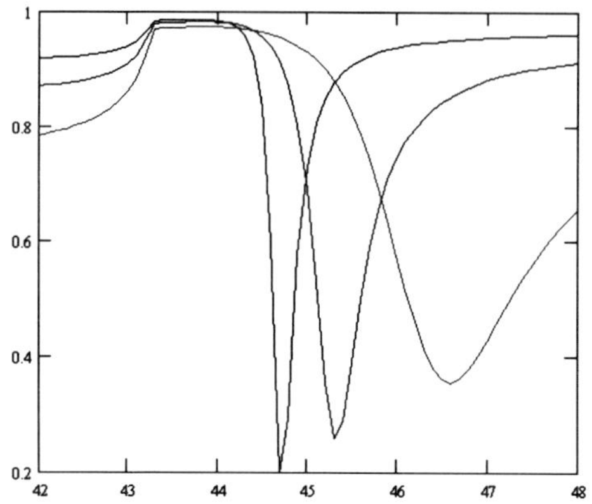

Figure 30 Drop and shift in reflectance induced by a change in refractive index near the chip surface

coverage of about 1×10^{12} molecules/cm^2. The area surrounding the DNA-probes is coated with, e.g., polyethylene glycol to prevent the nonspecific binding.

Instead of using traditional SPR in which the wavelength of the light is fixed and the angle of incidence is varied, or in which the angle is fixed and the wavelength varied, SPR imaging is employed. In this setup, both the wavelength and the angle of the light are fixed, and the reflectivity at various points on the surface is measured. Thus, a picture of the microarray can be obtained.

A detection limit of 10 nM for the 18 base nucleotides and of 2 nM for the 1500 base natural sequence has been reported. Although the fluorescence detection is more sensitive for small numbers of molecules, SPR imaging can detect adsorption online at nanomolar concentrations, having the big advantage of being a label-free method.

3 Sensors in bioprocess control

For optimal growing conditions of cells, knowledge about the physical, chemica, and biochemical characteristic parameters of fermentation is necessary. These parameters can be monitored with suitable probes in different ways:

- Online: measurement continuous or automatical in short intervals
- Offline: measurement discontinuous, often manual sample handling necessary

Sensors for fermentation control have to fulfill various criteria:

- Accuracy: difference between true and determined value
- Precision: variation of signals for repeated measurement of one sample
- Resolution: smallest signal change measurable
- Sensitivity: slope of calibration curve

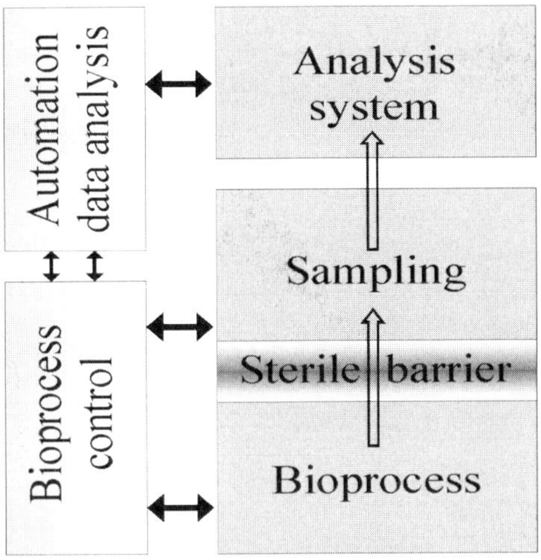

Figure 31

- Signal to background ratio: influence of unspecific background
- Specificity: signal only dependent on the concentration of analyte
- Measuring range: difference between maximal and minimal signal
- Delay: interval between change of the signal and detection of the change
- Reliability
- Maintenance
- Hysteresis
- Linearity
- Drift

Online measurement can be performed *in situ* (directly in contact with fermentation broth, inside of the sterile area) or *ex situ* (directly in contact with fermentation broth but outside of the sterile area). For *in situ* application, additional qualities are necessary:

- sterilization of the sensor must be possible (preferably in autoclave), and
- stability of the sensor or possibility of recalibration must exist without affection of the running fermentation process.

In bioprocess control, the physicochemical parameters (pH, pO_2, temperature) are usually monitored and regulated online and *in situ*.

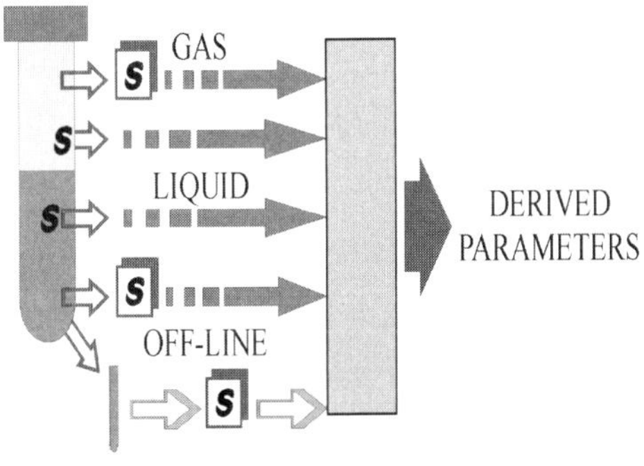

Figure 32

Table 4 *In situ* sensors for bioprocess control

Parameter	Sensing device	In-line	Sterile	Ex-situ	Online
Temperature	• Pt 100	*	*		*
	• thermo element	*			*
Pressure	• piezo-electric	*	*		*
PH	• potentiometric	*	*		*
pO_2	• amperometric	*	*		*
pCO_2	• potentiometric	*	*		*
Conductivity	• potentiometric	*	*		*
Weight	• piezo resistance	*	*		*
	• strain gauge	*	*	*	*
	• differential pressure				*
Flow Gas	• thermal mass flow	*			*
Flow Liquid	• magnetic inductive	*	*		*
	• turbine	*			*
	• corriolis	*			*
Level	• capacitive				
	• potential drop	*	*		*
	• fiber optic	*	*	*	*
	• ultrasonic	*			*
	• radar	*			*
	• radioactive				
Foam	• potential drop	*	*		*
Biomass -Cell Count	• optical-turbidity	*	*		*
	• flourometer	*	*		*
	• bugmeter	*	*		*
	• laser scatter	*	*		*
Agitation rate	• optical			*	*
	• proximity			*	*
Agitation power	• electrical power consumption			*	*
	• torque of shaft			*	*

In situ sensors are much more difficult to handle, but they are still more accepted in fermentation industry. The *in situ* application of biosensors is quite critical because they cannot be sterilized. There are some *in situ* systems that are in contact with the fermentation broth by a microfiltration membrane or sterile barriers. Quantification of most biochemical parameters is often still performed offline because there are a few sensor elements and sampling systems available and automation of analyzing systems is connected with great effort. Besides sample dilution, separation of interfering substances, addition of reagents, sample derivatization, etc., can be necessary and are most easily done externally.

Trend leads to online *ex situ* determination of biochemical parameters (substrates, metabolites, products of the cells). Sensors do not have to be sterile, recalibration becomes easier, and failures can be overcome without interrupting the fermentation process. Table 5 shows a list of published applications of biosensors for bioprocess monitoring [43, 44].

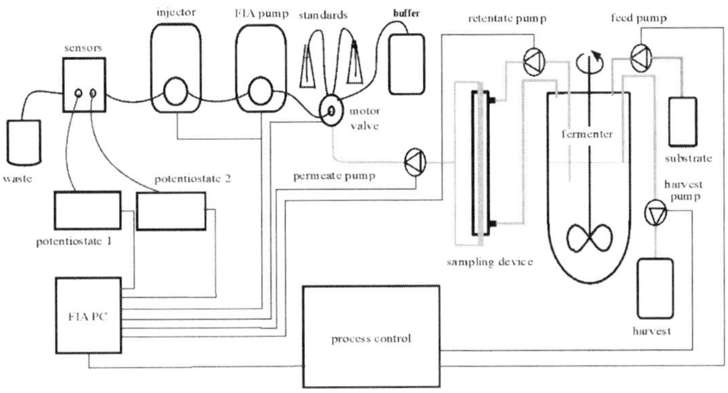

Figure 33 Automated feedback system using FIA and amperometric biosensors

Table 5 Examples for on-line *ex situ* bioprocess analysis

Analyte	Analytical System	Detection	Sampling
Salicylic acid	Spectro-photo-meter	optical 325 nm	filtration
Isoleucine, valine	HPLC		filtration
Glucose, lactate glutamate	FIA	immobilized oxidase electrochemical detection	filtration
Glucose, alcohol, acid phosphatase	FIA	immobilized oxidase, electrochem. detection, substrate dye, optical	filtration
Penicillin	FIA	thermal enzyme biosensor (thermistor)	tangential flow filtration
Microbial activity	FIA	bioelectrochemical, hexacyanoferrate as electron shuttle	sampling valve
Sugars, organic acids	HPLC		sampling valve system and microcentrifuge
Lipase	FIA	turbidimetric triolein emulsion test	filtration
Monoclonal antibody	FIA	turbidimetric immunocomplex formation with anti IG	cross-flow microfiltration
Glucose	HPLC		cross flow microfiltration
Beta-galactosidase	FIA	enzyme assay	sampler and automatic ultrasonic cell disrupter
Primary metabolites (acetone, acetic acid)	GC		cross flow microfiltration

Analyte	Analytical System	Detection	Sampling
NH_4^+		alkali addition and NH_3 detection by air-gap membrane pH sensor	sampler
Protease inhibitor (hirudin)	Zone sampling + FIA	thrombin activity by chromogenic substrate	microfiltration
Dehydrogenase	FIA	optical NAD cofactor detection	samples from downstream cell disruption step
Glucose	FIA	enzyme covered pH sensitive FET	microfiltration
Glucose, ammonia, total protein	double injection FIA	photo diode array, chromogenic substrates or reagents	microfiltration
Glucose	FIA	amperometric biosensor	hollow fiber cross- flow
Glucose, lactate, glutamine	FIA	amperometric biosensor	sterile filtration
Glucose, urea, cephalosporine	FIA	FET biosensor	sterile filtration
Glucose	FIA	immobilized oxidase, chemiluminescent detection of H_2O_2 with luminol	microfiltration membrane
Glucose	FIA	amperometric biosensor	cross-flow filtration
Glucose	FIA	amperometric biosensor	microfiltration
Glucose, lactate, glutamine, glutamate, ammonia	FIA	immobilized oxidase, chemiluminescent detection of H_2O_2 with luminol	cross-flow filtration

4 Flow injection analysis

The term 'Flow Injection Analysis', (FIA) used by Ruzicka in 1975 describes an analytical method that is very useful for a high turnover of samples.

An exactly defined volume of liquid sample is injected into a continuous flowing carrier stream and transported to a detector. On the distance between injection and detector, a mixing of sample and surrounding carrier solution takes place. The carrier solution either can be used for transport of the sample to the detector or it can contain reagents for a reaction with the analyte. A simple FIA system is setup with: pump 1 fills the sample into the injection loop, pump 2 transports the carrier buffer by an injection valve and a mixing loop to the detector. After switching off the valve, the carrier buffer flushes a very defined volume of the sample into the system.

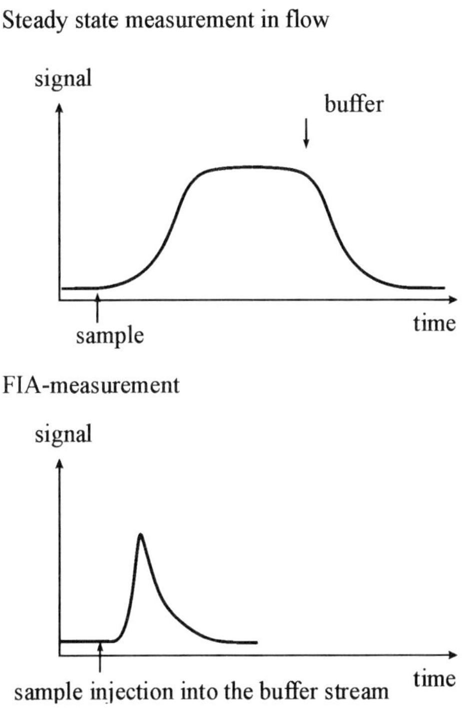

Figure 34 Peak shape *versus* injection volume

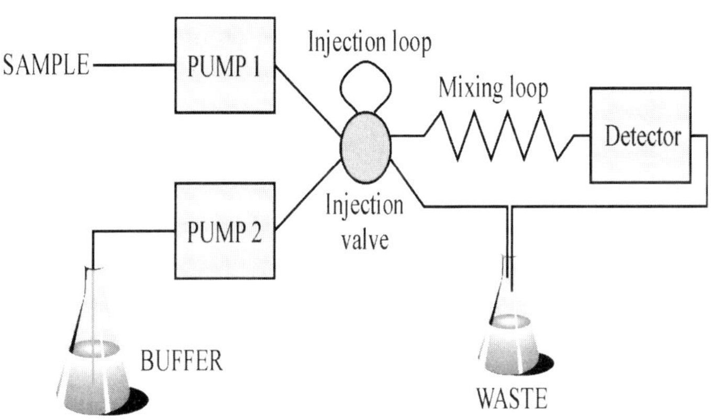

Figure 35 Schematic FIA system

According to chemical or biochemical requirements the dispersion of the sample zone can be controlled by various parameters, e. g., by type or length of the mixing loop between the place of injection and the detector, the length of the distance between injection and detector, and the diameter of the tubes or different injection volumes.

In contrast to other flow methods, the flow injection analysis leads to very reproducible signals in the form of sharp peaks, though there is no homogeneous mixing of sample and carrier and reactions taking place in the system do not reach equilibrium in many cases. FIA signals are so reproducible because

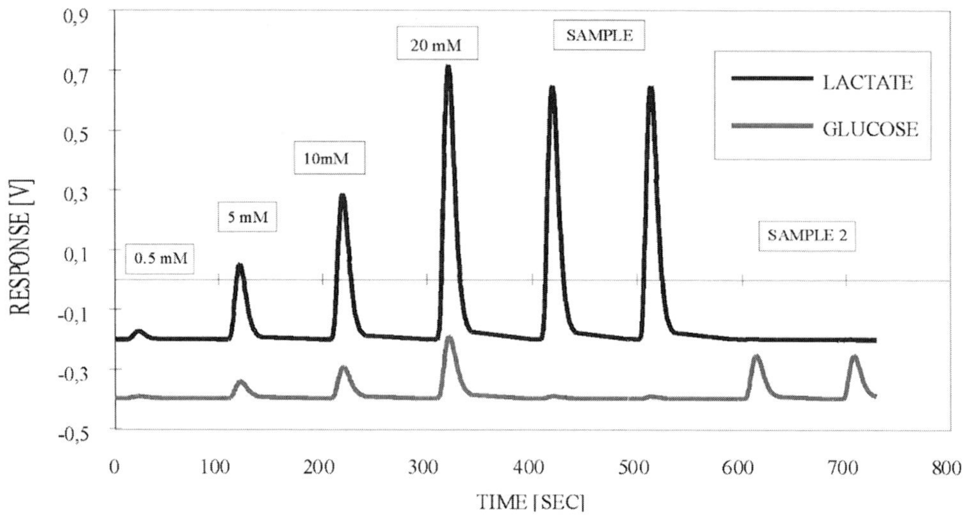

Figure 36 Plot of a dual sensor setup with calibration and sample data

the time samples stay in the system and all parameters influencing the dispersion of the sample zone are held extremely constant. In contrast to steady-state measurements where samples have to flow through the system until the signal reaches constancy, FIA signals are peaks.

Peaks usually are characterized by their height; other possible characteristics are the area under the peak, the height of the signal at a defined moment after injection, the width of the peak, or kinetic analysis for stopped flow techniques.

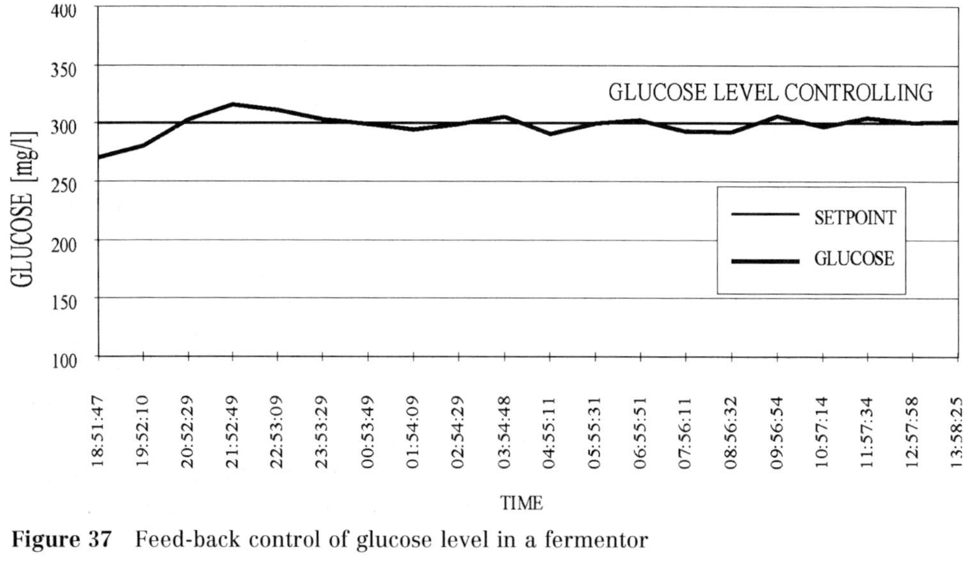

Figure 37 Feed-back control of glucose level in a fermentor

There are many advantages to flow-injection systems. The response time is short because it is not necessary to wait for equilibrium; therefore a high turnover of samples is possible. In addition, the sample volume needed is very low compared to other methods (a few microlitres).

Flow-injection analysis is very appropriate to automation, and a high number of samples therefore can be managed. Adaptability to various problems is possible because of the flexibility of the parameters of the system like dispersion, injection volume, type of detector, etc.

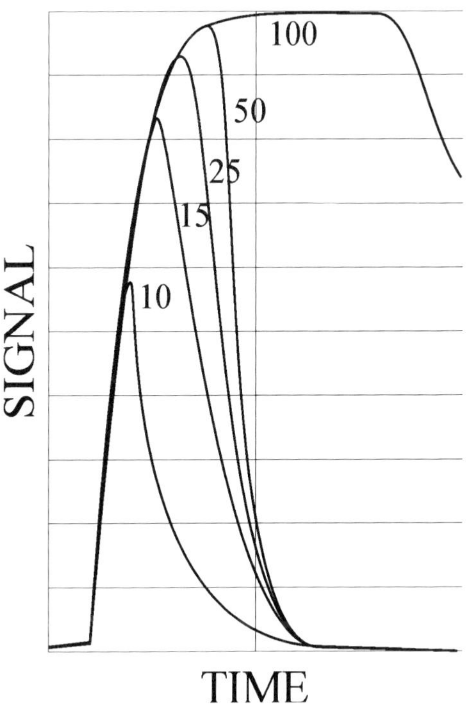

Figure 38 Peak shape *versus* injection volume

4.1 Controlled dispersion

The crucial parameter in a FIA system is the controlled dispersion of the sample. Controlled dispersion means the physical process of distribution of the injected sample volume in the FIA system. Without any flow, the zone of the sample would disperse regularly to both of the adjacent carrier zones by diffusion. Under ideal circumstances, in very narrow tubes there is a laminar flow. The carrier drags the zone of the sample along, immediately leading to an axial dispersion of the sample. In reality, the distribution of the zone of the sample is a combination of radial diffusion and axial convection. In general, axial diffusion can be neglected in FIA systems.

Molecules at the edge of the zone of the probe diffuse to the fringe with lower sample concentration and lower flow rate. Because of the radial concentration gradient, molecules in the fringe of the extreme end of the sample zone migrate to the center with a higher flow rate. Therefore, the expansion of the sample is finite and the signal returns to the baseline.

The dispersion of the zone of the sample in the carrier stream is described by the so-called coefficient of dispersion:

$\mathbf{D = C_0 / C_{max} = H_0 / H_{max}}$,

D	coefficient of dispersion
C_0	concentration of undiluted sample
C_{max}	concentration of the sample at the maximum of the peak
H_0	height of the signal of the undiluted sample
H_{max}	height of the peak

The equation is exact if the calibration curve has the same slope at C_0 and C_{max}. The coefficient of dispersion describes the dispersion of the sample in the part of the zone of the sample corresponding to the maximum of the peak. Knowledge of the coefficient of dispersion eases the optimization of a FIA system regarding sensitivity. Furthermore, optimal concentrations of reagents in carrier streams can be determined for FIA systems using the carrier not only for transport but also for chemical reactions.

Typical changes of the coefficient of dispersion and the form and basis width of a peak according to variations in injection volume or mixing loop are shown in figure 38: influence of the injection volume on height and form of the signal and

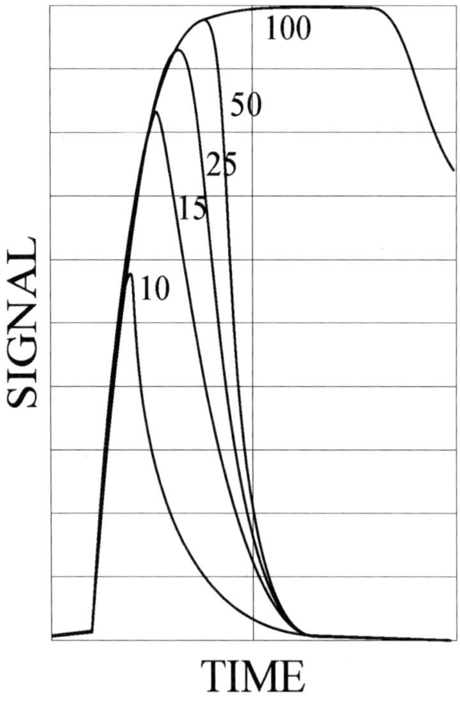

Figure 39 Peak shape *versus* mixing column length

on the coefficient of dispersion, and influence of the length of the mixing loop between the place of injection and the detector on signal and coefficient of dispersion.

The coefficient of dispersion changes proportionally to the square root of the distance between the place of injection and the detector. Other parameters like velocity of flow, type of the mixing loop, or turbulences in the system also influence the coefficient of dispersion. Systems with low coefficients of dispersion (D = 1 to 3) are interesting for applications using the carrier only for transport of a sample with a high local concentration. For methods needing one or even more chemical reactions for the detection of a certain substance, systems with medium coefficients (D = 3 to 10) are used. Very high coefficients of dispersion (D > 10) are interesting either for methods using the system also for dilution of the sample or for others evaluating by gradient techniques.

References

1 Urban G, Jobst G, Kohl F, et al. (1991) Miniaturized thin film biosensors using covalently immobilized glucose oxidase. *Biosensors and Bioelectronics* 6: 555–562

2 Schalkhammer Th, Lobmaier C, Ecker B, Wakolbinger W, Kynclova E, Hawa G, Pittner F. (1994) Microfabricated Glucose-, Lactate-, Glutamate- and Glutamine Thin Film Biosensors. *Sensors and Actuators B* 18–19: 587–591

3 Turner APF, Karube I, Wilson GS (1987) Biosensors-fundamentals and applications, Oxford Science Publications, London

4 Freitag R (1993) Applied biosensors *Curr Opin Biotechnol* 4: 75–79

5 Doblhoff-Dier O, Rechnitz GA (1989) Amperometric method for the determination of superoxide dismutase activity at physiological pH. *Analytica Chimica Acta* 222: 247–252

6 Karyakin AA, Karyakina EE, Schuhmann W, et al. (1994) New amperometric dehydrogenase electrodes based on electrocatalytic NADH-oxidation at poly(Methylene Blue)-modified electrodes. *Electroanalysis* 6: 821–829

7 Mann Buxbaum E, Hawa G, Schalkhammer Th, et al. (1992) Construction of electrochemical biosensors: Coupling techniques and surface interactions of proteins and nucleic acids on electrode surfaces. In: G Costa, S Miertus (eds) *Trends in electrochemical biosensors*: World Scientific, Singapur/New York/London/Hong Kong, pp 69–84

8 Pittner F, Schalkhammer Th, Mann-Buxbaum E, Urban G. (1990) Miniaturized thin film biosensors. *Microsystem Technologies* 90: 124–125

9 Moser I, Schalkhammer Th, Mann-Buxbaum E, et al. (1992) Advanced immobilization and protein techniques on thin-film biosensors. *Sensors and Actuators* B7: 356–362

10 Schalkhammer Th, Mann-Buxbaum E, Urban G, Pittner F. (1990) Electrochemical biosensors on thin film metals and conducting polymers. *J Chromatography* 510: 355

11 Williams RA, Blanch HW (1994) Covalent immobilization of protein monolayers for biosensor applications. *Biosens Bioelectron* 9: 159–167

12 Ahluwalia A, De Rossi D, Ristori C, et al. (1992) A comparative study of protein immobilization techniques for optical immunosensors. *Biosen Bioelectron* 7: 207–214

13 Schalkhammer Th, Mann Buxbaum E, Moser I, et al. (1992) New immobiliza-

tion techniques for the preparation of thin film biosensors. In: U Sleytr, P Messner, D Pum, M Sara (eds) *Immobilized Macromolecules* Springer Verlag Berlin, Heidelberg, New York p. 119–139

14 Lobmaier C, Schalkhammer Th, Hawa G, et al. (1995) Photo-structurized electrochemical biosensors for bioreactor control and measurement in body fluids. *J Mol Recognition* 8: 146–150

15 Schalkhammer Th, Lobmaier C, Pittner F, et al. (1995) A metal island coated polymer sensor for direct determination of chaotropic agents. *Mikrochimica Acta* 121: 259–268

16 Collioud A, Clemence JF, Sanger M, Sigrist H (1993) Oriented and covalent immobilization of target molecules to solid supports: Synthesis and application of a light-activatable and thiol-reactive cross-linking reagent. *Bioconjugate Chem* 4: 528–536

17 Schalkhammer Th, Mann-Buxbaum E, Urban G, Pittner F (1991) Electrochemical glucose sensors on permselective non-conducting substituted pyrrole polymers. *Sensors and Actuators* B4: 273–281

18 Bartlett PN, Cooper JM (1993) A review of the immobilisation of enzymes in electropolymerized films. *J Electroanal Chem* 362: 1–12

19 Heiduschka P, Göpel W, Beck W, et al. (1996) Microstructured peptide-functionalised surfaces by electrochemical polymerisation. *Chem Eur J* 2: 667–672

20 Lukachova LV, Karyakin AA, Ivanova YN, et al. (1998) Non- aqueous enzymology approach for improvement of reagentless mediator based glucose biosensor. *Analyst* 123: 1981–1986

21 Schalkhammer Th, Hartig A, Pittner F, Moser I. (1991) Surface modification of platinum based electrochemical thin film electrodes for DNA biosensors. *Microsystem Technologies* 91: 76–81

22 Karyakin AA, Bobrova OA, Lukachova LV, Karyakina EE (1996) Potentiometric biosensors based on polyaniline semiconductor films. *Sensors & Actuators B* 33: 34–38

23 Karyakina EE, Neftyakova LV, Karyakin AA (1994) A novel potentiometric glucose biosensor based on polyaniline semiconductor film. *Analyt Letters* 27(15): 2871–2882

24 Cornell BA, Braach-Maksvytis VLB, King LG, et al. (1997) A biosensor that uses ion-channel switches. *Nature* 387: 580–583

25 Weiss-Wichert C, Smetazko M, Saba M, Schalkhammer Th (1997) A new analytical device based on gated ion channels. *J Biomol Screening* 2 (1): 11–18

26 Schalkhammer Th, Weiss-Wichert C, Smetazko M, Saba M (1997) Ion channels in artificial bolaamphiphilic membranes deposited on sensor chips – optical detection in an ion channel based biosensor. *SPIE* 2976:117–128

27 Neumann-Spallart C, Pittner F, Schalkhammer Th (1997) Immobilization of active facilitated glucose transporters (GLUT-1) in supported biological membranes. *Appl Biochem Biotechnol* 68: 47–63

28 Berggren C, Johansson G (1997) Capacitance measurements of antibody-antigen interactions in a flow system. *Anal Chem* 69 (18): 3651–3657

29 Ngeh-Ngwainbi J, Suleiman AA, Guibault GG (1990) Piezoelectric crystal biosensor. *Biosens Bioelectron* 5: 13–26

30 Wirth M, Gabor F, Schalkhammer Th, Pittner F (1995) Determination of mitoxantrone via fiber-optic device. *Mikrochimica Acta* 121: 87–93

31 Trettnak W, Wolfbeis OS (1989) A new type of fiber optic biosensor based on the intrinsic fluorescence of immobilized flavoproteins. *Proc SPIE* 1172: 287–292

32 Moreno-Bondi MC, Wolfbeis OS, Leiner MJP, Schaffar BPH (1990) Oxygen optrode for use in a fiber optic glucose biosensor. *Anal Chem* 62: 2377–2380

33 Vogel H, Sanger M, Sigrist h (1995) Covalent attachment of functionalized lipid bilayers to planar waveguides for measuring protein binding to biomimetic membranes. *Prot Sci* 4: 2532–2544

34 Duschl C, Vogel h (1994) A new class of thiolipids for the attachment of lipid bi-

layers on gold surfaces. *Langmuir* 10: 197–210
35 Stelzle M, Weissmüller G, Sackmann E (1993) On the application of supported bilayers as receptive layer for biosensors with electrical detection. *J Phys Chem* 97: 2974–2981
36 Gustafson I, Artursson E, Ohlsson PA (1995) Retained activities of some membrane proteins in stable lipid bilayers on a solid support. *Biosens Bioelectron* 10: 463–476
37 Sambles JR (1991) Optical excitation of surface plasmons. *Contemporay Physics* 32: 3
38 Fägerstam LG, Frostell-Karlsson A, Karlsson R, et al. (1992) Biospecific interaction analysis using surface plasmon resonance detection applied to kinetic, binding site and concentration analysis. *J Chromatogr* 597: 397–410
39 Gershon PD, Khilko S (1995) Stable chelating linkage for reversible immobilization of oligohistidine tagged proteins in the BIAcore surface plasmon resonance detector. *J Immunol Meth* 183: 65–76
40 Harris RD (1996) Waveguide surface plasmon resonance biosensor, PhD Thesis, Southampton
41 Harris RD, Wilkinson JS (1995) Waveguide surface plasmon resonance sensor. *Sens Actuators B* 29: 261–267
42 O'Shannessy DJ (1994) Determination of kinetic rate and equilibrium constants for macromolecular interactions: A critique of the surface plasmon resonance literature. *Curr Opin Biotechnol* 5: 65–71
43 Loibner AP, Doblhoff-Dier O, Zach N, et al. (1994) Automated glucose measurement with micro structured thin layer biosensors for the control of fermentation processes. *Sensors and Actuators B* 18–19: 603–606
44 Mayer C, Frauer A, Schalkhammer Th, Pittner F (1999) Enzyme based flow injection analysis system for glutamine and glutamate in mammalian cell culture media. *Anal Biochem* 268: 110–116

6 Biochips

Norbert Stich

Contents

Methods and Tools in Biosciences and Medicine
Analytical Biotechnology, ed. by Thomas G.M. Schalkhammer
© 2002 Birkhäuser Verlag Basel/Switzerland

1 Introduction

The completion of sequencing of the genomes of plants, various eukaryotes, bacteria, and viruses has provided billions of basepairs in thousands of genes. Together with this development, not only new words have been created (genomics, transcriptomics, phenomics, proteomics, etc.) but also new types of biologists (bioinformaticians) and (biotech) companies have arisen. Moreover, new and more efficient software has been developed in order to fully analyze and display genomic information so that all genes involved in given metabolic, developmental, or oncogenic pathways can be identified as a basis to examine gene expression patterns on a genome-wide level. This paves the way for more detailed protein analyses, including protein-protein interactions, post-translational modifications, and three-dimensional structurization.

As a result, new perspectives have opened up in many fields, ranging from basic research and industrial biotech-applications to gene therapy. Faced with this large and rapidly increasing body of information, novel technologies are needed to take full advantage of this avalanche of data.

Novel devices provide the chance to highly parallize and fully integrate multiple reaction steps of complex analytical procedures. These novel methods form the basis of HT-analyzers and ultimately will be transformed into smaller devices for point-of-care testing.

1.1 From microtiterplates to biochips

Miniaturization offers a number of advantages. First, it offers the packing of considerably more reagents into the same volume. High numbers of different proteins can be placed on one protein chip instead of performing cumbersome Western blots or Sepharose affinity columns. Likewise, thousands of DNA hybridisation reactions can be done in parallel on a single glass slide and therefore replace Southern and Northern blots. An integrated laboratory can be provided on a simple CD-ROM-like disk, a lab-on-a-chip. Furthermore, miniaturization obviously can increase reaction kinetics by using smaller reaction volumes and significantly increased surface-to-volume ratios [1]. Reagents can be easily mixed and pumped just by spinning the CD-like lab. The experiments can be performed fast by using fewer reagents and smaller-sized reactions. In addition to all these advantages, the cost efficiency of this system also should be mentioned.

Microtiter plates represent well-established array formats that have been used for decades and are well known to the scientific community. To enable high-throughput screening, robots are employed, though the array density is still low. Moreover, assays usually take large amounts of reagents and can be laborious in terms of washing and analysing. So-called “nanoplates” with volumes between 100 and 1000 nl per well therefore have been produced by several universities (see chapter 5) and companies, such as GeSiM mbH (Dresden, Germany) and Orchid Biocomputer Inc (Princeton, NJ, USA). Nevertheless, the classical 96-well microtiter plate has been the starting point for low-density arrays on filter membranes (Fig. 1) [2].

DNA tests based on thin-film electrodes were among the first types of miniaturized DNA sensors developed. One of the first examples is a thin-film sensor with oligonucleotides attached to a metal electrode [3]. It was designed by using photolithography, and first radioactive labels as transducer and, in an optimised layout, glucose oxidase. The glucose oxidase that were attached to the oligonucleotides produced hydrogen peroxide in the presence of the substrate (glucose). The hydrogen peroxide formed was then quantified amperometrically by the sensor. The authors found that the electrode was capable of hybridising with analyte DNA from solution and that the reaction proceeds similar to blot-type membrane devices. Detailed studies on the chemistry of oligonucleotide immobilization and unspecific background adsorption have been done.

Much effort was invested in expanding the density onto the arrays, e. g., by designing a microarray as a glass plate that bears 96 wells formed by an enclosing hydrophobic Teflon mask [4]. Arrays of 144 (4 × 36) elements per well were spotted, employing a 36-capillary-based print head attached to a precise X-Y-Z robot. For detection, an enzyme-linked immunosorbent assay (ELISA) and a scanning CCD detector were used. This format represents a compromise between the high-density microarrays and the microtitre plates.

In a further approach, capture molecules are attached onto microscopic slides and flow chambers are applied in a cross-wise fashion. Detection can be done via fluorescent labels and CCD-based optical readout. Since the system combines automated image analysis and microfluidics, it has high throughput potential, though the array density still remains low (i. e., a 6 × 6 pattern). In order to considerably increase the sensitivity of measurements, three-dimensional arrays on a flat surface have been designed. By using a gel photo- or persulfate-induced copolymerisation technique, oligonucleotide, DNA, and protein microchips have been produced on polyacrylamide gel pads from 10 × 10 to 100 × 100 μm, separated by a hydrophobic glass surface. It has been proven that three-dimensional polyacrylamide gels provide > 100 times greater capacity for immobilisation compared with two-dimensional glass supports [5].

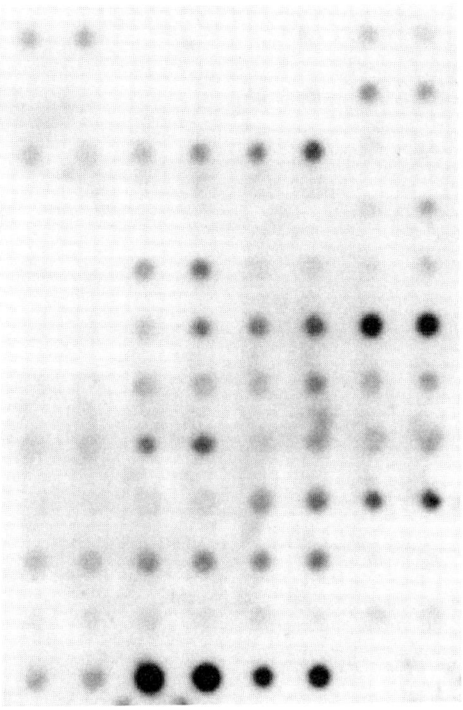

Figure 1 Low-density protein array for autoradialgraphical detection of protein-protein interactions.

In another approach, micromolded hydrogel “stampers” and aminosilylated receiving surfaces are used to array antibodies onto a chip [6]. A stamper deposits the antibody as a submonolayer, while the antibody retains its activity.

Other advances are based on the design of miniaturised immunoassays by arraying single proteins such as bovine serum albumin, kinases, avidin, or monoclonal antibodies, which are often gained by Phage Display or other selection techniques. These kinds of immunoassays include photolithography of silane monolayers [7] or gold [8] combining microwells with microsphere sensors [9] or ink-jetting onto polystyrene film [10]. Specific ligand-coated surfaces are booming nowadays and are provided by several companies in order to facilitate the immobilization of proteins and their analysis. Examples are chips from SELDI ProteinChip (Ciphergen Biosystems Inc., Palo Alto, CA, USA [Fig. 2]); INTERACTIVA Biotechologie GmbH, Ulm, Germany; and various BIAcore chips (Biacore AB, Uppsala, Sweden).

An example of medical applicability of a biochip is a microarray for high-throughput molecular screening of tumour specimens [11]. First, a robot (commercially available from Beecher Instruments, Silver Spring, MD, USA) was designed to punch cylinders (0.6 mm wide, 3 to 4 mm high) from 1000 individual tumour biopsies embedded in paraffin. Next, they were arrayed in a 45 × 20 mm paraffin block. On serial sections, tumours were then analysed in parallel by immunohistochemistry, fluorescence *in situ* hybridisation, and RNARNA *in situ* hybridisation. This system turned out to be useful for micro-

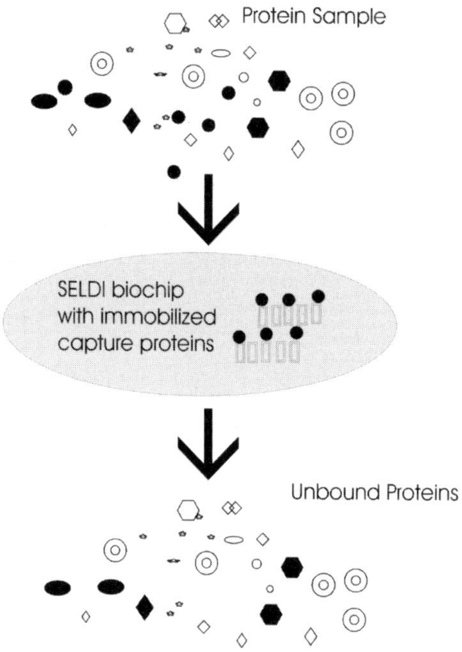

Figure 2 A high-density protein biochip from Ciphergen: The chip surface bearing immobilized capturing proteins is probed with a diverse protein sample. Unbound proteins are then washed away, while proteins that remain attached are analyzed by surface-enhanced laser desorption/ionization (SELDI).

scopic scanning of an immunohistochemistry array slide containing 645 specimens in less than two hours.

Within the past decade, classical devices, such as microtiterplates, have turned into smaller structures, thereby allowing faster and easier analyses at a high throughput level. These efficient analyzers represent mini-computers in a chip look-alike form that are able to scrutinize thousands of biological reactions within relatively short time periods. Moreover, these biochips more and more are replacing cumbersome equipment by miniaturized assays by providing ultra-sensitive detection methodologies with significantly lower costs.

All these obvious advantages have led to a real boom in the variety of biochip developments within the past few decades [12].

Biochips, representing more or less a special class of biosensors, bear various transducer elements and are based on integrated circuit microchips. The term "biochip" is derived from the word "computer chip". This chip is a silicon-based substrate, which is mainly used for the fabrication of miniaturized electronic circuits. With the tremendous development in the field of high-throughput screening in the course of the past few years, the term biochip has taken on a variety of definitions. In general, a biochip is now defined as a device that bears a high-density array of probes used for a variety of assays. Any device comprising a two-dimensional array and having biological materials on a solid substrate therefore has been named a biochip. Biochips offer the advantage of combining miniaturization (usually in microarray formats) and the possibility of low-cost mass production.

1.2 Protein and DNA biochips

Broadly, two kinds of chips can be distinguished, according to the biomolecules on which the chips are based. Protein-protein recognition is the basis for protein biochip design, whereas DNA chips rely on DNA hybridization. Both systems have in common that one of the interaction partners is immobilized on the chip surface. Subsequent probing of the chip surface with a solution containing a biomolecule with an affinity for the immobilized partner results in a recognition and binding event. This specific interaction can then be detected in various ways. Both of the two methods have advantages and drawbacks that have to be taken into consideration for individual experiments.

DNA microarrays are used widely nowadays for the analysis of both single-nucleotide polymorphism (SNP) and mutations and also for the detection of microorganisms. Moreover, DNA chips are applied broadly for gene-expression analyses [13].

As an example of the use of such a chip, it was employed by Schena et al. [13] to investigate the gene expression of patients suffering from rheumatoid arthritis (RA). Because of the wide spectrum of genes and endogenous mediators involved, microarray technology is needed to analyse this chronic disease. In RA, inflammation of the joint is caused by the gene products of many different cell types present in the synovium and cartilage tissues, plus those infiltrating from the circulating blood. Broadly, in inflammatory diseases such as RA, the expression patterns of diverse cell types contribute to the pathology. Schena et al. monitored the gene expression at the inflammatory disease state with a microarray of selected human genes of probable significance in inflammation as well as with genes expressed in peripheral human blood cells. Despite their obvious advantages for expression monitoring, DNA microarrays provide relatively little information about the final concentrations and presence of gene products in a cell. Even patterns of gene-expression profiling are incomplete, since the mRNA level (on which DNA arrays are based) within a cell does not necessarily correspond to its amount of protein. Hence, these kinds of arrays do not give evidence about protein activity, protein-protein interactions, and post-translational modifications. In contrast, protein chips offer the big advantage of directly detecting and analyzing proteins out of crude samples. However, there are no complete sets of affinity reagents with the required specificities and affinities available. This is caused mainly by two reasons. First, it is significantly more laborious to develop high-affinity reagents, such as antibodies, than to synthesize an oligonucleotide or to purify a PCR product. Second, one cannot know exactly which protein and which post-translational pattern is expressed differentially under defined conditions. Moreover, DNA can be thoroughly washed, dried, heated, and frozen and even partially degraded without losing the affinity for its complementary strand. Proteins are usually more sensitive to such treatments and often lose their binding capacity. Moreover, nucleic acids can attach to the surface of a solid support in any direction, whereas proteins must be correctly oriented and active sites must be accessible. In addition, unlike DNA hybridization, protein interaction cannot be performed under denaturing conditions, leading to additional difficulties with background false-positive staining. However, in addition to the already mentioned fact that protein arrays detect proteins directly, they offer more advantages. Protein chips are advantageous in comparison to other traditional approaches in proteomics, including two-dimensional gels and/or chromatography coupled with MS. Protein arrays have the potential to considerably improve the scale-up of analysis by running various affinity-recognition steps on a single chip and by reducing the time of analysis. Moreover, protein microarrays also can improve reproducibility and enable fully quantitative protein-expression analysis at high throughput. Still, it is necessary to further improve surface-immobilization techniques and to increase the affinity and density of the immobilized antibody. Whether the development of protein arrays will follow the way paved by DNA-array developers remains to be seen and is hardly predictable. There is growing variability in the style of protein arrays in all areas, including assays, arrayers, antibodies, surfaces, analyses, and readouts. More and more research interests today are focusing on the improvement of these protein microarrays' design [14, 15] in order to complement DNA microarrays.

Table 1 Comparison of DNA and protein biochips

DNA Chips	Protein Chips
Can be performed under denaturing conditions	Proteins lose their correcting folding upon denaturing
Resistent to harsh treatment, such as heating, freezing, drying, and tringent washing.	Cannot withstand most strong treatment
Immobilization at the chip surface can be done in any direction	Proteins have to be attached in an orientation such that the binding capacity does not get lost
The protein level is identified via the mRNAs	Proteins are directly detected as well as their modifications, such as phosphorylations
Probes are easily labelled, mostly by fluorophores	Labelling requires complex procedures (protein-detecting arrays)
DNA probes can be produced easily and fast	Each binding protein has to be synthesized via selection procedures, such as Phage display, taking considerably longer than the oligonucleotide synthesis

1.3 Protein biochips

Broadly, one can distinguish between two kinds of protein chips: A *protein-function array* comprises large numbers of native proteins attached to the surface in a defined pattern. Such arrays can be applied for high-throughput parallel monitoring of the activities of native proteins. However, for studying of a protein's function, it is necessary to array the protein itself in a way that preserves its native conformation. The other type of protein array is named *protein-detecting array*. It consists of arrayed protein-binding agents, allowing for expression profiling to be done at the protein level. In this way, proteins can be detected out of complex biological solutions, such as crude cell extracts. This is accomplished by arraying antibodies or antibody mimics.

Protein-function arrays

Protein-function arrays offer the advantage of being constructed more easily than protein-detecting arrays. Indeed, a lot of effort has been put lately into the development of useful protein-function arrays. MacBeath and Schreiber [16] described the immobilization chemistry and robotics that were needed to design high-density protein-function arrays. As a starting point, they chose three different kinds of kinase/substrate pairs. Then they produced thousands of proteins and spotted them by high-precision contact printing as nano-droplets down onto an aldehyde-coated microscope slide. By employing this system, spot densities of > 1600 spots cm^{-2} could be achieved with spot diameters of 150 to 200 μm. Each slide was then incubated with a different kinase in the presence of [γ-^{33}P] adenosine triphosphate. This technology could be applied efficiently for the identification of protein kinase substrates at a high throughput.

In another approach, Snyder and colleagues [17] evaluated the catalytic activity of a large number (119) of protein kinases from yeast with a wide array of substrates on a protein chip. This work represented the first genome-wide attempt to analyze protein kinase function using protein-chip technology. Covering nearly all kinases of yeast, the authors used disposable arrays of microwells in silicone elastomer sheets placed on top of microscope slides. Because of the high density and small size of the wells, high-throughput batch processing and highly simultaneous analysis of many individual samples were possible.

For complete protein-function arrays (i. e., covering the entire proteome of an organism), it is inconvenient to spot the whole set of proteins using standard methods of expression of tagged recombinant proteins in Sf9 cells or *E. coli*, followed by laborious affinity purifications. Perhaps this is possible for smaller protein arrays [18]. However, larger protein sets cannot be dotted by this system. In that case, *in vitro* transcription/translation methods are likely more useful, since only small amounts of proteins are required then. Many arrayed clones are commercially available today and could be employed as templates for PCR with primers that are designed to integrate a promoter and an affinity tag. In that way, the DNA substrate could be easily prepared for the following *in vitro* transcription. Then, the tagged protein could be synthesized by *in vitro* translation and selectively immobilized at a certain spot on the array by specific interaction with a capture agent. One possibility would be to tag a protein and to immobilize its corresponding antibody at the chip surface. Alternatively, an approach has been presented to incorporate biotin cotranslationally into a protein [19]. Another possible (but tedious) solution would be first to perform an automated affinity purification of the tagged protein and then to tether it covalently to a chip surface that is modified with lysine-reactive groups.

Protein-detecting arrays
The construction of protein-detecting microarrays is dominated mainly by the detection problem. Analogous to DNA microarrays, where the DNA probes are fluorescently labeled during the reverse transcription step, it is of course not possible to label a protein in the same way. However, a fluorescence label for proteins can be done chemically. For that purpose, e. g., *N*-hydroxysuccinimide ester of carboxyfluorescein can be used to activate the terminal amine of lysines of proteins. This solution then can be incubated with the chip onto which the binding partners of the proteins have been immobilized. The fluorescing proteins can be subsequently detected with a simple reader as is now used in the DNA microarray field. However, there are several problems with using this approach, including the differences of proteins in labelling efficiency and the requirement of calibration curves (otherwise the dot intensity of a particular spot would be meaningless).

In order to avoid different chemical efficiencies in labelling, one can, e. g., treat extracts derived from wild-type yeast and a deletion mutant with two different (coloured) dyes. Subsequent mixture of the extracts is then followed by incubation with the protein-detecting array. The colour intensities of each single spot on the array can then be compared. This provides a measurement of the change in the production levels of the various proteins. The main problem of chemical labelling of proteins is that it markedly influences their surface characteristics, leading to a significant protein denaturation or to the loss of the protein's affinity to its immobilized ligand. Therefore, chemical modification represents an unwanted and labor-intensive process and is not the optimal solution. The detection problem could be solved by avoiding the chemical modification and replacing it with a modification of the capture ligands with a sensitive reporter. This reporter can then uncover the analyte protein-ligand binding. One possibility is to attach the capture ligand to a material that can significantly change its conductive or emission property in response to a binding event. Alternatively, the capture ligand can be modified with a reporter that is capable of signalling the binding of the target.

Protein-detecting arrays also can be designed by a sandwich assay. In that approach, the selected protein is bound by both the immobilized capture ligand and a soluble sandwich ligand. This ligand can be labelled with a reporter (i. e., an enzyme) that is easily detectable. Only in the case of bound protein can the marked sandwich ligand subsequently bind to the chip surface. If the sandwich ligand is conjugated to an enzyme, this method also allows significant amplifications of the binding signal. This means that this approach is also highly useful for the detection of low-level proteins, such as cytokines, hormones, etc.

Huang and coworkers [20], presented a novel antibody-based protein array. This assay combined the advantages of the specificity of ELISA, the sensitivity of enhanced chemiluminescence (ECL), and the high throughput of microspotting.

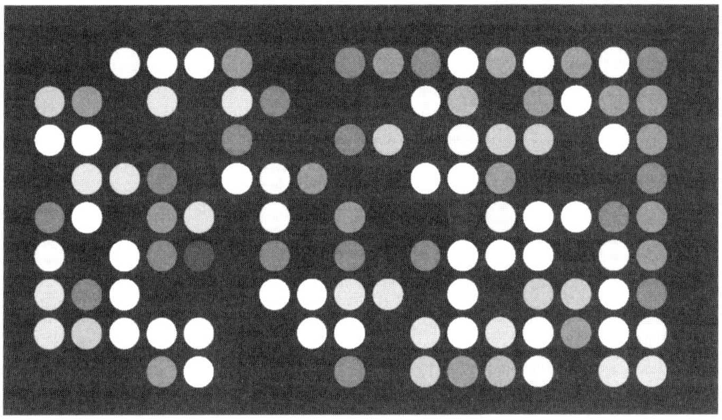

Figure 3 A typical presentation of gene-expression data gained by high-density DNA microarrays. The color-coded spots reflect the abundance of the mRNAs.

When applying this system, the authors first spotted the capture proteins, either antibodies or antigens, onto membranes in an array format (Fig. 3). Then they probed the membranes with biological samples. After binding of the antigens or antibodies in the probes to their corresponding immobilized targets, the membranes were exposed to horseradish peroxidase (HRP)-conjugated antibodies. For analysis, the signals were visualized with ECL system.

There are already some protein chips that are commercially available today and are used in research. Ciphergen(r) (Ciphergen Biosystems Inc., Palo Alto, CA, USA) provides its patented SELDI ProteinChip(r) [21] for detection and analysis of proteins at the femtomolar level directly from biological samples. The detection of the purified proteins is accomplished by Laser desorption / ionization Time-Of-Flight mass analysis (Fig. 2).

First, non-specifically bound proteins are washed off from the chip surface. Then the crystallized sample is ionized and accelerated into the MS tube.

Another example of broadly applied and commercially available protein chips is the chips provided by BIAcore(r) (Biacore AB., Uppsala, Sweden). The chips are flow-based instruments, which use the detection principle of surface plasmon resonance [22]. The surface of these chips consists of a glass slide coated with a thin (50 nm) gold film. On top of this film, a chemical matrix is deposited where one of the binding partners can be immobilized using well-defined chemistries. The flow cells are formed by interfacing the sensor chip with a thermostatically controlled integrated fluidic cartridge (IFC). Parallel channels (60 nl volume, 1.5 mm^2) are formed on the sensor surface, and microfluidics is used to deliver reagents and samples over the chip sensor surface.

Biacore's surface plasmon resonance technology combined with MS/MS represents a versatile system for the discovery of protein interactions [23].

1.4 DNA biochips

In general, the term DNA biochip refers to high-density arrays of oligonucleotide or complementary DNA (cDNA) sequences immobilized on a solid support. The technique is based on the principle that complementary sequences of DNA (probes) can be probed and bound to their immobilized DNA partners (targets). For that purpose, labelled sample DNA and/or RNA in solution is hybridised to a large number of immobilized DNA molecules that are arrayed on the chip surface (Fig. 4). By simultaneously hybridizing to all the sequences on the array, high-throughput analyses of gene-expression patterns can be performed (Fig. 5). The arrayed DNA fragments often represent expressed gene sequences that were obtained from various sources, including those identified in public databases or those from expressed-sequence tags (ESTs).

Two types of microarrays can be distiguished from each other: oligonucleotide-based (short nucleic-acid fragments) and cDNA-based (longer DNA-fragments) microarrays. Both of them are used successfully in industry (e.g. Affymetrix, Santa Clara, CA, USA; and Incyte, Palo Alto, CA, USA) and more and more in academic research. As substrates, glass, silicon, silicon/glass, quartz, or plastic devices are used mainly to provide a rigid surface and therefore often replace chips consisting of classical membranes, such as nitrocellulose and nylon. However, membranes offer the advantage of posses-

Figure 4 A typical scan of high-density arrays. The intensity of the spots reflects the abundance of the analyte.

Figure 5 Micro-dotting robot with pin and ring setup

sing excellent binding properties, whereas one has to make use of a variety of immobilization strategies with working with rigid surfaces.

On the other hand, glass slides offer the advantage of allowing effective immobilization of the probe and robust hybridization of the target with the probe. Moreover, glass is a durable material that sustains high temperatures and washes of high ionic strength. Since it is non-porous, the volume necessary for hybridization can be kept extremely small, leading to an enhanced kinetics of hybridising probes to their target strands.

Whereas nylon arrays are restricted to serial or parallel hybridisations, thousands of probes on glass arrays can be labelled with different fluorophores and subsequently incubated on a single microarray. Furthermore, glass slides show merely weak self-fluorescence and therefore do not contribute to background signals. Suitable methods for immobilization of oligonucleotides must efficiently and irreversibly attach a uniform probe density over the chip surface. Moreover, this method must not negatively influence the subsequent hybridisation process. Conversely, blocking and hybridisations should not be able to disturb the immobilization. The attachment method should be as easy to handle as possible and should be transferable from the laboratory to mass-production scale.

Usually, DNA is spotted onto activated glass surfaces coated with poly-L-lysine or aminosilane. The efficiency of this step of the assay depends on several factors, the most important of which is the homogeneity of the DNA spots and the chemical properties of the solution in which the DNA is dissolved. The spotting buffer of choice is mostly saline sodium citrate (SSC) buffer. However, binding efficiency and spot uniformity are often poor. Adding 50% dimethyl sulfoxide to SSC can reduce those kinds of problems. However, this leads to the drawback that it is toxic and a solvent for many materials.

Another critical step of DNA chip manufacturing is the processing of the glass surface after spotting, during which the remaining, unreacted amino residues of the poly-L-lysine polymer or aminosilane has to be deactivated (Blocking). This leads to decreases of the background signal. Therefore, an efficient blocking step represents one of the most critical steps and should be considered extremely carefully. Blocking usually is accomplished by probing the surface with succinic anhydride in aqueous, borate-buffered 1-methyl-2-pyrrolidinone (NMP), converting the amines into carboxylic moieties. However, during this process the immobilized DNA becomes influenced by the blocking solution, leading partially to its dissolving and spreading across the entire slide. In order to prevent this, Diehl and coworkers [24] developed a robust processing protocol that makes use of a non-polar, non-aqueous solvent and accelerates the blocking reaction by the addition of a catalyst. As far as high-density oligonucleotide arrays are concerned, two main strategies are applied to immobilize the DNA. The first approach, *direct spotting of cDNAs,* comprises the immobilization of pre-synthesized oligonucleotides and is being used for applications such as specific diagnostic tests and drug discovery. In this approach, pre-synthesized oligonucleotides are arrayed onto a chemically active surface (by spotting or printing technologies) and subsequently immobilized through either the 5′- or 3′-oligonucleotide terminus. In contrast, *in situ oligonucleotide synthesis* makes use of a parallel on-chip synthesis of oligonucleotide probes by photochemistries, thereby arraying up to 96,000 different oligonucleotides. The major disadvantages of this method are the high costs and the limited lengths of the synthesized oligonucleotides.

Direct spotting of pre-fabricated cDNA

In order to covalently immobilize probes, oligonucleotide has to be modified with a functional group that allows the covalent attachment to a reactive group on the chip surface. Several attachment strategies have been applied since the development of DNA chip technology, including (1) the modification of the oligonucleotide with an NH_2 group for immobilisation onto epoxy-silane-derivatised or isothiocyanate-coated slides, (2) succinylated oligonucleotides that are coupled to aminophenyl- or aminopropyl-derivatised slides via peptide bonds, (3) disulfide-modified DNA that is immobilised onto a mercapto-silanised support by a thiol/disulfide exchange reaction, (4) the use of unactivated microscope slides with activated, silanised oligonucleotides, and (5) a broad range of methods that are based on heterobifunctional cross-linking molecules, thereby providing many alternatives to the modification of the oligonucleotide and to the linking molecule. Chemically pre-activated microscope slides also have become commercially available lately, leading to considerably more attachment of the DNA.

Non-covalent oligonucleotide attachment also has been reported [25] and may represent an alternative approach. In that case, short oligonucleotide probes are tethered to a solid support by simple electrostatic adsorption onto a positively charged surface film. Attachment is then obtained by microfluidic application of unmodified oligonucleotides in water onto amino-silanized glass. By applying this method, an extremely stable monolayer of oligonucleotide is formed, reaching a high density of molecules.

Another non-covalent approach consists of binding to glass slides with a microporous polymeric surface [26]. This surface comprises a nitrocellulose-based polymer that binds DNA and proteins in a non-covalent but irreversible manner. Some data indicate that these slides have a much higher binding capacity for DNA and better spot consistency than traditional polylysine-coated slides.

In situ oligonucleotide synthesis

Generating DNA probes on chip requires considerable expertise and technical know-how. The DNA has to be applied to the chip surface quickly, yet with high fidelity and reproducibility. *In situ* (or "on-chip")synthesis requires a careful monitoring of the process in order to ensure fidelity of the growing oligonucleotide. In contrast, the setup of cDNAs or prefabricated oligomers is less difficult to perform and therefore more widely used. *Step yield*, meaning the amount of product correctly synthesized from a substrate, represents an important term for oligo synthesis systems, those which are based on prefabricated oligomers and those rested on *in situ* synthesis. A Perkin Elmer (ABI) synthesizer (Foster City, CA), allowing the synthesis of 40 to 60 base long oligomers, can achieve a step yield of, e. g., 0.99.

Piezoelectric printing (the basis of ink-jet printers) offers similar step yield values, whereas Affymetrix's photolithographic method is less efficient (step yield 0.95).

Piezoelectric printing makes use of a technology that is similar to that used for "ink-jet" printers. In short, at each spot of the array, electric current expands an adapter that encircles a tube, which contains reagents for one of the four bases. This leads to a drop of some millilitres of the reagent onto the coated surface, where it is attached via standard chemistry.

Using photolithography (Fig. 6), a mercury lamp first is shone through a photolithographic mask onto the chip surface, thereby removing a photoactive group.

That leads to the display of a 5′hydroxy group that is capable of reacting with another nucleoside. Further rounds of de-protection and chemistry result in the synthesis of oligonucleotides with a length of up to 30 bases. Apart from the low step yield, photolithography also has the disadvantage that its complex process requires an in-house synthesis. In contrast, contact-dotting or piezoelectric printing of DNA allows scientists at the bench top to perform tailor-made expression-monitoring experiments and to determine themselves both the length and type of the probes.

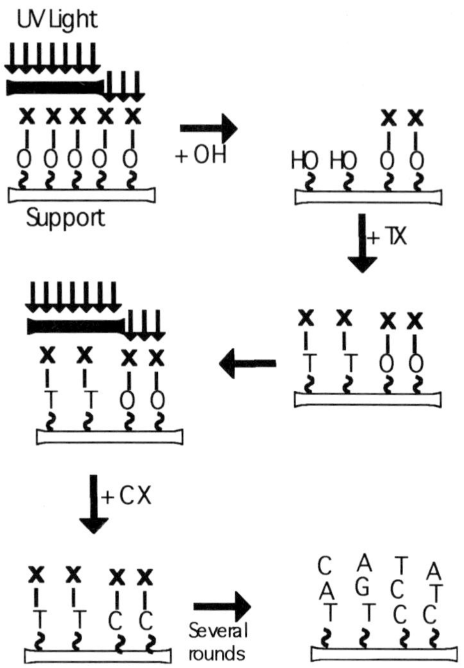

Figure 6 The principle of photolithography. Synthesis of oligonucleotides is initiated by excitation via UV light. Photoprotected hydroxyls (X-O) are illuminated through a photolithographic mask, giving rise to free hydroxyl groups. 5′ photodeprotected deoxynucleosides (e. g., Thymidine [T]) phosphoramidite can then be coupled to these hydroxyl groups. Second row: Another mask is used, leading to the coupling of a new photoprotected phosphoramidite (e. g., Cystein [C]). Subsequent rounds of illumination and coupling result in a growing oligonucleotide chain.

Moreover, array construction can be simplified by pre-fabricating oligonucleotides via traditional controlled pore-glass (CPG) synthesis and subsequently printing them onto the surface. *Nanogen* (San Diego, CA) varies this method by immobilizing pre-fabricated biotinylated oligomers to individual spots on the streptavidin-coated chip surface via controlled electric fields.

However, a cDNA library synthezised on-chip is usable merely for weeks, while prepared oligonucleotide arrays can be stored for much longer.

Summing up, both *in situ* syntheses have their advantages and drawbacks, and it is up to every scientist to decide which of the two systems is more advantageous for a particular experiment.

2 Methods

2.1 Protocols for protein-function arrays

Material

Aldehyde slides can be purchased from TeleChem International (Cupertino, CA). BSA-NHS slides, displaying activated amino and carboxyl groups on the surface of an immobilized layer of bovine serum albumin (BSA), can be fabricated as described in protocol 1. Proteins can be purchased from New

England Biolabs (Beverly, MA). EasyTides γ-^{33}P-adenosine 5′-triphosphate ([γ-^{33}P]ATP) is available at NEN Life Science Products (Boston, MA). NTB-2 autoradiography emulsion, Dektol developer, and Fixer are available from Eastman Kodak Company (Rochester, NY).

Protocol 1 Chemically derivatized glass slides

1. Dissolve 10.24 g N,N′-disuccinimidyl carbonate (100 mM) and 6.96 ml N,N-diisopropylethylamine (100 mM) in 400 ml of anhydrous N,N-dimethylformamide (DMF).
2. Immerse 30 CMT-GAP slides (Corning Inc., Corning, NY), displaying amino groups on their surface, in this solution for 3 h at room temperature.
3. Rinse the slides twice with 95% ethanol and then immerse them in 400 ml phosphate-buffered saline (PBS), pH 7.5, containing 1% BSA (w/v) for 12 h at room temperature.
4. Rinse the slides twice with ddH_2O, twice with 95% ethanol, and centrifuge at 200 × g for 1 min. to remove excess solvent.
5. Immerse the slides in 400 ml DMF containing 100 mM N,N′-disuccinimidyl carbonate and 100 mM N,N-diisopropylethylamine for 3 h at room temperature.
6. Rinse the slides four times with 95% ethanol and centrifuge as above to yield BSA-NHS slides.
7. Store the slides in a desiccator under vacuum at room temperature for up to 2 months (without noticeable loss of activity).

Protocol 2 Arraying of proteins on slides

1. Dissolve proteins in 40% glycerol, 60% PBS, pH 7.5, at a concentration of 100 µg/ml.
2. Spot proteins on aldehyde slides using a microarrayer, such as the GMS 417 Arrayer (Affymetrix, Santa Clara, CA, USA) or the SDDC-2 DNA Micro-Arrayer from Engineering Services Inc. (Toronto, Canada). (In order to avoid differences between pins, you can use a single SMP3 pin [TeleChem International Inc., Sunnyvale, USA]).
3. Incubate for 3 h in a humid chamber at room temperature.
4. Invert the slides and drop them onto a solution of PBS, pH 7.5, containing 1% BSA (w/v).
5. Turn the slides after 1 min. right-side up and immerse them in the BSA solution for 1 h at room temperature with gentle agitation.
6. Rinse briefly in PBS.

Protocol 3 Screening for substrates of protein kinases

1. Spot Kemptide, I-2, and Elk1 with the microarrayer on BSA-NHS slides and process as described above.
2. Wash the slides three times for 10 min. each with washing buffer (WB): 20 mM Tris, 150 mM NaCl, 10 mM EDTA, 1 mM EDTA, 0.1% Triton X-100, pH 7.5.
3. Wash the slides once for 10 min. with Kinase Buffer (KB): 50 mM Tris, 10 mM $MgCl_2$, 1 mM DTT, pH 7.5.
4. Incubate for 10 min. with KB supplemented with 100 µm ATP and wash for additional 10 min. with KB.
5. Incubate the slides for 1 h at room temperature with 200 µl of kinase solution that is applied to the slides under a PC200 CoverWell incubation chamber (Grace Biolabs).
6. The kinase solution is composed of the recommended buffer for each kinase supplemented with the recommended amount of ATP, 2 µl of [γ-^{33}P] ATP (20 µCi), and 2 µl of purified enzyme (10 units of cAMP-dependent protein kinase [catalytic subunit], 1000 units of casein kinase II, or 100 units of Erk2).
7. Wash the slides six times for 5 min. each with WB, twice for 5 min. each with WB lacking Triton X-100, and three times for 3 min. each with ddH_2O.
8. Centrifuge the slides at 200 × g for 1 min. to remove excess of water.

Protocol 4 Visualization of protein-function arrays

1. Melt the NTB-2 autoradiography emulsion at 45 °C for 45 min. in a dark room. Dip the slides in the emulsion for 2–3 s and allow to dry vertically at room temperature for 4 h.
2. Seal the slides in a β-radiation box with desiccant and incubate in the dark at 4 °C for 5 to 10 d.
3. Develop the slides by immersing them successively in Dektol developer for 2 min., ddH_2O for 10 s, Fixer for 5 min., and ddH_2O for 5 min. To visualize the slides, take successive images with a DeltaVision automated microscope (Applied Precision) in DIC mode and stitch the individual panels together in order to form a single larger image.

2.2 Protocols for protein-detection arrays

Material

Immunoglobulins (IgGs) and corresponding HRP-conjugated monoclonal antibodies can be purchased from different companies (see Tab. 2).

Table 2 Overview of antibodies and antigens and their providers

Bovine IgG	SIGMA	Anti-bovine IgG	JIRL
Chicken IgG	SIGMA	Anti -chicken IgG	JIRL
Goat IgG	SIGMA	Anti -goat IgG	SCB
Guinea pig IgG	SIGMA	Anti -guines pig IgG	JIRL
Human IgG	SIGMA	Anti -human IgG	JIRL
Mouse IgG	SIGMA	Anti -mouse IgG	SIGMA
Rabbit IgG	SIGMA	Anti -rabbit IgG	SIGMA
Rat IgG	SIGMA	Anti -rat IgG	Amersham
Sheep IgG	SIGMA	Anti -sheep IgG	JIRL
BSA	Roche		

Pairs of antibodies against cytokines can be obtained from BD PharMingen (San Diego, CA). Cytokines can be purchased from Peprotech (Rochy Hill, NJ).

All donkey anti-Igs against specific species can be purchased from Jackson ImmunoResearch Laboratories (JIRL, West Grove, PA). HRP-conjugated anti-donkey IgG are purchased from Rockland (Gilbertsville, PA). Agarose immobilized guinea pig IgG, goat IgG, human IgG and sheep IgG also can be purchased from Rockland.

Table 3 Different membranes and their properties provided by several companies

Membrane	Provider	Cat.no.	Absorption	Sensitivity of IgG detection	Sensitivity of cytokine detection
Biotrans	ICN	BNRQ3R	Excellent	Excellent	?
Colony-plaque screen	NEN	NEF-978X	Very good	Excellent	?
Hybond N+	Amersham	PRN303B	Very good	Excellent	Excellent
Hybond ECL	Amersham	RPN2020D	Excellent	Excellent	Excellent
Magnacharge	MSI	NBOHY00010	Poor	Excellent	Good
MagnaGraph	MSI	NJOHY00010	Excellent	Very good	Good
Zetaprobe	Bio Rad	162–0155	Good	Excellent	?

Protocol 5 Preparation of antibodies and cytokines

1. Prepare IgG antibodies as stock solutions at a concentration of 4 mg/ml and dilute with TBS to 100 μg/ml as working solutions before the experiments.
2. Prepare the cytokines as stock solutions at a concentration of 100 μg/ml and dilute them into suitable working concentrations before experiments.

Protocol 6 Preparation of array membranes

A template of 504 spots with 28 spots in width and 18 spots in length with a size of 6 × 8 cm can be generated from a computer. This template is used as a guide to spot solution onto membranes. In order to spot capture proteins onto membranes, the template is first placed on the top of a white light box.

1. Put the membranes on the top of the template. Through the light, dark spots in the template are clearly visible from the membrane and are used to guide spot solution onto membranes.
2. Load 0.25 µl of solution manually onto a single spot using a 2 µl Pipetman. HRP-conjugated or biotin-conjugated antibodies can be spotted onto membranes as positive control and identification of orientation of arrays.
3. Different IgGs (0.25 µl of 100 µg/ml) can be loaded onto membranes in that fashion.
4. Block the membranes with 5% BSA/TBS (0.01 M Tris HCl pH 7.6/0.15 M NaCl) for 1 h at room temperature and incubate individually or collectively with HRP-conjugated antibodies for 2 h at room temperature.
5. Wash arrays three times with TBS/0.1% Tween 20 and then twice with TBS. The signals can be imaged with enhanced chemiluminescence (ECL).

Protocol 7 Screening assay of multiple cytokines

A pair of antibodies that recognize different epitopes of the same antigen is used to capture and detect a certain antigen.

1. Spot 0.25 µl of an individual capture antibody at a concentration of 200 µg/ml onto the membranes as described above.
2. Block with 5% BSA/TBS and incubate the membranes with a single or a combination of different cytokines prepared in 5% BSA/TBS for 2 h at room temperature.
3. Wash out unbounded cytokines with TBS/0.1% Tween 20 and TBS.
4. Incubate the membranes individually or collectively with biotin-conjugated anti-cytokines antibodies.
5. Wash and image the arrays as done in ECL system.

Protocol 8 Screening assay of multiple antibodies

1. Immobilize different species of IgGs (100 µg/ml) onto membranes at a quantity of ~ 0.25 µl per spot.
2. Block with 5% BSA/TBS
3. Incubate the arrays individually or collectively with various anti-IgG antibodies.
4. Wash 5x with TBS.
5. Pre-absorb rabbit anti-donkey IgG with agarose-immobilized guinea pig IgG, goat IgG, human IgG, and sheep IgG (to remove cross-reaction components).
6. Incubate the membranes with these pre-absorbed anti-donkey IgG.
7. Carry out images with ECL.

2.3 Protocols for DNA chips

Protocol 9 Preparation of total human RNA from cultured cells

Materials

RPMI media (from Sigma, it is supplemented with 10% fetal bovine serum, 500 U/ml penicillin, and 100 µg/ml streptomycin

Phosphate-buffered saline (Sigma), GTC solution, Ultra-Clear centrifuge tube 13 × 51 mm (Beckman), low-speed ultracentrifuge (Beckman), CsCl (United States Biochemical)

1. Use a confluent flask containing 50 ml cultured human cells (e. g., Jurkat line J25).
2. Setup five flasks (162 cm^2) containing 100 ml per flask of rpmI media
3. Split the cells 1:10 and grow for 48 h.
4. Add phorbol ester for induction and grow for 24 h.
5. Transfer the cells (500 ml in total) to a 500 ml centrifuge bottle.
6. Pellet the cells by centrifugation for 5 min. at 3000 rpm in a JA-10 rotor.
7. Remove and discard the entire medium.
8. Resuspend the pellet in 10 ml phosphate buffered saline (PBS).
9. Transfer it to a 15 ml conical tube.
10. Pellet the cells again by centrifugation for 5 min. at 1000 rpm.
11. Remove and discard the PBS.
12. Freeze the cell pellet (app. 5×10^8 cells) in liquid nitrogen and store at –80 °C.
13. Take the cell pellet (app. 5×10^8 cells) from –80 °C.
14. Add 12 ml GTC solution (60 g guanidine thiocyanate, 0.5 g sodium N-lauroylsarcosine, and 5.0 ml 1 M sodium citrate are dissolved in 100 ml of H_2O, sterile filtered).
15. Add 0.5 ml β-mercaptoethanol.
16. Re-suspend the pellet by vortexing.
17. Draw the cell lysate through a 19-gauge needle at least 10 times (using a 12 cc syringe).
18. Setup six ultra-centrifuge tubes (Ultra-Clear centrifuge tubes) containing 3.0 ml each caesium chloride solution (dissolve 95 g CsCl and 20 ml 0.5 M EDTA, pH=8.0 in 100 ml H_2O. After sterilization by filtration, add 0.1% diethyl pyrocarbonate, then autoclave it).
19. Pipette 2 ml of the GTC-cell lysate onto the CsCl layer of each tube.
20. Pellet the RNA by centrifugation in an SW50.1 rotor at 35,000 rpm for 12 h at room temperature.
21. Discard the supernatant by inverting each tube (Note: Prevent the supernatant from draining back onto the RNA pellet).
22. Invert the tubes on paper towels and dry them for 1 h.
23. Add 100 µl of 1x TE (1 mM Tris, 0.1 mM EDTA, pH 8.0) to each tube (Note: Do not touch the inner sides of the tube; it might contain ribonuclease).

24. Heat it up for 1 min. at 65 °C in order to dissolve the pelleted RNA.
25. Pipet vigorously up and down to fully dissolve the pelleted RNA.
26. Combine all six 100 RNA samples.
27. Use 400 µl 1x TE (Note: Add 0.1% diethyl pyrocarbonate to the 1x TE before use in order to inactivate ribonucleases).
28. Extract the 1 ml RNA sample twice with 500 µl phenol/chloroform/isoamyl alcohol (25:24:1 v/v) for removal of residual protein
29. Extract twice with ether.
30. Divide the RNA sample into two 500 aliquots and transfer them into two Eppendorf tubes.
31. Add 50 µl (diethyl pyrocarbonate-treated) 2.5 M sodium acetate and 1 ml ethanol to each tube.
32. Pellet the RNA by centrifugation for 15 min. at room temperature in a micro-centrifuge.
33. Discard the supernatant.
34. Dry the pellet containing the RNA in a speedvac.
35. Resuspend this pellet in 400 µl 1x TE. For a total yield of 3 mg RNA per 500 ml cells, the concentration should be 7 mg/ml.

Protocol 10 Preparation of polyA and mRNA from total human RNA

Materials

Oligotex mRNA Midi Kit from *Qiagen.*

1. Thaw the total human RNA (~ 7 mg/ml) that has been stored at –80 °C.
2. Mix 150 µl RNA (~ 1 mg), 150 µl 2x binding buffer, and 55 µl Qiagen Oligotex-dT resin.
3. To denature the RNA, heat for 3 min. at 65 °C.
4. Cool for 10 min. at room temperature to allow the annealing of mRNA to resin.
5. Pellet the resin with its bound mRNA by spinning for 2 min. in a micro-centrifuge.
6. Resuspend the resin in 600 µl wash buffer by vortexing vigorously.
7. Pellet the resin by spinning for 2 min. in a microcentrifuge.
8. Resuspend the resin in 600 µl of wash buffer.
9. Transfer the solution to a Qiagen spin column.
10. Centrifuge for 30 s in a microcentrifuge for spinning out of the washing buffer.
11. Resuspend the resin in 33 µl elution buffer (heated to 80 °C).
12. Spin for 30 s and transfer the eluant to a new tube.
13. Elute the remaining mRNA in the spin column using two additional 33 µl of elution buffer heated up to 80 °C.
14. Combine all three (33 µl) mRNA-fractions.
15. Add to that solution 10 µl of 2.5 M sodium acetate and 220 µl of 100% ethanol.

16. Pellet the mRNA by centrifugation for 15 min. at room temperature (again in a microcentrifuge).
17. Dry the mRNA pellet using a speedvac.
18. Resuspend the pellet in 10 μl of 1x TE.
19. Use 1.0 μl for quantitation of the sample.
20. Dilute to1.0 μg/μl (The yield should be 20–25 μg mRNA, taken from 1 mg of total RNA).
21. Wash in an Eppendorf tube in order to remove particles and debris in the spin column.
22. Heat the tube containing the spin column to 80 °C for 30 s.
23. Treat the solutions with 0.1% diethyl pyrocarbonate (for inactivation of ribonucleases).
24. Mix 1.0 μl of that solution with 500 μl TE and measure the absorbance at 260 nm (OD of 1.0 = 40 μg/ml mRNA).

Protocol 11 Amplification and purification of cDNAs for microarray fabrication

Materials
PCR primers modified with a 5′-NH_2-modifier C6 can be purchased from Glen Research (#10–1906–90), 96-well thermal cyclers are provided by PCR system 9600-Perkin Elmer, 96-well PCR plates by MicroAmp 96-well Perkin Elmer (#N801–0560)
Taq DNA polymerase by Stratagene (#600139), PCR Purification Kit can be purchased from Telechem PCR-100, Flat-bottom 384-well plates from Nunc (#242765)
Micro-Spotting Solutions are provided by TeleChem (#MSS-1).

1. Add 1 μl of 10 ng/μl plasmid DNA into each well of a 96-well plate.
2. Add 99 μl of this PCR mix:
   10 μl of 10x PCR buffer (500 mM KCl; 100 mM Tris-Cl, pH 8.3; 15 mM Mg^{2+}; 0.1% gelatin)
   10 μl of dNTP cocktail (2 mM each)
   1 μl primer 1 (100 pmole/μl), 1.0 μl primer 2 (100 pmole/μl)
   1 μl plasmid DNA (10 ng/μl)
   76 μl H_2
   1 μl Taq Polymerase (5 units/μl)
3. Amplify cDNAs by 30 rounds of PCR (94 °C, 30 s; 55 °C, 30 s; 72 °C, 60 s).
4. Purify the products with a 96-well PCR Purification Kit.
5. Add 100 μl of 0.1x TE (pH 8.0) to elute the PCR products.
6. Dry the products in a speedvac.
7. Resuspend each product in 7.5 μl Micro-Spotting solution.
8. Transfer it to a flat bottom 384-well plate (e. g. Nunc) for microarraying.

Note: NH_2-linked cDNAs can be produced during PCR using 21-mers that contain a C6- NH_2 modifier (from Glen Research) on the 5′ end of each primer. Plasmid DNA can be prepared by alkaline lysis. The 96-well REAL prep (Qiagen #SQ811 and #19504) facilitates the preparation.

Protocol 12 Preparation of fluorescent probes from total human mRNA

Materials
100 mM dATP, dGTP, dCTP , dTTP (Pharmacia #27–2050, -2060, -2070, -2080)
StrataScript RT-PCR kit (Stratagene #200420)
Oligo-dT 21-mer (treated with 0.1% diethyl pyrocarbonate for inactivation of ribonucleases)
SuperScript II RNase H- Reverse Transcriptase (Gibco BRL , #18064–014)
1 mM Cy3-dCTP (Amersham #PA53021)
1 mM Cy5-dCTP (Amersham #PA55021)
1 mM fluorescein-12-dCTP (DuPont. #NEL-424)

1. In a microcentrifuge tube, mix 5.0 µl of total polyA+ mRNA (1.0 µg/µl) (the total mRNA has to be purified from total RNA using Oligotex-dT (Qiagen), 1.0 µl of control mRNA cocktail (0.5 ng/µl) (Control mRNAs from *in vitro* transcription are doped in at molar ratios of 1:1000, 1:10,000, and 1:100,000 for an average length of 1.0 kb for mRNAs) , 4.0 µl oligo-dT 21mer (1.0 µg/µl), and 27.0 µl of H_2O (diethyl pyrocarbonate-treated).
2. Denature the mRNA for 3 min. at 65 °C.
3. Add 10.0 µl 5x first strand buffer (250 mM Tris-HCL; pH 8.3, 375 mM KCl, 15mM KCl), 5.0 µl 10x DTT (0.1M), 1.5 µl Rnase Block (20 units/µl), 1 µl dATP, dGTP, dTTP cocktail (25 mM each), 2 µl dCTP (1 mM) (for labelling the mRNA with other fluorophores, substitute Fl12-dCTP or Cy5-dCTP to the reaction), 2 µl Cy3-dCTP (1 mM) and mix by gently tapping the microcentrifuge tube.
4. Add 1.5 µl SuperScript II reverse transcriptase (200 units/µl) for a total reaction volume of 50 µl.
5. Mix again by gently tapping the microcentrifuge tube.
6. Anneal Oligo-dT to mRNA for 10 min. at room temperature.
7. Reverse transcribe the polyadenylated RNA for 2 h at 37 °C.
8. Add 5.0 µl 2.5 M sodium acetate and 110 µl 100% ethanol at room temperature.
9. Centrifuge for 15 min. at room temperature in a microfuge in order to pellet cDNA/mRNA hybrids.
10. Remove and discard the supernatant and wash the pellet carefully with 500 µl 80% ethanol (Note: To avoid any loss of pellet, centrifuge 1 min. before removal of 80% ethanol).
11. Dry the pellet in a speedvac and resuspend in 10 µl 1x TE.
12. Heat sample for 3 min. at 80 °C for denaturing of cDNA/mRNA hybrids. Put the sample on ice immediately thereafter.

13. Add 2.5 µl 1 N NaOH and incubate for 10 min. at 37 °C for degration of the mRNA.
14. Neutralize the cDNA mixture by adding 2.5 µl 1 M Tris-HCl (pH 6.8) and 2 µl 1 M HCl.
15. Add 1.7 µl 2.5 M sodium acetate and 37 µl 100% ethanol.
16. Centrifuge for 15 min. at full speed in a microfuge (to pellet the cDNA).
17. Discard the supernatant and wash the pellet with 500 µl 80% ethanol.
18. Dry the pellet in a speedvac and resuspend it thoroughly in 13 µl H_2O.
19. Add 5 µl 20x SSC (3 M NaCl, 0.3 M sodium citrate, pH 7.0) and 2 µl 2% SDS.
20. Heat at 65 °C for 30 s.
21. Centrifuge for 2 min. in a microfuge at high speed.
22. Transfer the supernatant to a clean tube.

Note: The final cDNA concentration should be ~ 250 ng/µl per fluor in 20 µl of 5x SSC and 0.2% SDS.

Protocol 13 Spotting of cDNA

Materials

5- NH_2-Modifier C6 (# 10-1906), from Glen Research, Sterling, VA
ChipMaker2 (# CMP2), ChipMaker3 (#CMP3), or Stealth Micro-Spotting Device
Wash Station (#AWS-1), SuperAldehyde Substrates

1. Attach covalently a 5′ amino linker to oligonucleotides or PCR products either directly during oligonucleotide synthesis or by enzymatic incorporation of amino-modified PCR primers into cDNAs during PCR amplification-modification of the oligos. (Use of 5′ amino-modification promises the best success.) The $NH_2(CH_2)_6$ linker from Glen Research is broadly employed. Once bound to the chip surface, the amino-modification is resistent to a wide range of conditions.
2. Re-suspend the DNA samples containing a 5′ amino modification in dH_20 at the respective concentration. For PCR products in gene-expression monitoring, an amino-modified DNA concentration of 0.2–1 µg/µl gives best results. For ~ 15-mer oligonucleotides in mutation detection, a DNA concentration of 10–100 pmole/µl is useful.
3. Transfer the DNA solutions into 96-well or 384-well plates via multichannel pipettes. (Note: Purification of cDNAs with a conventional purification kit results in a 96-well or 384-well format for the cDNA samples. By using 96-well formats, oligonucleotides can be obtained commercially in a 96-well or 384-well format).
4. Add 4 µl per well of the spotting solution with a multi-channel pipetting device when the amino-modified DNA samples are transferred to a 96-well or 384-well formats.

5. Mix the spotting solution and the DNA thoroughly several times by pipetting vigorously up and down. (Note: The Micro-Spotting Solution contains a concentrated mixture of ionic and polymeric components. Therefore, mix well in order to achieve an efficient DNA printing).
6. Print the DNA samples with a microarrayer onto silylated microscope slides. (Note: Evaporation can be avoided by using wetted filter discs on the underside of the microplate lid and sealing the microtiterplates with parafilms before and after printing). These sealed plates can be stored for several weeks at 4 °C without detectable evaporation or DNA degrade.
7. Upon printing, leave the slides at room temperature for 24 h to permit efficient attachment of the DNA onto the surface. Mark the region where the DNA arrays have been applied with a diamond pencil on the underside of the slide.
8. Put several printed and dried slides into a washing rack and place to a 600 ml beaker with a stir bar.
9. Wash vigorously with the following solutions: 2x with 0.2% SDS at 25 °C for 5 min. each, 2x in dH_20 at 25 °C (5 min. each), 1x in dH_20 at 95 °C for 2 min.
10. Chill to 25 °C for 5 min. in 1x in sodium borohydride solution (1.3 g $NaBH_4$ dissolved in 375 ml phosphate buffered saline, then add 125 ml pure ethanol) at 25 °C for 5 min., 3x in 0.2% SDS for 1 min. each, 2x in dH_20 at 25 °C for 1 min. each. Air dry the slides.
11. The slides are now ready for hybridization.

Protocol 14 Hybridization

1. Place the microarray in a hybridization chamber.
2. Add 5 µl 5x SSC + 0.2% SDS to the slot chamber for keeping it wet.
3. Pipette 6 µl fluorescently labeled DNA probe along the edge of a 22 × 22 mm lifter cover slip. Place the cover slip onto the microarray so that the sample forms a thin monolayer by capillary forces between the cover slip and the spotted microarray.
4. Seal the hybridization chamber (containing the microarray) and incubate the device in a water bath at 62 °C. Hybridize then for 6–12 hrs at 62 °C.
5. Upon hybridization, remove the microarray from the hybridization chamber and place it immediately into a washing rack .
6. Wash the microarray in the rack for 5 min. at room temperature in a 600 ml beaker containing 400 ml 1x SSC + 0.1% SDS.
7. Transfer the washing rack to a second beaker with 400 ml 0.1 × SSC + 0.1% SDS.
8. Wash the microarray for 5 min. at room temperature with 0.1 × SSC and 0.1% SDS.
9. Rinse the microarray briefly in a third beaker containing 0.1 × SSC to remove the SDS. Allow the microarray to dry at air.
10. For detecting the fluorescence emission, scan the microarray in a Fluoroscanner (i. e. ScanArray 3000).

Note:

1. Take care that the cover slips are free of oil, dust, and other contaminants.
2. Prevent drying of the sample under the cover slip.
3. Lower the cover slip onto the microarray from left to right, thereby avoiding air bubbles. (Small air bubbles under the cover slip can be eliminated by placing the microarray at 62 °C for several minutes.)
4. While 62 °C is well suited for cDNA-cDNA hybridizations, lower temperatures are useful for hybridizations to oligonucleotides.
5. Transfer the microarray as quickly as possible from the cassette to the washing rack. Do not leave the microarray too long at room temperature. It will lead to an elevated background of fluorescence.
6. Slid off the microarray during the wash step. If the cover slip does not slid off within one minute, remove it gently from the microarray surface.

Troubleshooting after scanning:

1. *Poor printing quality*
Incomplete mixing of DNA samples and Micro-Spotting Solution.
2. *Poor DNA attachment to the surface*
Did not use silylated slides and/or amino-modified DANN.
3. *High background fluorescence*
Poor slide processing, bad blocking, use higher hybridization temperatures. For PCR products, use a PCR purification kit.
4. *Contaminations in the labeling reactions*
Use a Fluorescent Probe Purification Kit.

Protocol 15 Gene expression analyses/Overview

Materials

SuperAldehyde Substrates (from TeleChem), fluorescently labelled samples derived from mRNA, cDNA microarrays

ScanArray 3000, 4000, or 5000

1. Fabricate a cDNA microarray bearing genes of interest.
2. Prepare a fluorescent sample from total mRNA.
3. Hybridize the fluorescent sample to the microarray as described above.
4. Wash and scan as above.
5. Quantitate the fluorescent emission for each spot on the microarray.
6. Monitor the gene-expression values by comparing the experimental data with the controls.

Note: cDNAs of known or unknown sequence can be used. They can represent either a subset of genes or a whole set of the genes in a genome. Synthetic oligonucleotides made from expressed sequences (e. g., ESTs) also can be used.

Protocol 16 Preparation of poly-L-lysine-coated glass slides of 75 × 25 mm

1. Place microscope glass slides (e. g., Corning) in slide racks and the racks in chambers.
2. Prepare a cleaning solution: Dissolve 70 g NaOH in 280 mL ddH_2O and add 420 mL 95% ethanol. Stir until it is completely mixed (the total volume is 700 mL (2x 350 mL); Add ddH_2O, if the solution remains cloudy (Note: the solution has to be totally clear).
3. Pour that solution into chambers with slides.
4. Cover the chambers with glass lids and mix well on an orbital shaker for 2 h at room temperature (Note: Once the slides have been cleaned, they should be exposed to air as shortly as possible. Avoid dust particles, otherwise it will interfere with coating and printing.)
5. Quickly transfer the racks to chambers filled with ddH_2O. Plung the racks vigorously up and down. Repeat the rinses four times, each with fresh ddH_2O. Remove all traces of NaOH-ethanol.
6. Prepare a polylysine solution: Mix 70 mL poly-L-lysine + 70 mL tissue culture PBS in 560 mL water. Use plastic cylinder and beaker (1x tissue culture PBS is 8 g sodium chloride, 0.2 g potassium chloride, 1.44 g sodium phosphate, dibasic anhydrous, and 0.24 g potassium phosphate, monobasic).
7. Transfer the slides to this polylysine solution and shake gently on an orbital shaker for 15 min. to 1 h.
8. Transfer the rack to a fresh chamber with ddH_2O. Plunge the rack five times up and down.
9. Centrifuge the slides for 5 min. at 500 rpm on microtiter plate carriers (first place paper towels below the rack for absorption of liquid).
10. Transfer the slide racks to empty chambers.
11. Dry the racks in a 45 °C (vacuum) oven for 10 min.
12. Store the slides in a closed clean plastic slide box.
13. Check before spotting that the polylysine coating is not opaque. Use print-, hybridization-, and scan-sample slides to check the slide quality.

Protocol 17 Spotting and blocking of poly-L-lysine-coated chips

1. Adjust the DNA spotting solution to 45 mM sodium citrate, pH 7.0, 450 mM NaCl (3x SSC). Alternatively, this same composition, but supplemented with 1.5 M betaine (*N,N,N*-trimethylglycine; Sigma), can be used.
2. Spot the DNA with a microarrayer and print the samples in an adequate format.
3. Leave the slides at room temperature overnight.
4. The next day, heat the slides on a metal block at 80 °C for 5 s.
5. Cross-link the DNA to the chip surface by UV irradiation using a total energy of 60 mJ in a Hoefer UV-cross-linker (Amersham Pharmacia Biotech, Freiburg, Germany).

6. For blocking, freshly dissolve 1 g succinic anhydride (Fluka, Deisendorf, Germany) in 200 ml anhydrous 1,2-dichloroethane (DCE; Fluka) and add – as soon as the succinic anhydride is completely dissolved – 2.5 ml *N*-methylimidazol (Fluka). Pour this blocking solution immediately into the slide chamber.
7. Incubate for 1 h at room temperature on an orbital shaker with slight agitation.
8. Wash the slides briefly in 200 ml fresh DCE and incubate in boiling water for 2 min. Rinse briefly in 95% ethanol, then leave them to dry at room temperature.

Protocol 18 Hybridization of labelled samples and scanning

1. Mix for each hybridization 0.2 µg Cy3- or Cy5-labelled and 1.8 µg unlabelled PCR product and precipitate it with ethanol.
2. Dissolve the pellet in 15 µl hybridization buffer (50% formamide, 3x SSC, 1% SDS, 5x Denhardt's reagent, and 5% dextran sulfate) (50x Denhardt's reagent is 5g Ficoll [type 400, Pharmacia], 5g polyvinylpyrrolidone, and 5g bovine serum albumin [Fraction V, Sigma] in 400 ml dH_2O).
3. QS with dH_2O to 500 ml and mix overnight.
4. Filter through a filter and freeze in 10 ml aliquots.
5. Denature the sample at 80 °C for 10 min.
6. Place the sample on a microarray that is covered by a lifter cover slip of 22 × 22 mm.
7. Hybridize for 6–16 h at 62 °C in a humid hybridization chamber (TeleChem International Inc.).
8. Wash the slides in 2x SSC, 0.1% SDS for 2 min., then in 1x SSC for 2 min.
9. Rinse briefly in 0.2x SSC and dry by centrifugation at 500 rpm for 5 min.
10. Read out the signals by a ScanArray5000 unit and analyse them with the QuantArray1.0 software package (GSI Lumonics, Billerica, USA).

3 Results

3.1 Protein-function arrays

As Schreiber et al [16] showed, it was possible to selectively phosphorylate substrates for a chosen enzyme at a high throughput. They applied a sophisticated method to identify the phosphorylation and to detect the radioactive decay. Since neither x-ray films nor conventional PhosphorImagers offer sufficient spatial resolution to visualize the spots, they exploited a technique that is well known from isotopic *in situ* hybridization. Dipping the slides first in a photographic emulsion, they developed them afterwards manually, leading to the deposition of silver grains directly onto the glass surface. They subsequently visualized the slides by an automated light microscope and stitched the individual frames together.

In their approach, Snyder and colleagues [17] overexpressed 119 of the 122 known and predicted yeast protein kinases and analysed them using 17 different substrates and protein chips. They revealed novel activities and found out that many protein kinases were capable of phosphorylating tyrosine. The tyrosine-phosphorylating enzymes often share common amino-acid residues that are near the catalytic region. The protein-function array thus allowed them to identify several novel features of protein kinases and proved that protein-chip technology is a useful tool for high-throughput screening of protein biochemical activities.

3.2 Protein-detecting arrays

The results of the work of Huang and coworkers [20] demonstrated that multiple cytokines and antibodies could be detected simultaneously with this new approach. Their system comprising (1) the spotting of capture proteins onto membranes, (2) the subsequent incubation with biological samples, (3) the addition of antibody-associated enzymes, and (4) the evaluation of the resulting chemiluminescent color change allowed for the simultaneous detection of multiple cytokines and antibodies. The procedure was so simple that no sophisticated equipment was required. Starting with low-density array formats, it could be shown that high-density arrays with a total of 504 spots also could be simultaneously detected with specificity and sensitivity similar to that of lower density arrays.

The authors expect future applications of this approach for many applications, including direct protein expression profiling, immunological disease diagnostics, and discovery of new biomarkers.

However, the authors also admit that protein arrays are still in the infant state. The cross-reaction among different antibodies is one of the most challenging problems we in the development of antibody-based protein arrays face. A solution to this problem might be overcome by more careful selections of antibodies, pre-absorption of antibodies with antigens, and application of protein-interacting peptides that are selected by phage libraries.

3.3 DNA chips

Effectiveness of DNA binding

In their attempt to improve both the binding efficiency of the spotted DNA and the homogeneity of the spots, Diehl and coworkers [24] achieved a considerable improvement of spot quality and a significant reduction of background by adding betaine to the spotting solution.

In principle, an important factor in microarray analysis is the amount of probe material attached to the surface. This factor can directly influence the sensitivity and the dynamic range of measurements. Diehl and coworkers investigated the effectiveness of different conditions to the attachment of DNA to the support. In order to determine the influence of the buffer condition to the spotting solution, and thus to the binding efficiency of the spotted DNA, they produced PCR products of 500 bp in length from individual randomly picked clone inserts from a subtractive human clone library. The DNA then was diluted serially to concentrations of 500, 250, 100, 50, and 25 ng/μl and applied to glass slides in four replica spots each. As a negative control, they spotted a solution without DNA. In addition to 3x SSC and the same buffer supplemented with 1.5 M betaine, they investigated a commercial ArrayIt™ micro-spotting solution (TeleChem International Inc.).

For hybridisations, labelled PCR products were used as target DNA. Analysis of the signals revealed that irrespective of the buffer, hybridizations were specific to the complementary probe molecule. Moreover, all signal intensities increased with increasing concentrations of the spotted DNA probe solution. However, quantification uncovered that at a DNA concentration in the spotting solution of up to 100 ng/μl, the signal intensities were 2.5-fold higher when betaine was present in the spotting buffer. Accordingly, the binding capacity of the glass surface was nearly saturated at a DNA concentration of 250 ng/μl, while without betaine this level was already reached only at a concentration 500 ng/μl.

Spot homogeneity

The homogeneity of the spots was dependent on the variation of the DNA concentration across a spot. There were distinct, frequently occurring patterns that could be observed upon hybridisation, such as a higher DNA concentration at the edges ("doughnut" effect) or the aggregation of the DNA at few points within a spot. The former effect was seen on slides printed with DNA in pure SSC buffer, while the latter occurred when the ArrayIt™ micro-spotting solution was used. Adding 1.5 M betaine to SSC led to much more homogenous spots. This effect was evaluated by calculating the variation coefficient of signal intensity across all pixels that represented a spot. At a DNA concentration of 100 ng/μl during spotting, for example, the variation coefficient was found to be 7% with the commercial buffer, 14% if SSC was used, and only 5% for SSC supplemented with betaine.

Spot-specific background signal

The spotting solution also had a strong effect on the background signal produced at the spots in absence of a complementary target DNA. The authors took particular care to avoid any carry over of DNA from other samples by extensive washing steps and by spotting the buffer probe before proceeding to samples containing DNA. The signal/noise ratio of each feature was calculated by dividing the mean signal intensity of the four spot areas by the mean of the background signal in between spots. A ratio of 0.7 (± 0.2) was found for 3x SSC supplemented with 1.5 M betaine, while much higher ratios of 5.1 (± 0.8) and 10.5 (± 1.5) were determined for SSC without betaine and the TeleChem ArrayIt™ micro-spotting solution.

Suppression of overall background
The first protocol of slide post-processing with succinic anhydride was introduced by Schena et al. [27] and is widely used at the moment for the blocking of aminated surfaces by acylating the unreacted primary amines. In this process, succinic anhydride is first dissolved in NMP before sodium borate buffer pH 8 is added; the final concentrations are 164 mM succinic anhydride, 96% (v/v) NMP, and 4% (v/v) aqueous sodium borate buffer.

In their approach, Diehl and coworkers [24] assumed that incubation in this solution re-dissolved a part of the DNA deposited on the glass surface, which could then spread across the slide, causing additional background. In an effort to avoid this effect, they substituted the non-polar, non-aqueous solvent 1,2-dichloroethane (DCE) for NMP. The concentration of succinic anhydride was decreased to 50 mM. Furthermore, no aqueous buffer was added to the solution. Instead, the acylating catalyst *N*-methylimidazol was added for acceleration of the process. In addition, they compared slides that were produced and processed in parallel but acylated by either the NMP method or DCE protocol. Notably, they achieved a significantly reduced background with the latter blocking reaction. Since using the DCE-based process as a routine blocking procedure, the authors are no longer faced with background problems that could be caused by inefficient blocking. However, when using NMP the authors struggled with a lot of known problems, such as inverted signal phenomena or a higher background around DNA spots.

Discovery and gene-expression monitoring of inflammatory disease-related genes using cDNA microarrays
Messenger RNA from cultured macrophages, chondrocyte cell lines, primary chondrocytes, and synoviocytes provided a variety of expression profiles for cytokines, chemokines, DNA-binding proteins, and matrix-degrading metalloproteinases [13]. In this work of Heller and coworkers, comparisons between tissue samples of rheumatoid arthritis and inflammatory bowel disease verified the involvement of many genes and revealed novel participation of the cytokine interleukin 3, chemokine Groα, and the metalloproteinase matrix metalloelastase in both diseases. From the peripheral blood library, tissue inhibitors of metalloproteinase 1, ferritin light chain, and manganese superoxide dismutase genes were identified as being expressed differentially in rheumatoid arthritis compared with inflammatory bowel disease. These results successfully demonstrate the use of the cDNA microarray system as a general approach for scrutinizing human diseases.

Arrays of proteins (antigens, antibodies, enzymes), oligodeoxynucleotides, DNA, and cDNAs represent valuable analytical tools in biological and clinical sciences. The range of techniques adapted to a microarray format includes simultaneous multi-analyte immunoassays , mutation analysis, expression assays, tumor cell analysis, and sequencing. The array format provides a simple way of performing tens to hundreds of thousands of tests simultaneously using a relatively small analytical device. These properties make the technology of microarrays one of the most promising tools for the future.

4 Troubleshooting

For reasons of simplicity, the troubleshooting information is given in the text where it fits to its context.

References

1 Ekins RP, Chu FW (1991) Multianalyte microspot immunoassay - microanalytical "compact disk" of the future. *Clin Chem* 11: 19551–967

2 Ge H (2000) UPA, a universal protein array system for quantitative detection of proteinprotein, proteinDNA, proteinRNA and proteinligand interactions. *Nucleic Acids Res* 28:e3, I–VII

3 Schalkhammer T, Hartig A, Pittner F, Moser I (1991) Surface modification of platinum based electrochemical thin film electrodes for DNA biosensors *Microsystem Technologies* 91: 76–81

4 Mendoza G, McQuary P, Mongan A, et al. (1999) High-throughput microarray-based enzyme-linked immunosorbent assay (ELISA). *Biotechniques* 27: 778–788

5 Parinov S, Barsky V, Yershov G, et al. (1996) A DNA sequencing by hybridization to microchip octa-and decanucleotides extended by stacked pentanucleotide. *Nucleic Acids Res* 24: 2998–3004

6 Martin BD, Gaber BP, Patterson CH, Turner DC (1998) Direct protein microarray fabrication using a hydrogel "stamper". *Langmuir* 14: 3971–3975

7 Mooney JF, Hunt AJ, McIntosh JR, et al. CT (1996) Patterning of functional antibodies and other proteins by photolithography of silane monolayers. *Proc Natl Acad Sci USA* 93: 12287–12291

8 Jones VW, Kenseth J.R, Porter MD, et al. (1998) Microminiaturized immunoassays using atomic force microscopy and compositionally patterned antigen arrays. *Anal Chem* 70: 1233–1241

9 Michael KL, Taylor LC, Schultz SL, Walt DR (1998) Randomly ordered addressable high-density optical sensor arrays. *Anal Chem* 70: 1242–1248

10 Ekins RP (1998) Ligand assays: From electrophoresis to miniaturized microarrays. *Clin Chem* 44: 2015–2030

11 Kononen J, Bubendorf L, Kallioniemi A, et al. (1998) Tissue microarrays for high-throughput molecular profiling of tumour specimens. *Nat Med* 4: 844–847

12 Soloviev M (2001) EuroBiochips: spot the difference! *Drug Discovery Today* 6 (15): 775–777

13 Heller RA, Schena M, Chai A, et al. (1997) Discovery and analysis of inflammatory disease-related genes using cDNA microarrays. *Proc Nat Acad Sci USA* 94 6: 2150–2155
14 Kodadek Th(2001) Protein microarrays: prospects and problems. *Chemistry & Biology* 8: 105–115
15 Walter G, Buessow K, Cahill D, et al. (2000) Protein arrays for gene expression and molecular interaction screening. *Current Opinion in Microbiology* 3 (3): 298–302.
16 MacBeath G, Schreiber S (2000) Printing proteins as microarrays for high-throughput function determination. *Science* 289: 1760–1763
17 Zhu H, Klemic JF, Chang S, et al. (2000) Analysis of yeast protein kinases using protein chips. *Nat Genet* 26: 283–289
18 Marten MR, McCraith SM, Spinelli SL, et al. (1999) A biochemical genomics approach for identifying genes by the activity of their products. *Science* 286: 1153–1155
19 Tsao KL, DeBarbieri B, Michel H, Waugh DS (1996) A verstile expression vector for the production of biotinylated proteins by site-specific enzymatic modification in *Escherichia coli. Gene* 169: 59–64
20 Huang RP (2001) Detection of multiple proteins in an antibody-based protein microarray system. *Journal of immunological Methods* 255 1–2: 1–13
21 Davies HA (2000) The ProteinChip® System from Ciphergen: A new technique for rapid, micro-scale protein biology. *J Molecular Medicine* 78: B29
22 Rich RL, Myszka DG (2000) Advances in surface plasmon resonance biosensor analysis. *Curr Opin Biotechnol* 11: 54–61
23 Natsume T, Nakayama H, Jansson, et al. (2000) Combination of biomolecular interaction analysis and mass spectrometric amino acid sequencing *Anal Chem* 72: 4193–4198
24 Diehl F, Grahlmann S, Beier M, Hoheisel JD (2001) Manufacturing DNA microarrays of high spot homogeneity and reduced background signal. *Nucleic Acids Res* 1: E38
25 Belosludtsev Y, Iverson B, Lemeshko S, et al. (2001) DNA microarrays based on noncovalent oligonucleotide attachment and hybridization in two dimensions. *Anal Biochem* 15: 250–6
26 Stillman BA, Tonkinson JL, (2000) FAST slides: a novel surface for microarrays. *Biotechniques* 29(3): 630–635
27 Schena M, Shalon D, Davis RW,Brown PO (1995) Quantitative monitoring of gene expression patterns with a complementary DNA microarray. *Science* 270: 467–470

7 Nano-clusters and Colloids in Bioanalysis

Georg Bauer, Norbert Stich and Thomas G. M. Schalkhammer

Contents

Methods and Tools in Biosciences and Medicine
Analytical Biotechnology, ed. by Thomas G.M. Schalkhammer
© 2002 Birkhäuser Verlag Basel/Switzerland

1 Introduction

In chapter 4 application of metal clusters for immunechromatography is covered. Some of these and related applications have been successful for decades, while others are novel and have to prove their usefulness. It has been described how clusters can be applied in test strips and as efficient quenchers for fluorescent molecular beacons. The fundamental properties in that respect are their absorption high coefficient, their plasmon resonance, and their high density.

These unique properties have been exploited in various ways. The enormously high coefficient of absorption makes metal clusters perfectly suitable for direct optical labels, even visible to the naked eye. Their extremely dense matrix is used to obtain a high contrast in electron microscopy. Both these applications use isolated clusters, without considering particle-particle or particle-surface interactions.

In the past ten years, understanding of the latter, more complex phenomenon has grown strongly. Physicists are now able to describe these interactions and predict the extent of non-linear effects. It is these effects that are made use of in surface-enhanced raman scattering (SERS) [1, 2], surface plasmon resonance (SPR) [3–5], resonance light scattering (RLS) [6–8], 3D particle arrays, and surface enhanced absorption (SEA) [9–13] and surface enhanced fluorescence (SEF) [14].

We will show how with minimum effort each of these techniques can be implemented in a classical laboratory and will strongly discriminate between solution-based techniques and those where vacuum coating is indispensable. We will also point out which (especially analytical) techniques are available as service.

2 Methods

2.1 Materials and equipment

Chemicals

- Solution based-methods: Distilled water, $HAuCl_4$, sodium citrate, tannic acid, $AgNO_3$, PAS, PEI, alginate, chitosane, silanes, oligonucleotides, proteins,
- Vacuum-based methods: targets or corns of gold, silver, copper, tin, aluminium, Al_2O_3, ITO

Equipment

- Solution based methods: hood, heating and stirring plate with contact thermometer, spin coater (i. e. from LOT), UV-Crosslinker (i. e. RS-Components), microscope with at least 50fold magnification, AFM, electron microscope, Biacore or other device for SERS measurements, power supply, highly cleaned glassware.
- Vacuum-based methods: all of the above and vacuum coater with at least two different targets, or one target and one evaporation source. Preferentially, you should choose a coater suitable for reactive sputtering. This technique allows optimal process control and does not heat the probes. However, you will also get along with DC-sputtering or an evaporation source. As an evaporation source, an electron-beam evaporator is preferred, because this reduces the heat impact on the probes. Such vacuum coaters can be obtained from Leybold, Edwards, Pfeiffer, and many others. Most of them have availabilities of three to six months. For some applications, conventional coaters for electron microscopy can be used.

Solutions, reagents and buffers

1% $HAuCl_4$ in distilled water
1% sodium citrate in distilled water
1% tannic acid in distilled water
5 M NaCl in distilled water
1M HCl
1M NaOH
1M Tris pH 7.4
1% PAS adjusted to pH 7 with 1 M NaOH, 10 mM NaCl
1% PEI adjusted to pH 7 with 1 M HCl, 10 mM NaCl
1% alginate adjusted to pH 7 with 1 M NaOH, 10 mM NaCl
1% chitosane adjusted to pH 7 with 1 M HCl, 10 mM NaCl
1% silane in EtOH with 1% distilled water (fresh!)

2.2 Methods

In order to allow the reader to get into the matter, the methods described follow a line from simple to more complex. We will start with particles and their synthesis and how to attach them to surfaces. We will modify the particles and bind them selectively to each other as well as to specifically crafted thin-film structures.

General methods

Colloidal synthesis

Colloidal gold has an orange, red, or red-violet colour, depending on the preparation method, ultimately resulting from the species adsorbed to its surface. Colloidal gold is prepared by condensation of metallic gold produced by reducing gold salts with numerous reagents, the most common being phosphorus, tannic acid, ascorbic acid, and sodium citrate. Depending on the method used, colloidal gold preparations vary in particle size (expressed as the average particle diameter) and size variability (expressed as the coefficient of variation [CV]) of the gold particle diameter in a sol. Such a sol is called monodisperse when the CV is smaller than 15%. The colloid preparations applied in this work are the two major methods employed for purposes of cytochemistry, especially staining of electron microscopic samples.

Protocol 1 Gold colloid preparation according to Frens

The method provides monodisperse colloidal gold with a size range from 14 to 50 nm. To prepare a 14 nm gold solution:

1. 1 ml of a solution of 1% $HAuCl_4$ (in double distilled water) is added to 100 ml of double distilled water in a very clean Erlenmeyer flask covered with aluminum foil with a small hole in the center.
2. The solution is heated to boiling.
3. As soon as it starts to boil, 5 ml of filtered 1% trisodium citrate (in double distilled water) is quickly added (i. e. blow the solution out of a 5 ml glass pipette).
4. The solution is gently boiled until an orange red color develops (usually, the color changes from translucid to blue, and to red within minutes).
5. All parameter, cited here have been varied, and it was found that the concentration of the 1% gold solution and the temperature and mixing rate of the heated water are most crucial.
6. It is also possible to bring 100 ml double-distilled water to boil alone, add the 1% trisodium citrate, and finally add the gold-solution under vigorous stirring. The size of the clusters can be adjusted by adding different amounts of reducing agent.

Table 1 Relation of nano-particle size and amount of trisodium citrate to be added

Average particle diameter in nm	Amount of 1% trisodium citrate added (ml)
14	5
24	2
30	1,6
46	0,8
65	0,6

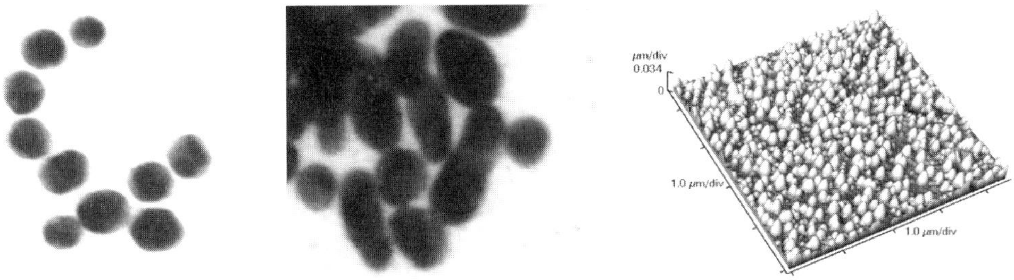

Figure 1 Nano-cluster: crystalline and amorphous via chenical synthesis (left, middle) and sputter coated (right)

Protocol 2 Gold colloid preparation according to Slot-Geuze [15]

If smaller clusters are desired, an additional reducing agent has to be added (in order to prevent further condensation), and the temperature is lowered until the cluster formation is complete. Slot and Geuze [15] have developed such a method that combines tannic acid with trisodium citrate as reducing agents. They describe the optimal conditions to obtain monodisperse sols of different size (from 3 to 15 nm), varying the amount of tannic acid to be added.

1. To make 100 ml of a sol, two solutions are made:
2. Solution A: Gold chloride solution (250 ml Erlenmeyer flask with a stirring bar), 79 ml double distilled water and 1 ml 1% $HAuCl_4$ in double-distilled water
3. Solution B: Reducing mixture (25 ml Erlenmeyer flask) 40 ml filtered 1% trisodium citrate in distilled water, a variable volume (x) of 1% tannic acid and 25 µmolar K_2CO_3 (to correct the pH of the reducing mixture) if the tannic acid volume is above 1 ml.
4. Finally, use double distilled water to make 20 ml total.
5. Solutions A and B are heated to 60 °C.
6. Solution B is quickly added to solution A under vigorous stirring.
7. A red sol is formed very rapidly, depending on the amount of tannic acid added.
8. The colloid is then heated until boiling and gently boiled for another 5 min.
9. Sometimes the red colour develops only upon boiling, which does not imply a lower quality of this preparation.

Protocol 3 Silver cluster synthesis

1. 45 ml 0,1 M $AgNO_3$ solution is diluted in 220 ml double-distilled water.
2. After adding 1 ml 1% (w/v) tannic-acid solution, the mixture is heated to 80 °C.
3. 5 ml 1% (w/v) Na-Citrate is added under rapid stirring; the solution is kept at 80 °C all the time until the colloidal colour becomes visible (yellowish).

4. Colloid formation can be followed spectroscopically, an absorption peak develops with a maximum at 400 nm.
5. The colloids prepared in this way are polydisperse, with an average particle diameter of 30 nm and a standard deviation of 20 nm, and they adsorb readily to most surfaces.
6. However, unwanted adsorption has to be avoided by the measures cited in the troubleshooting section.

Protocol 4 Preparation of CoreShell-Cluster

Au colloids can be used as seeds for nucleation of Ag:
1. 50 ml 17 nM, 12 nm Au colloids (as cited above) are diluted with water to 200 ml and heated with rapid stirring.
2. Upon rapid addition of 5 ml of 10 mM $AgNO_3$ with a Gilson pipette, 1 ml 1% trisodium citrate is added at this time.
3. At 5 min intervals, additional 5 ml aliquots of 10 mM $AgNO_3$ are added, to a total of 30 ml.
4. At the addition of the fourth aliquot of $AgNO_3$ solution, an additional 1 ml of 1% trisodium citrate is added.
5. The colloid solution is boiled and vigorously stirred throughout the additions and for another 15 min., after which it is removed from heat and stirred until cooled.
6. The resulting solution has an absorbance maximum at 329 nm, with an absorbance of 1.7 after dilution 1:10 in water.
7. TEM analysis of these particles shows the presence of large (14–25 nm diameter) spheroid particles and tiny particles (4–8 nm diameter).
8. The absence of 12 nm particles indicates that all the Au nanoparticles have been coated with Ag, yielding the larger particles.
9. The two types of particles can be separated readily by centrifugation for 5 min. at 14,000 rpm in an Eppendorf centrifuge. The Ag-coated Au particles will sediment quantitatively.

Silanization

With a few exceptions, a silane layer can coat nearly any surface. Thus, a silane layer serves as an adhesion layer for a variety of polymer and nano-cluster films. For vapour-phase coating techniques, the silanes need to have a considerable vapour pressure in order to guarantee a sufficient adsorption rate.

Protocol 5 Vapour coating

In order to functionalize surfaces so that DNA, proteins, or thin-films can be covalently attached to the respective surfaces, silane derivatives can be adsorbed to the surface and cross-linked to form a stable monolayer or thin-film by backing the resulting chips at 105 °C.

1. In order to achieve homogenous coverage, the chip surfaces are cleaned by subsequent washing with isopropanol, ethanol, and water or under reactive oxygen plasma.
2. Both the chips and the silane or silane solution are placed in a vacuum chamber, onto which a vacuum is applied with a membrane pump for several minutes.
3. After the pump is turned off, silane fills the evacuated chamber and attaches to all the surfaces therein.
4. The adsorption process takes about 4 to 6 h, but is usually done overnight. Subsequently, the chips are heated for 1 h to 60 °C for 10 min. at 105 °C.

Heating is necessary in order to cross-link silane molecules with each other and the surface by condensation.

Protocol 6 Coating from solution

1. The surfaces to be coated are cleaned by washing with isopropanol, ethanol, and water.
2. The silanes are diluted in ethanol at concentrations from 0.01% to 5%, and an equivalent amount of distilled water is added to the solution.
3. Depending on how fast the reaction should occur (max.: few minutes in 5% solution) and how much a monolayer is desired (a few hours in 0.01% solution), the concentrations and periods of immersion into the silane solutions are chosen.
4. Thereafter, the chips can be dried directly, strongly increasing the tendency to form multilayers or the chips are washed with isopropanol to remove access silane.
5. The chips are then heat-treated as given above.

Adsorption of polymers

Polymer adsorption is a result of the interplay between ionic, hydrophobic, and more complex interactions. Hydrophobic adsorption is the most widely applied technology and will not be discussed in further detail, since it is very well understood and applied in immuno-adsorption assays (i. e., ELISA). However, especially with charged polymers (polyelectrolytes) there is a very good possibility to control adsorption to an extent impossible with hydrophobic interactions. A simple shift in pH can do what in the case of hydrophobic interaction will only be achieved with large amounts of detergent.

Protocol 7 Coating of polymers onto surfaces

In order to render a glass or plastic surface hydrophilic, you can coat it with two polyelectrolyte layers.

1. If you immerse a glass surface at neutral pH in water, it will appear acidic or negatively charged.

2. In order to coat it with a positively charged electrolyte (polyethyleneimine or chitosane), you have to dissolve the electrolyte at concentrations between 0.001 M and 0.1 M in buffer at pH 7 or higher.
3. To increase adsorption, add 0.1 M of salt for reduction of repulsive interactions of polyelectrolytes already bound to the surface and polyelectrolytes in solution.
4. After 10 min. to 1 h, the glass will be covered with polyelectrolyte, and, as a result of overcompensation of surface charges, a positively charged surface will appear alkaline in solutions of neutral pH.
5. If you repeat the same procedure with a negatively charged polyelectrolyte (polyacrylic acid, or alginate), you will come up with a surface that again appears negatively charged in solutions of neutral pH (like uncoated glass), but this surface hydrophilicity will have strongly decreased, minimizing the effect of undesired hydrophobic adsorption.
6. The same processes are applicable for the adsorption of DNA and proteins.
7. However, it should be kept in mind that these molecules are charged only as far as necessary from their isoelectric point. Also, many of the proteins are not stable at the pH most suitable for their coating to different surfaces.
8. Large molecules will adhere more slowly but more stably to surfaces than will low-molecular-weight compounds.

SERS

In surface-enhanced raman scattering (SERS), substances in close proximity to nanometer-scale roughened noble metal surfaces exhibit large (usually up to 10E6-fold) enhancements in vibrational spectral intensities. When a laser used to excite SERS is in resonance with an electronic transition of the substances (surface enhanced resonance raman scattering or "resonant SERS"), an additional 10E3 fold enhancement is seen. Accordingly, SERS has been utilized in a wide variety of applications, including detection of molecules; elaboration of

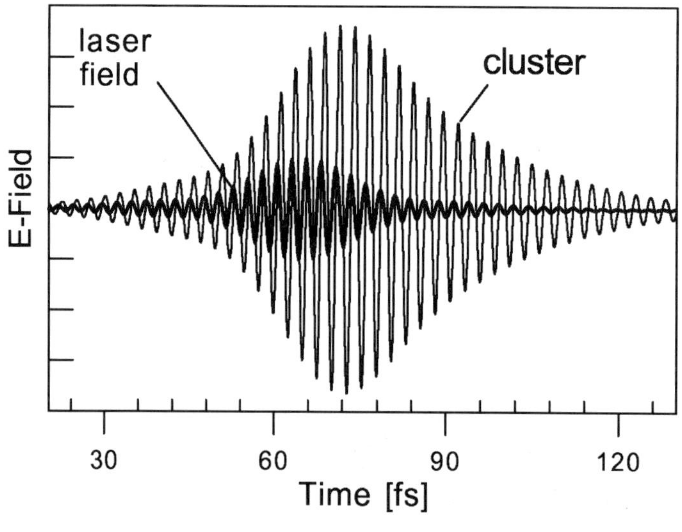

Figure 2 Metal nanocluster excited by a laser pulse

structure and function of large biomolecules; and elucidation of chemistry occurring at metal, metal oxide, and polymer surfaces.

The assay format proposed consists of three steps:

1. binding HemeC to modified colloidal particles
2. reacting the modified particles with the SERS active surface
3. incubating the SERS resulting surface with analyte
4. reading out the SERS-Signal

This assay format is particularly well suited for sandwich-type immunoassays and hybridisation assays. The readout can be done only with suitable instrumentation.

The literature gives a wide variety of approaches for setup, fabrication, and assembly of structures with the required nano-scale roughness. These include evaporated rough metal films, aggregated colloidal metal sols, electronically roughened macroscopic electrode, and others. In all these approaches, the substance being studied by SERS is placed in direct or close contact with the surface. For uncoated SERS substrates, there is typically biomolecule denaturation. In contrast, biomolecules conjugated to colloidal Au nano-particles often retain their biological activity. The SERS intensities for the colloids themselves are weak. The inability to observe SERS signals from even resonantly enhanced chromophores within proteins results from the exponential drop-off in electromagnetic fields away from the substrate surface. Placing a strong chromophore, like HemeC, near the surface of a colloidal particle generates a more suitable SERS active particle. The colloids are then adsorbed to a gold surface, preferably in a sandwich type assay. In the following, it is first described how to adsorb protein to the surface of any of the colloidal particles cited above, then how these particles can be bound in a sandwich-type assay to SERS active gold surfaces. Ideally, an analyte is sandwiched between two SERS active particles.

A special case of the principles described in the general method of coating surfaces with polyelectrolytes is the coating of colloidal gold sols with proteins. Generally, colloidal gold sols are unstable in the presence of electrolytes. When electrolytes are added, the color turns blue and clusters aggregate. However,

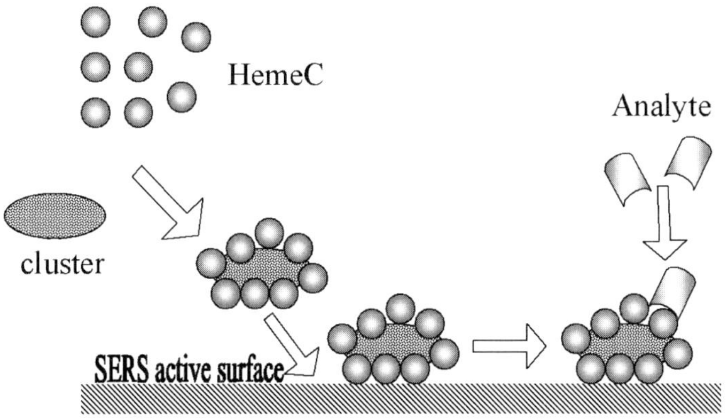

Figure 3 SERS assay using colloidal clusters with HemeC enhancer bound to the SERS chip surface

when proteins are added under proper conditions, these bind spontaneously to the gold particles and the sols are rendered hydrophilic and remain stable in the presence of added electrolytes. The binding of proteins to gold is irreversible, and the proteins maintain their biological activities, at least in part. This property makes nanoparticles highly useful as transducers, thereby considerably facilitating protein separation and detection. That enables reproducible, robust and high-throughput profiling of the proteome [16]. The degree to which biological activity is lost depends, as in all immobilization reactions, upon the stability of the protein, the degree to which conformational changes are vital for the proper function of the protein and how tightly the protein is bound by how many interactions to the surface. Furthermore, the pH during immobilization is critical as well as the isoelectric point of the protein to be immobilized. No quantitative measurement can be obtained easily of the degree to which biological function is retained, but it is assumed to be better than or at least comparable to other covalent immobilization techniques. On the average, one 14 nm colloid is covered with 50 to 150 proteins of a size of 3 nm in diameter. Therefore, even poor immobilization results in at least some biological activity of the cluster reagent. Nevertheless, the higher the success of immobilization, the faster and with higher affinity subsequent reactions of the cluster reagent will take place.

The binding of protein to the gold clusters is pH-dependent. In general, stable complexes are achieved at a pH 0.5 pH units higher than the isoelectric point of the protein involved. The pH of the colloidal gold solution can be adjusted by the addition of 0.1 M NaOH or by addition of 0.1 N HCl. The pH is measured directly with pH paper or with a gel-filled electrode, since the pores of normal electrodes can be plugged by the colloid.

If the protein employed is available only in very limited amounts, it is important to estimate the amount of protein needed to stabilize a certain volume of the sol. In order to provide optimal conditions for complex formation, the ionic strength of the protein-containing solution must be very low, since salts facilitate cluster aggregation. When a critical distance between unprotected colloids is reached, the clusters aggregate and flocculation occurs (a red gold sol turns blue). The easiest method is the salt flocculation test, where the colour change of the gold sol is observed.

Protocol 8 Labelling of gold clusters

1. The protein is dissolved in 200 μl of water at 1 mg/ml.
2. Serial dilutions of the protein in distilled water are prepared with 100 μl of volume each.
3. 500 μl of the pH-adjusted gold sol is added to each tube, and after 10 min. 100 μl of 10% NaCl in distilled water is added.
4. Tubes that contain enough protein to stabilize the gold sol maintain a red color even in the presence of electrolytes.

5. Flocculation takes place immediately if the gold sol is unstable and the red color turns to violet and blue.
6. The amount of protein considered sufficient corresponds to the second tube containing more protein than the one whose color changes to blue.
7. The success of conjugation can of course be followed spectrophotometrically. The maximum absorption of the colloidal solution is at $\lambda = 520$ nm.
8. Once the optimal pH and amount for adsorption has been determined, large scale protein-gold complex preparation (50–500 ml) is carried out.
9. The optimally stabilizing amount of protein plus about 10% in excess is dissolved in distilled water and filtered through a 0.2 micron filter.
10. The pH of colloidal gold is adjusted and then the sol is added to the protein solution under fast stirring.
11. The mixture is incubated overnight and washed by centrifugation the next day, thereby removing the excess of protein.
12. After pelleting of any of the colloidal particles, they can be resuspended in almost arbitrarily small volumes to prepare more concentrated solutions. This is particularly helpful for coated colloids that are to be adsorbed to various surfaces.

Most protein-gold conjugates retain their biological activity for months in the refrigerator at 4 °C with 0.02% sodium azide added to the buffer solution. Good bioactivity preservation for a longer period is also obtained upon storage at –20 °C in the presence of 45% glycerol.

The HemeC is perfectly suited for adsorption to particles with gold at its outside surface. However SERS enhancement is at least a factor of 10 higher with gold-coated silver particles. The SERS active surface is coated with, e. g., a first antibody. The SERS active particle is created by coating gold colloids first with the second antibody and secondly with the HemeC (both as given above). In flow-through chamber of an SPR device, you can consecutively add various analytes (e. g., antigens) and use the functionalised SERS active particle as a detection reagent. Devices providing two-dimensional resolution of the signal-generating surface can be used to apply the principle to array-type assays. (see application by Surromed).

Further reading on SERS devices and detection is given in physical handbooks. The techniques cited above focus on the chemical procedure required for optimal sample-handling, enhancement, and thus signal.

DNA-coated nano-particles

The potential utility of DNA for the preparation of biomaterials and in nano-fabrication methods has been recognized. A variety of methods have been developed for assembling metal and semiconductor colloids into nano-materials. They involve well-established thiol adsorption chemistry using linear alkanedithiols and conjugates thereof, as well as phosphorothioate groups, substituted alkylsiloxanes, cyclic disulfides, and resorcinarene tetra thiols. Especially with the increased application of these particles in biological assays, it has been found, that multivalent interactions of the linking group with the particle are beneficial, since often enough you want to apply elevated temperature in combination with hybridisation assays. At about 55 °C a single thiol binding to the surface may detach from the surface because of increased surface mobility of the gold atoms. The rate of hybridisation also can be increased by heating the solution containing the nucleic acid to be detected and the nanoparticle-oligonucleotide conjugates to a temperature above the melting temperature for the complex formed between the oligonucleotides on the nano-particles and the target nucleic acid and allowing the solution to cool.

The assay format proposed consists of three steps:

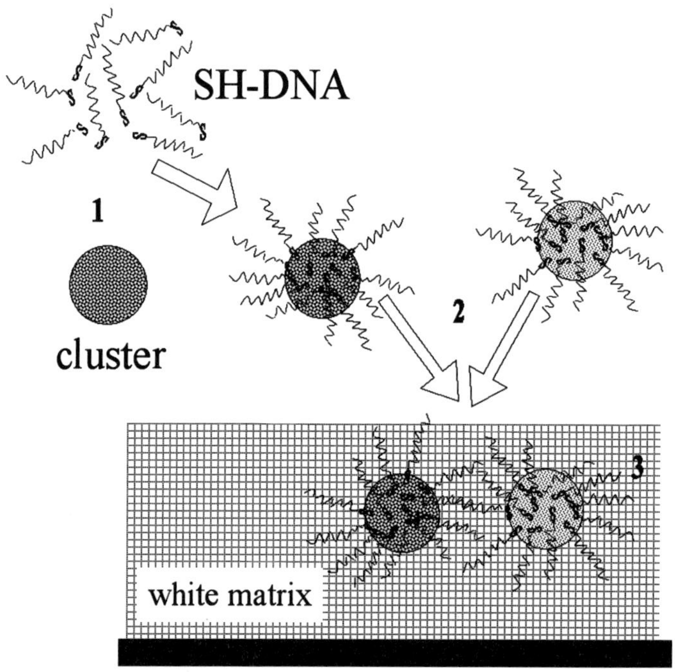

Figure 4 Color assay using assemblies of DNA-conjugated nano-clusters

1. coupling thiol-modified DNA to colloidal particles
2. reacting the modified particles with the sample
3. spotting the reaction mixture on porous white support

This assay format is particularly well suited for sandwich-type hybridisation assays. The readout can be done with the naked eye or, more sensitively, with a flat bead scanner and suitable densitometric software.

Protocol 9 Attachment of oligonucleotides to gold nanoparticles

The attachment of short synthetic oligonucleotides to gold nanoparticles follows slightly different rules than the attachment of proteins or larger polyelectrolytes, which is due to the linear nature of the species.

1. A preparation as cited above (Au colloids, 12 nm in diameter, 17 nM in particle concentration) is reacted with a 200-fold access of a thio-functionalized oligonucletoide (3.5 µm) in water.
2. The reaction is allowed to stand for 24 h at room temperature in order to allow reorientation at the surface and provide optimal space filling.
3. About 80 molecules are assumed to immobilize to the surface of a 12 nm particle.
4. The excess of oligonucleotides is removed by centrifugation.
5. The particles show very good stabilities at high salt concentrations.
6. However, in order to provide temperature stability at 70 °C and higher, the linking group has to be modified. Either a multivalent thiol linker is applied, or the particles are derivated with an alkyl-siloxane. The oligonucleotides can then be bound covalently to functional groups in the surface of the coated particle.
7. The test itself leads to the aggregation of nanocluster arrays. Two types of particles are coated with two sequences specific for the same target molecule.
8. Upon addition of the target, the particles bind to the target. Because any particle bears a large number of identical sequences aggregation of particles will occur.
9. The formation of these three-dimensional arrays gives a detectable change that occurs upon hybridisation of the oligonucleotides on the nano-particles to the nucleic acid, colour change from red to blue.

A colour change results from the formation of aggregates of the nano-particles or from the precipitation of the aggregated nano-particles. This colour change may be observed with the naked eye or spectroscopically. The formation of aggregates of the nano-particles can be observed by electron microscopy or by nephelometry. The observation of a colour change with the naked eye can be made more readily against a background of a contrasting colour, e. g., when gold nanoparticles are used, the observation of a colour change is facilitated by spotting a sample of the hybridisation solution on a solid white surface of an RP-18 TLC plate. Initially, the spot retains the colour of the hybridisation solution, upon drying at room temperature of 80 °C, a blue spot develops if the nano-particle conjugates had been linked by hybridisation. Otherwise, the spot appears pink. The results can be stored over long periods of time in this format. It has been found that a spacer region greatly contributes to the speed of the reaction. An optimum value is between 10 and 30 nucleotides. Applying conventional silver enhancement kits to the surface of the PR-18-plates where low concentrations of analytes are suspected can greatly increase the sensitivity of the reaction.

Any standard hybridisation test can be done in this manner; however, it should be noted that the formation of particle arrays give a very sharp signal because of the large number of interactions that form upon successful hybridisation between different particles. However, this limits the dynamic range of the reaction strongly.

Surface-enhanced absorption (SEA)

So far colloids or clusters have been applied to surfaces using the intrinsic properties of noble metal particles (strong optical absorption, high surface, good handling and modification), eventually in combination with chromophores. In a novel assay format, the colloid properties are altered by application in a specific thin-film setup. This setup is created by coating nanometric thin-films on top of a mirror from gold, silver, or aluminium. Upon binding of the colloids, their spectral absorption is strongly modified by a resonance with the mirror dipole of particle excitation. This optical property for the analytical application of metal cluster films is the so-called anomalous absorption (SEA). Thus, dependent on the thickness of the polymer film, the system can almost be tuned to any colour and used for reading out the resulting complex of e. g., an immunoassay. Bringing in this defined spatial orientation improves signal quality, assay stability, and handling. The feedback mechanism cluster-mirror strongly enhances the effective absorption coefficient of the cluster, increasing the signal by one order of magnitude compared to all standard colloid-based assays.

The assay format consists of four steps:

1. Coating with capture antibody
2. Incubation with a sample containing the antigen.
3. Detection with a secondary antibody coupled to any of the metal particles cited above. Gold particles are preferred due to their high chemical stability.
4. Readout in a standard flat-bead scanner with a diffracting foil to allow readout of the reflecting surface.

This assay format is particularly well suited for sandwich-type immunoassays and hybridization assays. The readout can be made more sensitive with a high resolution direct reflection (not scatter) scanner or a high resolution CCD-chip and suitable filters.

The SEA setup enables sensitivities about a factor of 10 higher than classical vial-based immunoassays. It is compatible with chip-based assays and gives spatial resolutions down the optical resolution of your detection device. In combination with very large particles, the scattering of light can be used as a further parameter of analysis (Genicon, RLS).

Protocol 10 Polymer spinning

1. A polymer is dissolved in a solvent in a concentration so that the resulting solution shows viscosity considerably higher than the solvent alone.
2. The polymer solution is spread over the reflecting chip (e. g., a gold mirror coated on a glass slide, or highly polished aluminum), and the chip is brought into rotation of 2500 to 5000 rpm.
3. During the increase of rotational speed, a major part of the polymer is removed from the surface of the chip.
4. If the indices of hydrophilicity of the surface and the polymer solution match, some of the polymer solution will attach to the surface and the remaining solvent will evaporate from the solution, so that a defined amount of the polymer remains attached to the surface.
5. Unless the film is cross-linked in a second reaction step, the thin-film can be removed easily by washing with the solvent in which the polymer has been originally dissolved. However, a hydrophobic film will remain intact over long periods of time in water-based solutions.
6. For SEA applications, polymers are applied at thickness typically between 20 and 200 nm (dry).

Protocol 11 Oxygen plasma-assisted adhesion

1. The surfaces of the chips are functionalized by oxidation in plasma.
2. The latter step greatly increases hydrophilicity of the surface and allows better handling in combination with any standard assay format.
3. The chip is inserted in the substrate holder inside the vacuum chamber of the sputter coater.
4. The chamber is evacuated, and, after washing the chamber with oxygen at a pressure of 1 mbar, the oxygen pressure is adjusted to 0.1 mbar.
5. A maximum electrical field of around 5 kV is applied, which results in the ignition of a plasma.
6. A plasma is formed between the target and the surface of the substrate. The activated oxygen atoms bombard the chip surface and perform an oxidation reaction there. As a result surfaces can be cleaned from organic substances and polymers can be rendered hydrophilic.

In the case of polymers, the oxidation reaction results in a wide variety of oxidized forms of carbon: hydroxyl, carbonyl, and carboxyl groups.

Carboxyl groups are generated only at a minor percentage of the whole number of reactive groups formed at the surface of such plasma activated polymers.

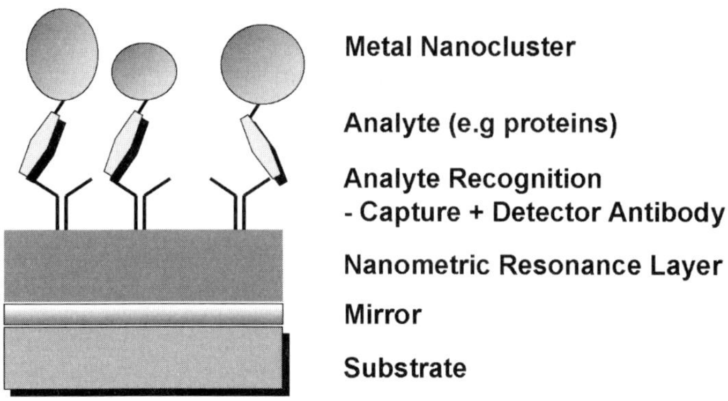

Figure 5 Setup of a surface enhanced nano-cluster protein assay

Table 2 A selection of polymers that can be spin-coated

Polymer	Mol. Weight [g/mol]	Solvent(s) applied:	Concentration for optimal spin films
poly-styrene	280000	toluene	2 to 7%
poly-hexylmetacrylate	320000	decane, AZ 1500	2 to 19%
poly-(4-tert-butylstyrene	300000	hexane, decane, AZ 1500	3 to 10%
poly-styrene bromated	480000	cyclohexanone	5 to 10%
poly-methylmetacrylat	350000	acetone	2 to 8%

Instead of using metal sheets for medium-or low-performance detection systems, a high-quality setup requires depositing a high-quality metal mirror by evaporation or sputter techniques. Using, e. g., a K675X sputter-coater (Emitech, Kent, U.K.), mirrors of up to 60 nm thickness with aluminum, titanium, chromium, tungsten, tin, platinum, gold, iron, silver, copper, and combinations

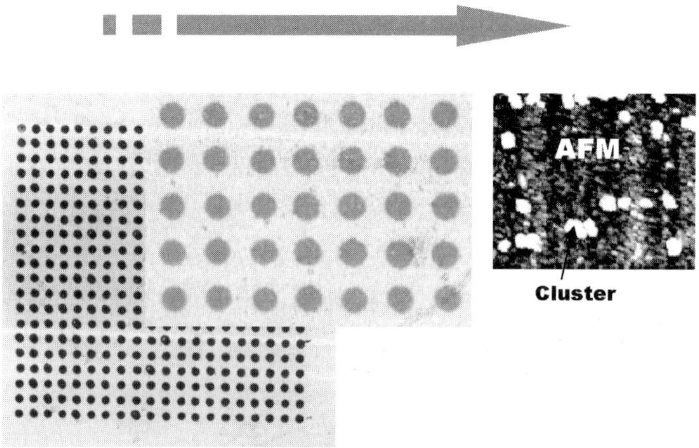

Figure 6 Zoom into SEA chip: Microdots (100 μm in diameter) with metal nano-clusters (see AFM)

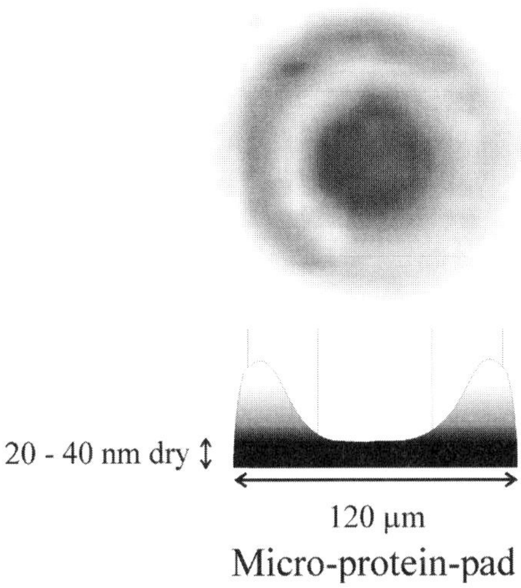

Figure 7 High resolution scan of a micro-protein-pad coated with metal nano clusters used for HT-protein-conformation screening

of these materials (as well as alloys and sandwich layers) might be coated on glass or plastic substrates. Some basic requirements of thermal, chemical, and mechanical stability are fulfilled for most materials, but, after considering the best reflection properties, only silver or aluminum should be used.

To setup an appropriate resonance interlayer, a variety of metallic glasses (tin oxide, aluminum oxide, ITO) can be coated for optimal chemical stability and mechanical performance. The most effective and therefore preferred method is to sputter a film of tin nitride by reactive sputter coating. (Tin is used as a target and sputtered via nitrogen plasma. Exact parameters depend on the machine used. Keep in mind that tin target melts at a very low temperature).

Protein conformation biochips

Instead of binding the colloids in classical assay formats to the surface of a specific thin-film setup, the behaviour of a smart polymer or protein interlayer can be studied by making use of the same phenomenon. New molecular-scale parameters become a matter of investigation without the need for expensive instrumentation.

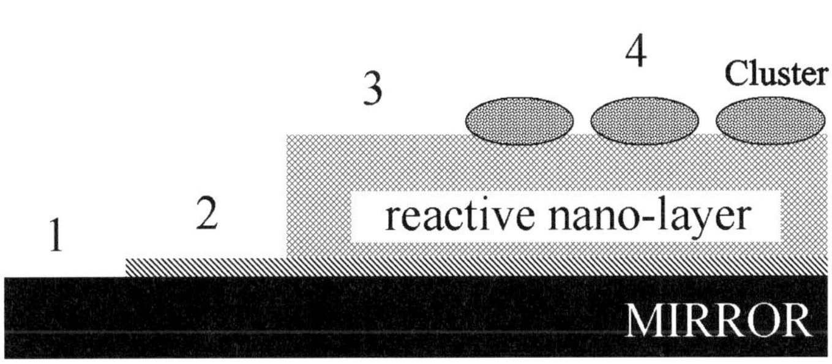

Figure 8 Setup of an reactive interlayer chip: 1: chip, 2: adhesion layer, 3: protein or polymer pad and 4: nano cluster

The assay format proposed consists of four steps:

1. activating the surface of a reflecting chip
2. spin-coating or dotting (see chapter 6) of the conformation-reactive material under investigation
3. adsorbing or sputtering clusters to the surface of the bio- or polymer layer or dot
4 reading the signal (even in real time) with a reflection mode spectrophotometer and a flow-through chamber holding the chip.

This assay format is also applicable for low- and high-molecular-weight analytes that interact with proteins or DNA and generally for the detection of any analyte that produces a volume or conformation effect in a smart material. In this case, a reflecting surface like in Fig. 8 is functionalised in order to allow covalent attachment of a thin-film.

Protocol 12 Setup of a reactive interlayer/protein conformation SEA chip

1. The surface is coated with an amino-silane as given above. The maximum thickness must not exceed 5 nm. Thus, dip-coating is preferred.
2. Thereafter, the chip is coated with a material under investigation, i. e. 2 μg of a protein is dissolved in 50 μl of water.
3. The solution is spin-coated to the chip's surface at 3000 rpm.
4. Because the reactive polymer usually performs large volume changes, it is necessary to cross-link the thin-film subsequently to the spinning procedure.
5. Conformational changes in the response to certain analytes are also the reason that it is necessary to covalently link the polymer to the surface of the substrate. As a cross-linker, the disodium tetrahydrate salt of diazostilbene-disulfonicacid (DIAS) might be used.
6. The protein and the cross-linker are dissolved in water. An ideal response *versus* stability is obtained with a mixture of 4% protein and 0.4% DIAS.
7. The thin-film is then cross-linked by UV-light at a wavelength of around 340 nm. The time for cross-linking depends strongly on the intensity and wavelength of the light source.
8. Under dry conditions, the resulting thin-film is less than 60 nm optical thickness (geometrical thickness multiplied with the refractive index of the distance layer). It swells to about 200 nm optical thickness upon insertion in water.
9. The surface of the reactive polymer is coated with large silver or gold clusters by adsorption. Noble metal clusters have a strong tendency to adsorb to proteins directly after surfaces. Therefore, clusters are synthesized at diameters of 35 nm and higher.

10. The cluster sols are concentrated by a factor of 10 or more and applied to surfaces of the reactive protein. After a few minutes to several hours, the unbound clusters are washed off. If the analyte is less sensitive to denaturation, e. g., synthetic polymers, sputter coating or evaporation also can apply the clusters. It is necessary, that the heat impact be as small as possible.

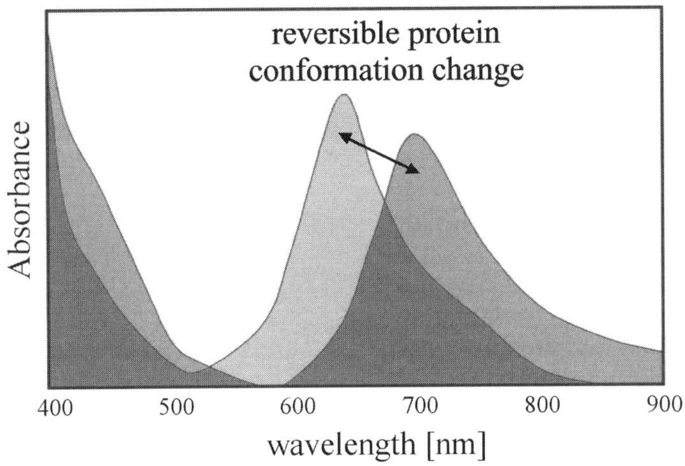

Figure 9 Spectral response of a protein SEA chip

Surface-enhanced fluorescence
In all chip-based devices, the need for ultra-sensitive signal transducers or novel amplification techniques of molecular binding is essential to achieve sufficient sensitivity and an optimized signal-to-noise ratio. By using fluorescence transducers, it is possible to directly visualize the binding of biomolecules at a given surface by a bound layer of fluorophores using flying object or camera-type readout. Nevertheless, both detectors suffer from the optimization compromise of detection time and angle of photon harvesting to achieve a sufficient number of photons per time well about the noise level.

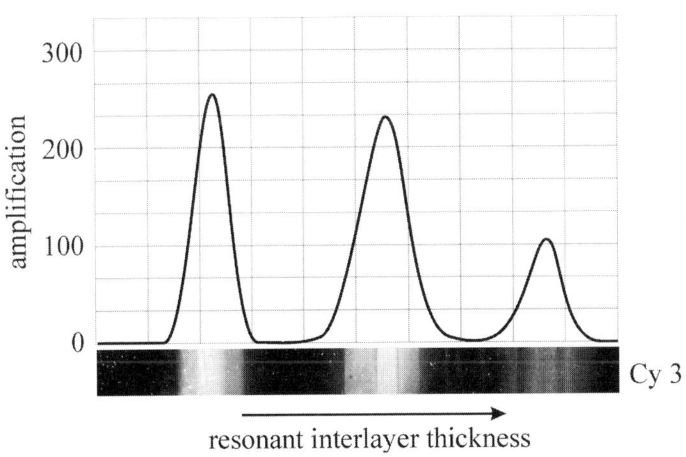

Figure 10 Fluorescence amplification of a nano-cluster–interlayer SEF–chip *versus* interlayer thickness for Cy 3 (530 nm) or Rhodamine. A similar amplification is observed for Cy5 (630 nm); but the curve is slightly shifted to the right.

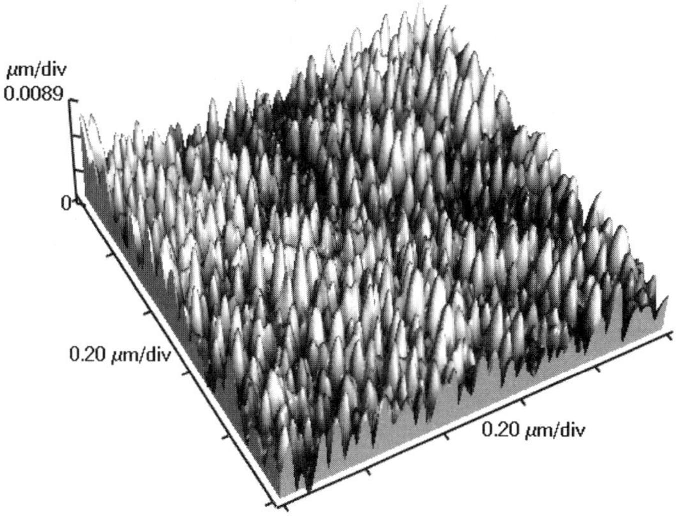

Figure 11 Atomic force microscopy of a SEF nano-cluster layer designed for optimal coupling to a He-Ne laser beam.

Within the past few years, novel approaches of signal boosting have been proposed based on surface enhancement of the field and/or of the excitation-de-excitation pathways. Among them are nano-particle-based effects (first described by Schalkhammer, Aussenegg, Cotton), rough surface effects (first described by Sheehy), surface plasmons over thin metal films (first described by Attridge, Knoll), and local waveguides with a microstructured surface (e. g., Novartis).

All these techniques strike for the development of new devices and are focused on increased sensitivity and simplified procedures.

The straightforward approach is the use of metal island layers as field-enhancing elements in fluorescence detection. The excitation of surface plasmon modes, and subsequently of a resonant interlayer at the interface to the analyte solution, increases the field strength of the incoming light, mainly but not exclusively laser light, by a factor of about 200 to 300. This enhanced field now drives fluorophores within this enhancement zone. Fluorophores feel more laser light than actually irradiated on the chip and, if not already on their maximal efficiency, thus emit more photons per unit time. Nevertheless, the effect needs to be tuned to gain an optimal signal because in the local near field of the metal surface a rapid de-excitation pathway for the fluorophore also is provided.

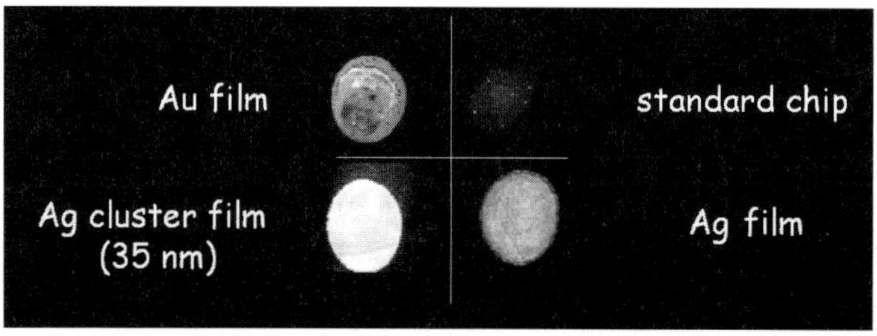

Figure 12 SEF effect for Cy 3 for various enhancer assemblies *versus* non-enhanced stanard glass slides

Contrary to flat metal films, rough particulate matter enables the coupling energy into the plasmon modes of an island layer. Cluster-enhanced fluorescence occurs when fluorophores are positioned at a distance of 10 to 100 nanometers (well apart from the Foerster-quenching distance) from the surface of a nano-particle layer.

A cluster enhanced resonant chip is setup via:

1. a layer consisting of a plurality of nanometric particles (clusters, islands, or colloids) of electrically conductive material, in particular metal, which is applied to the surface of a substrate,
2. an inert interlayer tuned to the appropriate wavelength (optical thickness allows a resonance of the desired wavelength; thus, thickness and refractive index are critical!),
2. a biorecognitive layer on that interlayer and
3. biorecognitive molecules that are labeled with fluorophores.

A method for measuring the concentration of an analyte in a sample comprises the following steps:

1. contacting the sample with a biorecognitive sensor layer,
2. contacting the sample with an analyte-specific fluorescent compound,
3. binding the analyte-specific fluorescent compound to the analyte, which in turn is bound by the biorecognitive layer,
4. radiating excitation radiation that is suitable for excitation of the analyte-specific fluorescent compound, the quantum yield per molecule and per time of the analyte-specific fluorescent compound increasing strongly in the vicinity of the cluster layer,
5. determining the fluorescence radiation emitted by the bound analyte-specific fluorescent compound as a measure for the analyte concentration.

Protocol 13 SEF chip calibration

1. As chips, use standard microscope slides.
2. The slides are coated with silver or gold clusters by absorption (see part on adsorption) or with a silver or gold cluster layer deposited by a sputter process. Nitrogen (Ag) or argon (Au) is used as sputter gas.
3. By sputter coating an inert and hard distance layer is applied onto the metal film using tin as target material and nitrogen as sputter gas, resulting in a tin nitride layer on top of the chips.
4. Aging in air for several days produces an adsorptive surface with good adhesion for protein binding.
5. Atomic force microscopy may be used to determine layer composition and thickness.
6. For chip enhancement calibration, bovine serum albumin (BSA, SIGMA) is used to coat the entire surface at a concentration of 1 mg/ml in phosphate buffered saline (PBS), pH 7.4 . Alternatively, 0.1% polylysine can be used.
7. After 20 min. of incubation, the chip is washed with MQ water and the surface is dried with air.

8. A dilution series (1:10, 10 x) of 0.3 µl cyanine 3 (FluoroLinkTM, Amersham Pharmacia Biotech) is applied onto the chip either by hand or preferably, by using a micro-dotting robot.
9. After drying for at least 10 min., the chips are read out without washing, using a GSI Lumonics Fluoroscanner (ScanArray Lite, MicroArray Analysis System).

Protocol 14 SEF chip

1. Setup chip as described in protocol 13.
2. Chips are covered with BSA or polylysine, then washed and dried as described.
3. Single-strand DNA, amino-modified at its 3 prime end, is diluted to 5 pmol/µl in PBS, pH 5.4, and 0.1 mg/µl N-(3-dimethylaminopropyl)-N-ethyl- carbodiimide hydrochloride (EDC.HCl) is added. Polylysine is able to bind DNA without using EDC activation via a strong ionic interaction (see chapter 6).
4. After micro-dotting, the chips are kept wet in a humidity chamber for at least 20 min. (e. g., inverted over a bath of warm water, small droplets should form where the microdots are!)
5. Subsequently, the surface is washed using MQ water and then dried.
6. The analyte DNA is hybridised in SSC or Dig Easy hybridisation buffer (1–16 h).
7. Complementary DNA, cyanine-5-labelled at its 5 prime end, is diluted in Dig Easy hybridisation buffer to 2 pmol/µl and incubated for 15 min. for 3 h.
8. The chip surface is washed with 10 mM $MgCl_2$, 5% glycerol.
9. The chip is dried and used for scanning.

3 Discussion

Nano-clusters are a state of matter in between a bulk, a surface, and a solution. Treating them asks for optimising strategies of surface chemistry or nano-colloidal solutions of particles. Detergents and wetting properties play a vital role. Important steps toward introducing these novel techniques have been made in several companies. Nanosphere (www.nanosphere-inc.com) uses gold colloids for a wide range of applications [17–19] including as a label for monitoring pathogen contamination on the basis of DNA traces.

Genicon (www.geniconsciences.com) applies large gold particles as labels in high-density array drug screening. Biological assay formats that support information generation and sophisticated instrumentation have been setup. By using "nano-sized" particle labels that specifically bind to targeted molecules, minimal amounts of samples of targeted proteins or nucleic acids can be measured by simple white light source-based instrumentation. The ultra-high sensitivity of resonance light scattering technology allows us to access novel biological information and avoids laborious amplification procedures such as PCR.

ThermoBiostar (www.thermobiostar.com) produces thin-films in order to enable immobilization platforms for immunoassays to detect infectious diseases.

Surromed (http://www.surromed.com) uses nano-barcodes (specifically designed nanoparticles) as identification tags in homogenous drug-screening assays. With Nano-barcode particles, the concept of the barcode and the barcode reader has been taken to the nanometer scale. Moreover, nano-barcode particles can be functionalized with nucleic acids or proteins. If the metals used are super-paramagnetic, nano-barcode particles can be magnetically recovered, thereby conserving the sample volume and simplifying the analysis of complex probe mixtures. Therefore, these particles allow infinite multiplexing of assays in homogenous or heterogeneous media. This technology has the potential to complement DNA microarray technology by functionalizing thousands of nano-barcode particles with oligonucleotides. These experiments can then be performed more efficiently in solution.

However, there is plenty of room for novel techniques, and the goals remain to develop high-resolution, low-cost screening on the one hand and simple, fast, point-of-care testing on the other.

After the successful use of nano-particles as markers for highly paralleled systems, the next step will comprise the design of new devices, such as complete chip detection setups, with metal nano-bead markers. For that purpose, novel low-cost readers and labelling protocols will be required also. By using nano-clusters as transducers, new possibilities will be opened, thereby paving the way for improved development for a variety of assays, including DNA and protein biochips [20–27].

Some of the techniques presented here have been developed in companies, while others are public domain.

The biggest challenge, however, is that with all these novel technologies, novel materials and methods have to be combined with the experiences in university-based institutes and clinical laboratories in order to build new standards in analytics. This will be achieved close only through cooperation of public institutions and private companies. The reward for this task will be data of molecular-scale interactions becoming obtainable on-site or in large scales, both for diagnostic and screening applications.

4 Troubleshooting

• *What if the minimum amount of protein required is not available?* In this case, the available amount of the desired protein is added and after about 30 min. a secondary protein like bovine serum albumin or horseradish peroxidase is added, which binds to the remaining free sites; the secondary protein is added in large surplus. In any case, the amount of the protein that should be attached to the gold colloid has to be at least 10% of the amount usually required for stabilization; otherwise, it cannot be assumed that sufficient protein bound to the colloid will retain its biological activity. Addition of too much electrolyte together with the first protein added will cause problems because the colloids will have a strong tendency to aggregate until they are fully protected by colloid.

• *What if the protein activity is lost by dissolving in distilled water?* The general rule is that as little salt as possible should be added to the gold sol until full protection of the sol has occurred. If a buffer is necessary for preserving the biological activity, organic buffer substances are preferred. If salt is absolutely necessary, the pH should be as high as possible without destroying the biological activity because colloids are more stable at pH above 8.5 and have practically no tendency to aggregate above pH 9.5. On the contrary, colloids bind proteins and also reactive groups like thiols much faster and tighter above pH 9.5.

• *What if the colloid adsorbs to the flask during synthesis?* Too much water may have been lost during synthesis, or the pH of the solution might have shifted. The former leads to increased hydrophobic adsorption, the latter increases ionic adsorption. Generally, large particles have a stronger tendency to adsorb than do smaller ones.

• *What if the colloid adsorbs during centrifugation?* The concentration increases dramatically, so hydrophobic interactions will dominate. If the addition of detergent is not an option because the colloids should be used for adsorption later on, you can revoke the adsorption by shifting the pH dramatically to the alkaline (pH 13 and higher). This will largely neutralize the negative charge of the particles.

• *What if the colloid adsorbs during stabilization?* Modify the concentration of the stabilizing agent, increase mixing, and add a large surplus of stabilizing proteins like BSA after the first few minutes of stabilization.

• *What if clusters do not adsorb to the desired surfaces of reactive polymers?* In the case that no cluster adsorption occurs, the pH of the cluster solution has to be shifted so that the surface appears positively charged. The cluster coming from citrate-based reduction will always appear negatively charged in solutions of neutral pH. If this does not help, the chips might be modified by polyelectrolyte adsorption so that they bind readily to the surface of the reactive polymer at neutral pH.

• *Stability and shelf-life of clusters* Noble metal clusters have the same chemical stability as macroscopic material. However, it has to be noted than clusters exhibit a much larger surface that the same amount of bulk material. This leads to an increased tendency of adsorption. These particles have high densities, which do not lead to strongly increased sedimentation until diameters of 30 nm or higher. Generally, clusters should be stored in dilute solutions.

• *Metal surfaces* Many metal surfaces have the tendency to adsorb protein and react with thiols, amines, and other entities. Less noble metal surfaces show aging as an effect of such adsorption. If such adsorption does not take place as desired, it is usually connected to impurities adsorbed to the surface of the metal. For these reasons, either use metal surfaces directly coming from vacuum deposition or clean the surfaces in piranha solution or under other highly oxidizing conditions.

References

1 Shadi IT, Chowdhry BZ, Snowden MJ, Withnall R (2001) Semi-quantitative trace analysis of nuclear fast red by surface enhanced resonance Raman scattering. *Analytica Chimica Acta* 450: 115–122

2 Toshio I, Uchida T, Teramae N (2001) Analysis of the redox reaction of 9,10-phenanthrenequinone on a gold electrode surface by cyclic voltammetry and time-resolved Fourier transform surface-enhanced Raman scattering spectroscopy. *Analytica Chimica Acta* 449: 253–260

3 Gomes P, Andreu D (2002) Direct kinetic assay of interactions between small peptides and immobilized antibodies using a surface plasmon resonance biosensor. *Journal of Immunological Methods* 259: 217–230

4 Kugimiya A Takeuchi T (2001) Surface plasmon resonance sensor using molecularly imprinted polymer for detection of sialic acid. *Biosensors & Bioelectronics* 16: 1059–1062

5 De G, Kundu D (2001) Silver-nanocluster-doped inorganic-organic hybrid coatings on polycarbonate substrates. *Journal of Non-Crystalline Solids* 288: 221–225

6 Cong X, Guo ZX, Wang XX, Shen HX (2001) Resonance light-scattering spectroscopic determination of protein with pyrocatechol violet. *Analytica Chimica Acta* 444: 205–210

7 Huang CZ, Li YF, Liu XD (1998) Determination of nucleic acids at nanogram levels with safranine T by a resonance light-scattering technique. *Analytica Chimica Acta* 375: 89–97

8 Pasternack RF, Collings PJ (1995) Resonance light scattering: a new technique for studying chromophore aggregation. *Science* 269: 935–939

9 Mayer C, Verheijen R, Schalkhammer T (2001) Food-allergen assays on chip based on metal nano-cluster resonance. *SPIE* 4265: 134–141

10 Mayer C, Stich N, Bauer G, Schalkhammer T (2001) Slide format proteomic biochips based on surface enhanced nanocluster-resonance. *Fres Anal Chem* 371: 238–245

11 Mayer C, Stich N, Palkovits R, et al. (2001) High-throughput Assays on the chip based on metal nano-cluster resonance transducers. *Journal of Pharmaceutical and Biomedical Analysis* 24: 773–783

12 Bauer G, Pittner F, Schalkhammer T (1999) Metal-nano-cluster biosensors. *Mikrochimica Acta* 131: 107–114

13 Schalkhammer T, Bauer G, Pittner F, et al. (1998) Optical nanocluster plasmonsensors as transducers for bioaffinity interactions. *SPIE* 3253: 12–19

14 Stich N, Mayer C, Alguel Y, et al. (2001) Phage display antibody-based proteomic device using resonance-enhanced detection. *Journal of Nanoscience and Nanotechnology*. Submitted.

15 Slot JW, Geuze HJ (1985) A new method of preparing gold probes for multiple-labeling cytochemistry. *European Journal of Cell Biology* 38: 87–93

16 Zhou H, Roy S, Schulman H, Natan MJ (2001) Solution and chip arrays in protein profiling. *Trends in Biotechnology* 19: S34–S39
17 Taton TA, Lu G, Mirkin CA (2001) Two-color labeling of oligonucleotide arrays via size-selective scattering of nanoparticle probes. *J Am Chem Soc* 123: 5164–5165
18 Taton TA, Mirkin CA, Letsinger RL (2000) Scanometric DNA array detection with nanoparticle probes. *Science* 289: 1757–1760
19 Elghanian R, Storhoff JJ, Mucic RC (1997) Selective colorimetric detection of polynucleotides based on the distance-dependent optical properties of gold nanoparticles. *Science* 277: 1078–1081
20 Mirkin CA, Letsinger RL, Mucic RC, Storhoff JJ (1996) A DNA-based method for rationally assembling nanoparticles into macroscopic materials. *Nature* 382: 607–609
21 Möller R, Csáki A, Köhler JM Fritzsche W 2000) DNA probes on chip surfaces studied by scanning force microscopy using specific binding of colloidal gold. *Nucleic Acids Res* 28: e91.
22 Niemeyer CM (2001) Semi-synthetic nucleic acid-protein conjugates: applications in life sciences and nanobiotechnology. *Rev Molec Biotechnol* 82: 47–66
23 Reichert J, Csáki A, Köhler JM, Fritzsche W (2000) Chip-based optical detection of DNA-hybridization by means of nanobead labeling. *Anal Chem* 72: 6025–6029
24 Fritzsche W 2001 DNA-gold conjugates for the detection of specific molecular interactions. *Reviews in Molecular Biotechnology* 82: 37–46
25 Horisberger M (1989) Quantitative aspects of labeling colloidal gold with proteins. In: Verkleij AJ, Leunissen JLM (eds) *Immuno-gold labeling in cell biology*, CRC Press, Boca Raton, FL, 49–60
26 Mayer C, Palkovits R, Bauer G, Schalkhammer T (2001) Surface enhanced resonance of metal nano clusters: A novel tool for Proteomics. *Journal of Nanoparticle Research* 3: 361–371
27 Stich N, Gandhum A, Matushin V, et al. (2001) Nano films and nano clusters – energy sources driving fluorophores of biochip bound labels. *J Nanoscience and Nanotechnology*, 1: 397–405

8 Atomic Force Microscopy

Yilmaz Alguel and Thomas G. M. Schalkhammer

Contents

1 Introduction

As a member of the scanning probe microscope (SPM) family, atomic force microscopy (AFM) became an efficient tool for imaging surfaces in nano and micro-scale. It is widely used for imaging thin- and thick-film coatings, ceramics, composites, glasses, synthetic and biological membranes, metals, and polymers. AFM not only enables imaging with a resolution down to an atom size scale, but also force measurements with nano- or pico-Newton resolution in between two molecules can be performed, which makes the AFM an essential tool in a nanotech laboratory. Citations of AFM experiments have grown exponentially since its birth. The applicability of AFM for imaging biological molecules [1] in their native environment was demonstrated shortly after the invention of this technique [2]. Through time AFM has become a successful tool, which complements other structural techniques such as NMR, x-ray crystal-

Methods and Tools in Biosciences and Medicine
Analytical Biotechnology, ed. by Thomas G.M. Schalkhammer
© 2002 Birkhäuser Verlag Basel/Switzerland

lography, and electron microscopy. These results are the basis for novel techniques using AFM not only for the visualization and manipulation of biomolecules and surfaces [3] but also for monitoring the change of conformational structures of proteins. In order to make use of the AFM for monitoring biological molecules, a number of techniques have been developed. The three main modes of operation are "contact mode" (CM), "non-contact mode" (C) and "tapping or intermitted contact mode" (IC). Data gained in non-contact-mode are processed either in "phase mode" or "amplitude mode". All these modes will be described in the following subsection more in detail. Both fundamental techniques, contact and non-contact, are necessary to study biomolecules on surfaces. In C-mode the biomolecules need to be bound to the surface and thus are not easily moved away from their place by the force applied through the AFM tip. Biomolecules, which are bound weakly to a surface, are wiped from the chip by C-mode AFM imaging. Thus, for soft bio-structures NC- or IC-mode imaging is vital.

Another field of application of AFM technology is its ability to measure binding forces between biomolecules such as epitopes and their antibodies, which is done in "force-mode". The forces determined by the AFM correlate well with calculated binding forces as was proven, e. g., in antibody-antigen [4] or streptavidin-biotin [5] systems.

2 Setup

The principles on how the AFM works are very simple. An atomically sharp tip is scanned over a surface with feedback mechanisms that enable the piezoelectric scanners to maintain the tip at a constant force (to obtain height information), or height (to obtain force information) above the sample surface. Tips typically are made from Si_3N_4 or Si, and extended down from the end of a cantilever. The AFM head uses an optical detection system in which the tip is attached to a reflective cantilever. A diode laser is focused onto the back of the reflective cantilever.

The beam is focused into a four-quadrant photodiode. As the tip scans the surface of the sample, guided by the contour of the surface, the laser beam is deflected from the position in the detector. The signal is the difference in light intensities between the four photosensors. Feedback from the photodiode difference signal enables the tip to maintain either a constant force or constant height above the sample.

In the constant force mode, the signal of the piezo-Z-axis reflects the height deviation. In the constant height mode (often used at ultra high resolution), the force on the sample is recorded. In order to give absolute force data this mode of operation requires calibration parameters of the scanning tip inserted in the AFM head.

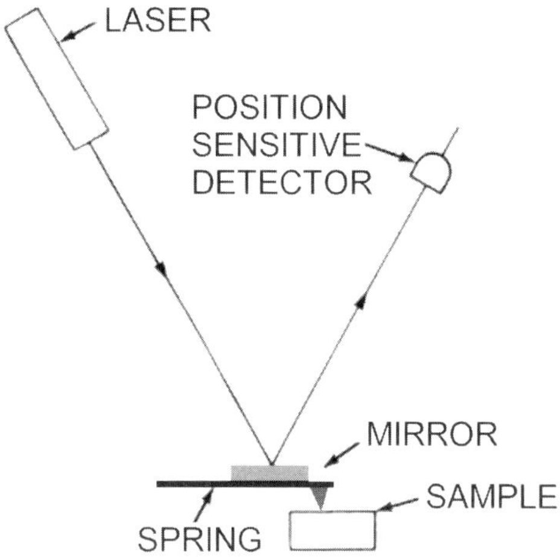

Figure 1

Figure 1 recorded by the AFM, represents a topographical map of the scanned surface. The data set gained by the scan encodes the height and width information of the scanned area. With image processing software, this information needs to be filtered and processed to obtain high-quality charts. Three-dimensional topographical maps of the surface are then constructed by plotting the local sample height *versus* horizontal probe tip position. In addition, height, width, distance profiles, and surface inhomogeneities of even single molecules can be measured.

Some AFMs can accept full 200 mm wafers, but most instruments use a 10 to 100 μm scanner table. In chip fabrication the primary purpose of these instruments is to quantitatively measure surface roughness with a nominal 5 nm lateral and 0.01 nm vertical resolution on all types of samples.

Depending on the AFM design, scanners are used to translate either the sample under the cantilever or the cantilever over the sample.

3 The resolution of AFM

Contrary to electron-microscopic techniques AFM has other resolution criteria. The criteria from radiation-based imaging cannot be applied to scanning microscopes and their ways of image processing. An AFM operates mechanically which is distinct from wave optical techniques. The optical microscopy depends on diffraction whereas the probe and sample geometry are the limiting factors for the AFM.

The double strand DNA in the B form has a known diameter of around 2.0 nm and can be used as reference of testing the quality of the used tip. The measurements found in literature describe the height of DNA between 1.4

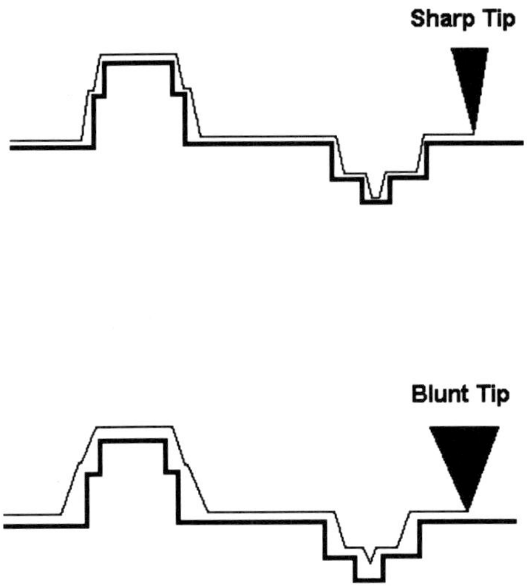

Figure 2

and 2.5 nm [6]. The measured height is within the range of precision expected and relates well to the data from molecular modeling.

Contrary to this, the width of any structure is significantly larger because of the lateral size of the tip used for the scan. The radius of curvature of the tip influences the resolving ability of the AFM. Figure 2 models the influence of the form of a chosen tip for the scan of a surface with a given profile. The sharp tip gives an image with high resolution whereas the blunt tip gives a smoothed profile of the scanned surface. The sharp tip is able to access the groove in more detail and trims better to the height changes of the sample. The use of a blunt tip during the scan leads to loss of detailed information of the surface of the sample. The grooves with small width are unable to be accessed by the blunt tip. Therefore, the height in small grooves also will be under-estimated. Thus, AFM images, except for small grooves, give an accurate measure of Z-profiles of any sample. The XY-profile is significantly influenced by the sharpness of the operating tip and cannot be better than the tip used for scanning for rough surfaces. Ultra-smooth surfaces can be scanned at higher XY resolution because of local force effects.

Atomic force microscopists generally use C-mode, NC-mode, tapping mode, and, today, sometimes, carbon-nano-tube tips. The "normal tip" [7] is a 3 μm pyramid with ~ 20 to 30 nm end radius.

4 Vibrations and AFM

Vibrations are a common source of trouble during AFM measurements. Vibrations of the desk or even of the whole building are registered by the AFM and noise the image. Special platforms are available to help avoid additional noise. A cheaper, but also effective method, is to dampen the vibrations by the use of synthetic rubber [8] as a platform. The synthetic rubber has as a property of a low resonant frequency, which results from the rubbing of the synthetic fibers inside the rubber against the outside matter. The resonant frequency of the synthetic rubber depends on the stretch of the synthetic fibers. Contrary to that, the AFM has a high resonant frequency from its hardware, which is greater than 10 kHz. This leaves the gauging station to become a band pass filter. To increase the filter efficiency, a sandwich of rubber pads and steel plates may be used. Another trick, often used in ultra-sensitive STM imaging, is a steel platform hanging from the ceiling mounted on three or four rubber bands.

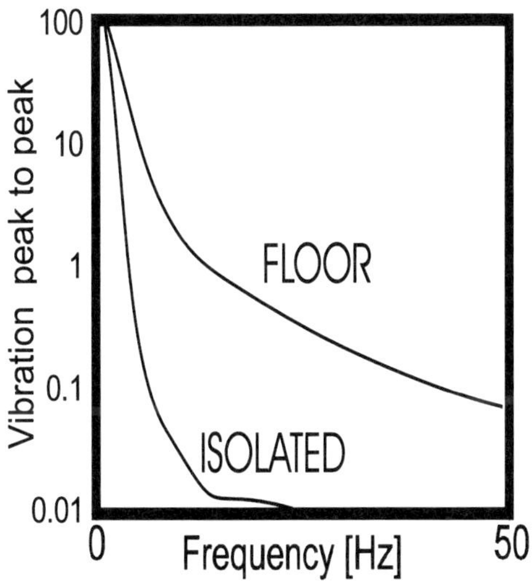

Figure 3

5 Techniques

5.1 Contact mode

Contact mode is the standard mode of AFM [9], where the tip is in direct contact with the scanned sample. A repulsive force with ~ 10 nN (nano Newton) is applied by the tip. The piezo tube pushes the sample against the tip with this force, which leads to the bending of the cantilever in different angles through the scan. The deflection of the cantilever is detected optically in the change of the laser focus.

The voltage output of the four-quadrant photodiode changes and monitors the surface height profile. A feedback amplifier uses the deflection signal of the cantilever and sets the force on the tip to the desired set point. The piezo tube is driven by the applied voltage and is navigated through the feedback loop. The feedback loop controls the tube to raise or lower the sample relative to the cantilever. This sets the laser beam deflection to the desired angle. The Z-axis signal of the piezo reflects the height change of the scanned sample surface and is monitored as a function of the XY-position of the sample. The applied force may vitally influence the picture gained for the real structure of the scanned surface.

In order to minimize imprinting and damage of the surface, the force of the probe has to be minimized. The applied force is limited if operating under ambient environment. A layer approximately 20 monolayers thick covers the sample during operating in ambient conditions. The layer mostly consists of water vapor and adsorbed gas, which contaminates the surface of the sample. After the approach of the tip at the contaminated surface a meniscus between the sample and the tip is formed, which pulls the cantilever by surface tension toward the sample surface.

The maximal force applied to the tip and thus to the sample is around 100 nN the minimal force is in the pN range and is more or less limited by the feedback of the system. A strategy to overcome the meniscus force and other undesirable attractive forces is to scan the sample in an immersed liquid where the cantilever also will be immersed from the used liquid. This leads to capillary forces that cannot effect the measurements anymore and the "Van der Waals" forces will be reduced to a very low level. The ability to image biological molecules immersed in liquids gives the advantage of varying the fluid compositions and study molecules in a biological microenvironment.

Molecular folding induced by pH [10] or microenvironment or conformational changes of membrane proteins after the binding of the ligand can be displayed.

The contact mode is limited in its application to biomolecules, which are not weakly bound to the surface and consequently are wiped away and partially destroyed through the scan. To avoid these negative effects on the sample the non-contact-mode has been developed.

5.2 Non contact mode

For cases of the contact mode where the tip affects the scanned sample in destroying or altering it, the non-contact mode is superior. In this mode the tip moves with a distance of ~ 5 to 15 nanometers above the surface of the sample. An attractive "Van der Waals" force is present between the tip and the sample during the scan, which is the feedback signal to register the topographical map of the sample surface. The attractive force between the sample and the tip in NC mode is significantly less than the force in the C mode.

To register these small forces, the cantilever is mounted on an oscillating piezo, which drives the tip up and down. The drive of the tip piezo occurs through applying an AC voltage on the tip-piezo with a constant frequency where the tip oscillates over the surface of the scanned sample. The AFM instrument registers the resonant frequency map of the tip (Fig. 4).

The user can choose the optimal frequency for the tip. As a response to the interaction with the sample surface, the applied resonance curve modifies its amplitude, phase, and frequency. This shift is used for generating and, afterwards, for displaying the samples surface map. The distance between sample surface and tip can be decreased to 1 nm for getting a high resolution. Nevertheless, this approach fails in most cases because of the contaminant layer of water vapor and adsorbates, which covers the "Van der Waals" surface force of the scanned sample surface.

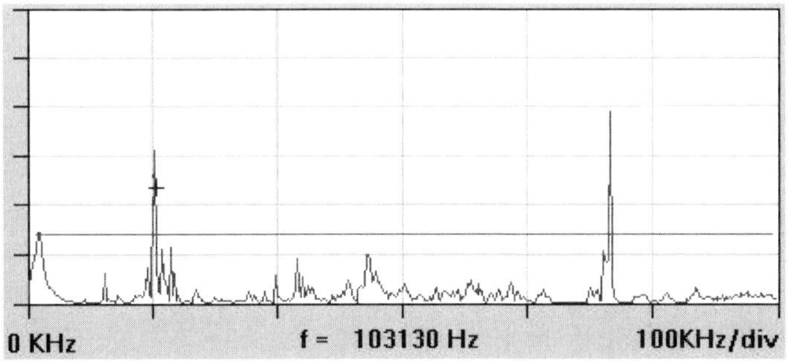

Figure 4

5.3 Tapping mode

Tapping mode, developed as a mixture of NC and C mode, reveals new possibilities for high resolution of topographic maps [11] of the scanned sample surface. The tapping mode is used to study biomolecules otherwise destroyed or weakly bound to the surface and thus wiped off. The conventional AFM accommodates problems associated with adhesion, electrostatic forces, and friction between the scanned sample surface and the tip. These problems are circumvented in tapping mode [12]. As in the non-contact mode, the tip is mounted on a piezo, which resonates at 50 to 600 kHz. The chosen frequency in the tapping mode to drive the cantilever is slightly left of the resonant frequency peak maximum, which is demonstrated in Figure 5 and indicated with a cross on the peak. A second parameter (shown in the same figure) is the amplitude visible as a thick black line. The amplitude of Z-movement of the cantilever while oscillating is around 10 to 25 nm. Parameters are adjusted while the tip is off the surface.

After the start of the approach, the oscillating tip is moved toward the surface until it contacts the chip. It still oscillates and taps (with a high frequency, typically around 100 kHz) the surface of the scanned sample. This intermitted tapping leads to a damped oscillating amplitude of the cantilever, which is used as the signal change from which the surface features can be derived.

The high-frequency tapping reduces the adhesion forces between the tip and the sample. In contrast to the contact mode, there is no shear force on the sample by the tip because of the vertical movement of the cantilever. The spring constant of a tapping-mode cantilever is around 1 to 100 mN. When applying the tapping mode in fluid [13], different spring constants are needed because of the damping of the oscillation in a liquid medium.

A feedback loop controls the distance in the tapping mode by regulating the amplitude and the force of the oscillating cantilever at the scanned sample. In depressions the amplitude will be increased, whereas in passing higher structures the amplitude will be decreased.

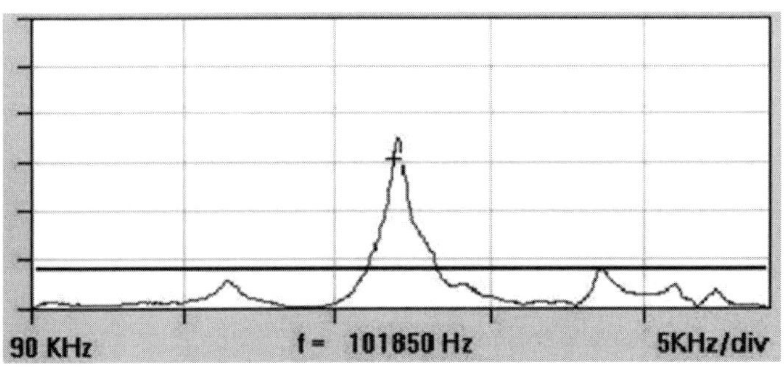

Figure 5

5.4 Phase recording

The phase mode is often applied for biomolecules [14], which are weakly bound to a surface or destroyed during a C-mode scan. After taking the image in amplitude and phase mode and comparing them, the phase mode appears as a contrast-enhanced form of the amplitude-mode image.

In comparison to the tapping mode, where the cantilever oscillates with the resonance frequency of its piezo crystal and the change of the amplitude is detected as a response to the scanned surface, in the phase mode the phase shift [15] of the oscillating piezo and the driving signal is used for feedback. This phase shift is induced by the sample surface via adhesion and viscoelasticity.

Many additional properties to the topographical map, such as compositional variation, adhesion, viscoelasticity and friction of scanned sample surface, are transduced through the phase mode of the tapping mode AFM.

The results of images taken give a better possibility to distinguish between higher and lower adhesions, hardness of the scanned surface, or identification of individual molecules in a composition, which leads to image resolution in nanoscale. These favorable characteristics of the phase mode are missing in other scanning probe microscopy techniques, and therefore the phase mode AFM is an essential tool in the field of sample surface scanning. For the phase mode, the same tip can be used as for the tapping mode described before. Most instruments enable the measuring and displaying the phase and amplitude mode in parallel.

6 Force curves

The ability to measure the force between the tip and the sample is an additional important feature of the atomic force microscope. Studies about force measurements are increasing, as seen by the count of citations.

The cantilever on which the tip is mounted is moved toward the sample and, after touching it, the tip is pulled back from the surface of the sample. The attractive force of the sample, which holds the cantilever back through pulling it away from it, is recorded and displayed as a curve diagram on the monitor.

The graph reflects either a single interaction cycle or a series of successive interaction cycles from tip to surface. This mode allows getting information about local force and elasticity of the sample.

The function of the voltage curve applied to the piezo for the force measurement has a triangle shape. Depending on the frequency and amplitude of the triangle voltage function, the distance and speed of the force measurement are set.

6.1 The basis of the distance and force curve

At point 1 the distance between the cantilever and the sample reaches its maximum and is at the same time the point of starting before the curve is taken. The tip is not in contact with the sample, and the cantilever is undeflected. The net force at the cantilever is set to zero. Now the tip is moved toward the sample until it reaches point 2, where an attractive interatomic force between the tip and the sample is experienced. The cantilever is pulled toward the sample, leading to a deflection of the cantilever called a “snap-in” point (Attractive).

The scanner continues to decrease the distance until the cantilever is bent away from the surface. The net force on the cantilever changes now to a positive value (repulsive).

After the endpoint is reached, the retraction begins at point 3. The force on the cantilever follows another path. The horizontal offset between the initial and the return paths is due to scanners hysteresis. The additional part of the curve that shows a negative (attractive) force on the cantilever is attributed to a thin layer of water that is usually present on the sample surface when the

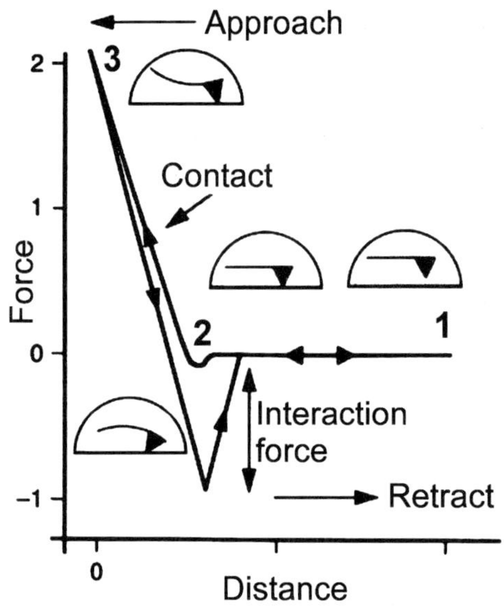

Figure 6

surface is exposed to air. This water layer induces a capillary force on the cantilever tip, which is strong and attractive. The water layer holds the tip in contact with the surface. This deflection is described in Figure 6 as an interaction force, where the net force of the cantilever is strongly negative.

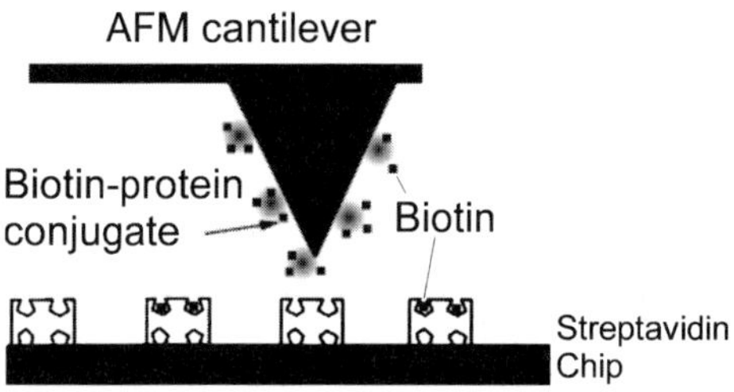

Figure 7

The scanner retracts far enough for the tip to spring free of the water layer. This point is called the "snap-out" or "snap-back" point. Beyond the snap-back point, the cantilever remains undeflected, and the net force on the cantilever is zero.

If the experiment is done in a fluid cell, the capillary force is suppressed and the direct interaction force in between two surfaces and finally in between two molecules, one from the tip, one from the surface is measured. A wide variety of setups has been used with, e. g., a biotin conjugate at the tip and a streptavidin on the chip (see Fig. 7). Calibrating the instrument and measuring the hysteresis enables one to quantify molecular interaction forces.

7 Surfaces and biomolecules imaged via AFM

As described, scanning of surfaces is the field of application for AFM. In order to achieve high-quality images and to avoid artifacts, all surfaces that are to be covered with proteins or DNA need to be scanned previously. This pre-scan is necessary to evaluate height fluctuations, roughness, and dust particles on the support. Thin layers of a few molecules to a few nanometers are necessary to ensure optimal images. For that, a nano-flat substrate is desired to deposit the biomolecules of interest. Moreover, AFM is a useful tool to study all materials in close contact with biofluids (steel, aluminum, glass, etc.).

7.1 Images of inorganic surfaces

• Steel

Figure 8 represents an AFM of steel used in a bio-fermenter imaged in contact mode. The two images show the changed surface of steel after corrosion. Figure 8A with a smooth surface is uncorroded steel, where as Figure 8B is steel surface etched by intense corrosion.

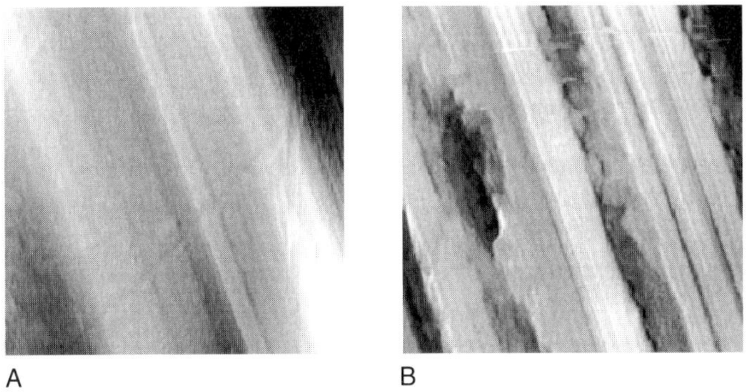

Figure 8

• Aluminum

Figure 9 represents an AFM of high-quality rolled aluminum used in optical reflectors and even as a matrix for biochips (imaged in contact mode). The images clearly indicate the directed growth of nano-crystals of a few hundred nanometers in size. After bio-corrosion, aluminum oxide or other deposits coat the sharp crystals.

• Glass

As a frequently used support for imaging and in the biochip field, two types of standard glass slides are scanned (1x1 μm) with the AFM in contact mode. Figure 10A (a trhee-dimensional image) and Figure 10B (a two-dimensional image) reflect a highly flat surface where the roughness is within 0.7 nm (7 angstroms). In contrast to this slide Figure 10C (also scanned in contact mode)

Figure 9

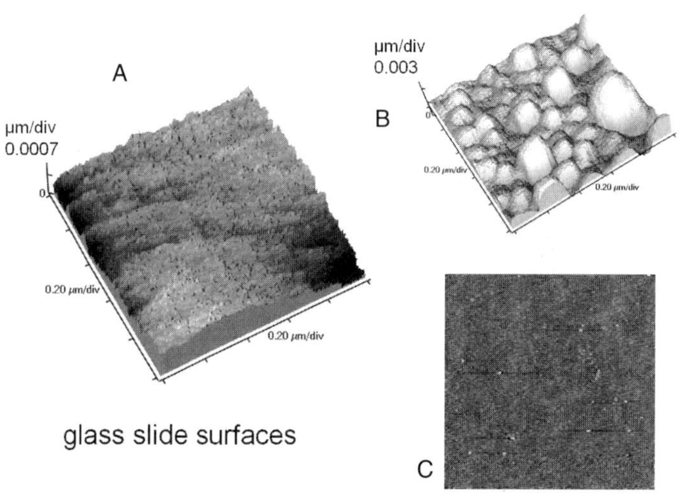

Figure 10

shows a rougher surface with nano-crystals of around 3 nm (30 angstroms) in size. Replacing Figure 10B with Figure 10A would lead to unexpected artifacts in bio-nano experiments or even biochip applications.

• Channel in sputter-coated Ag-nano-layer

Figure 11 represents a two- and three-dimensional image of a silver covered slide, with a nano-channel of 3 µm width. Mechanical engraving was used to imprint the channel. The metal removed from the channel is visible left and right of the channel as a nano well.

• Mica

Mica, as an atom-flat crystalline silicate material, is often used for immobilization of a wide variety of biomolecules down to single molecule resolution. Because it is very homogeneous, the image has a high contrast between molecule and its background, enabling good recognition of any structure

Figure 11

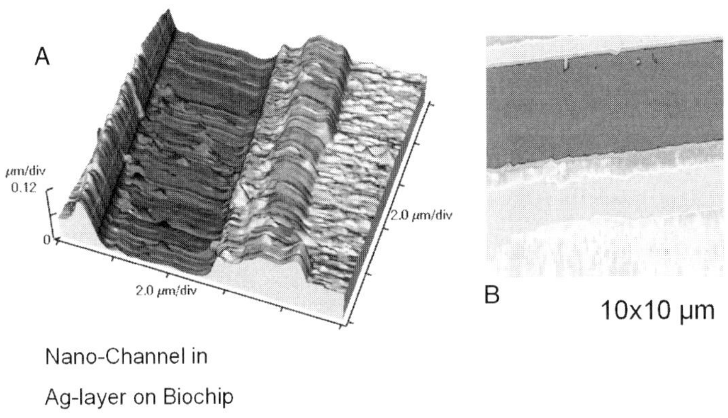

deposited onto the surface. The crystal lattice of the mineral is imaged in Figure 12A. After treating the mica with NaOH, single imperfections in the crystal lattice emerge (Fig. 12B). Mica is a layered silicate, which can be cleaved manually by peeling the silicate layer by layer. Thus, with a minimum of handling a new, clean and atomically flat surface is prepared. A scratched surface of mica shows these layers imaged as steps in the scan (Fig. 12C).

Figure 12

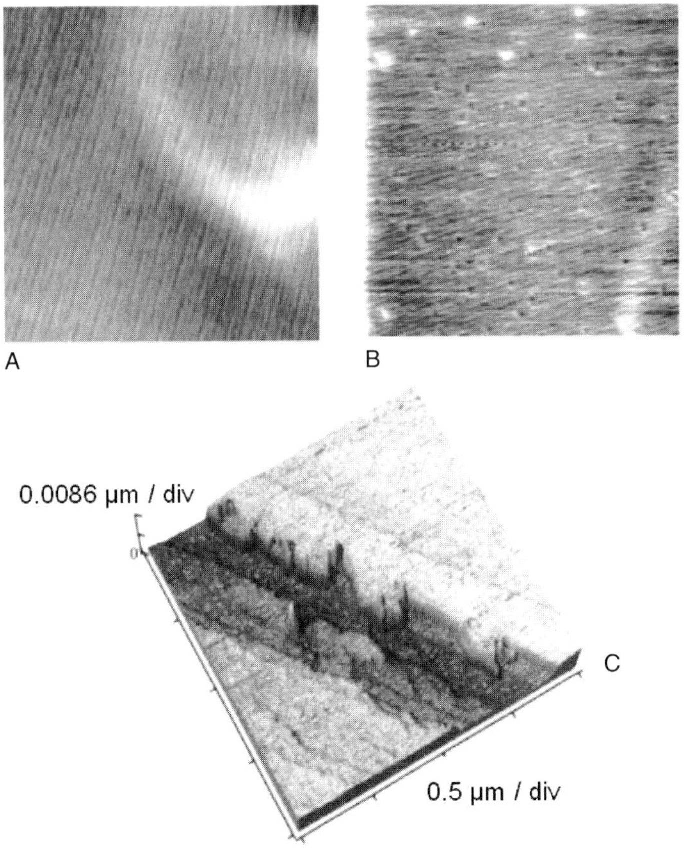

7.2 Images of whole cells

• Yeast (*Saccharomyces cerviseae*)

A surface of a slide is coated with yeast cells. Figure 13 is easily obtained by drying a drop of the cell suspension onto glass and then, by a quick move through a flame, attaching the cells to the chip by heat.

Whereas vital cells are round as micro-balls, old or dried cells are collapsed.

Because of the good adhesion, Figure 13 is done in contact mode without a risk of wiping the cells off the chip.

Figure 13

• *Escherichia coli*

Figure 14 shows the scanned surface of mica with coated *E. coli* cells on its surface. Three sizes of fields were scanned with increasing resolution. Figure 14A has a scan size of 50 μm, the right 20 μm, and Figure 14C 8 μm.

The *E. coli* culture was adsorbed on the surface of freshly cleaved mica. After, the surface was washed with deionized water, the chip was scanned with AFM. Because of the adsorption of *E. coli* cells to the surface of mica, an AFM operating in the contact mode just wipes the cells away and destroys them partially. Thus, these pictures were taken in non-contact tapping mode.

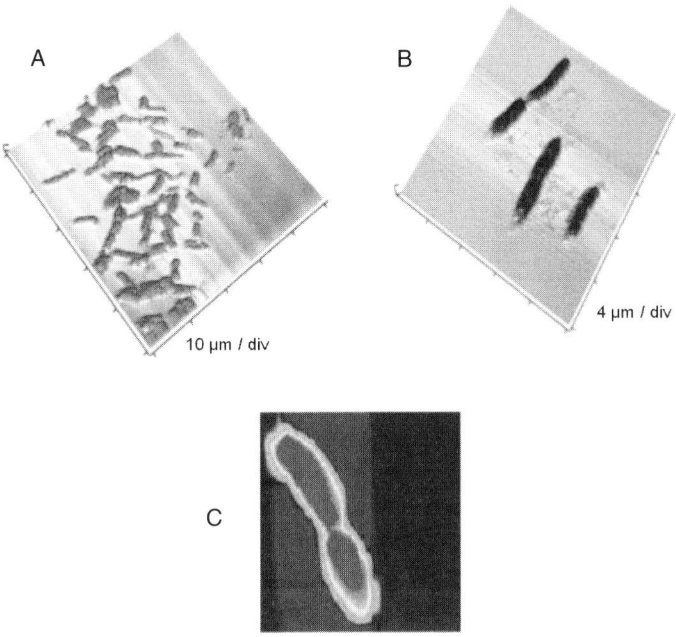

Figure 14

Protocol 1 Imaging of single *E. coli* cells

1. Cleave fresh mica.
2. Drop a small volume of about 20 µl of *E. coli* suspension on its surface.
3. Leave it for 15 min. to be adsorbed on the mica's surfaces.
4. Spin the mica with 2500 rpm and wash during spinning with deionized H_2O.
5. Dry by spinning.
6. Take AFM image.

7.3 Single-molecule images

• DNA

Imaging of single Lambda-DNA molecules, cut with *EcoRI* restriction enzyme is accomplished in two ways. Figures 15A and 15B are immobilized on the surface of mica [16] using magnesium as a divalent cation to mediate the binding of DNA.

In Figures 15C and 15D, the DNA was adsorbed to a poly-L-lysine-coated mica surface. Because of the positive charge of poly-L-lysine, the DNA binds as any other negatively charged molecule.

In the experiment given below, the DNA was heated to 65 °C and bound to the mica surface, mediated through magnesium ions. Figure 16A clearly visualizes a DNA fork with its single- and double-strand part of the DNA molecule. Figure 16B shows the double-strand DNA height at ~ 1.5 nm, where each single-strand DNA has a height of ~ 0.75 nm. These measurements correspond accordingly to

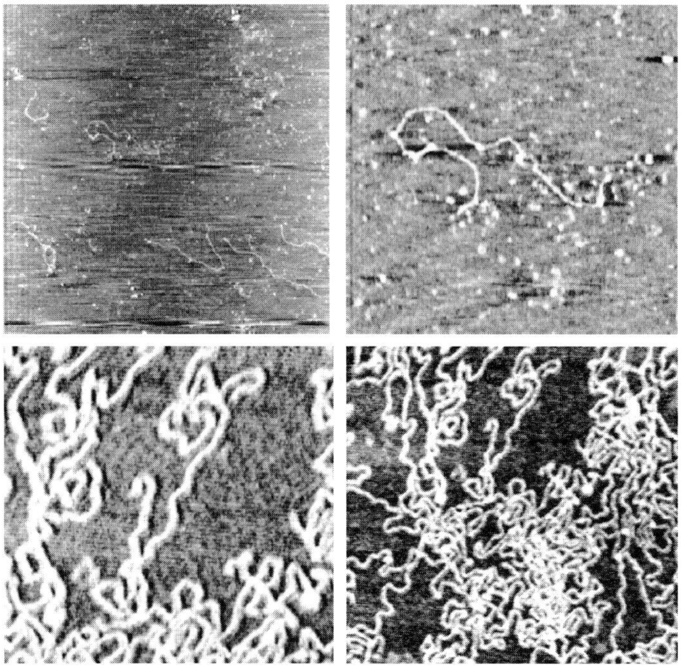

Figure 15

A B C D

those described in the literature. The precision of height measurement can be increased by calibration of the instrument prior to depth profiling.

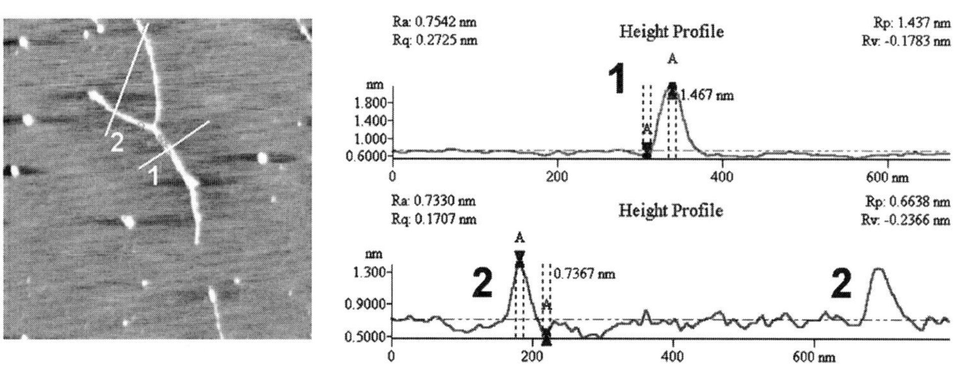

A B

Figure 16

Protocol 2 Imaging of DNA molecules

1. Cleave fresh mica.
2. Add 25 µl $MgSO_4$, 100 mM on its surface.
3. Leave for 10 min. to be adsorbed on the mica surface.

4. Spin the mica with 2500 rpm and wash during spinning with deionized H_2O.
5. Drop 20 µl DNA, 40 mg/ml solution onto surface.
6. Leave for 10 min. to be adsorbed to the divalent cations sites.
7. Spin the mica with 2500 rpm and wash during spinning with deionized H_2O.
8. Dry by spinning.
9. Take AFM image.

• Proteins

Proteins can be imaged easily on any surface with sufficient flatness. Figure 17 shows a MICA surface with a few partially aggregated bovine serum albumin proteins on its surface. The chip was covered by a protein monolayer and thereafter etched with 5% NaOH. The small dark dots are removed silica from the surface of the mica.

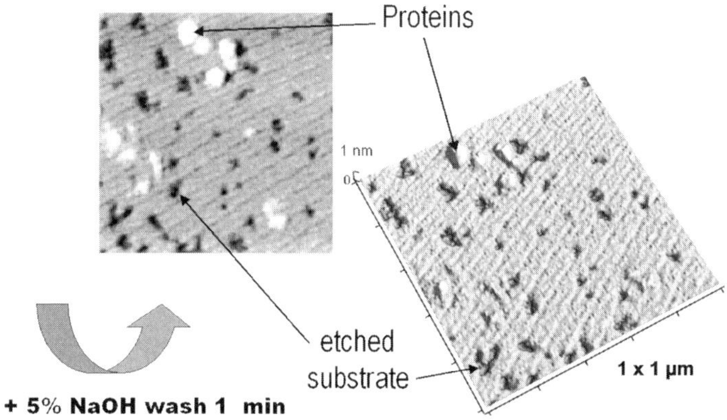

Figure 17

Figure 18 is a microdot of a protein deposited by a micro-dotting robot. The scan zooms into the border region of the 100 µm spot. Single protein molecules can be resolved easily and the surface coverage of the microdot can thus be determined.

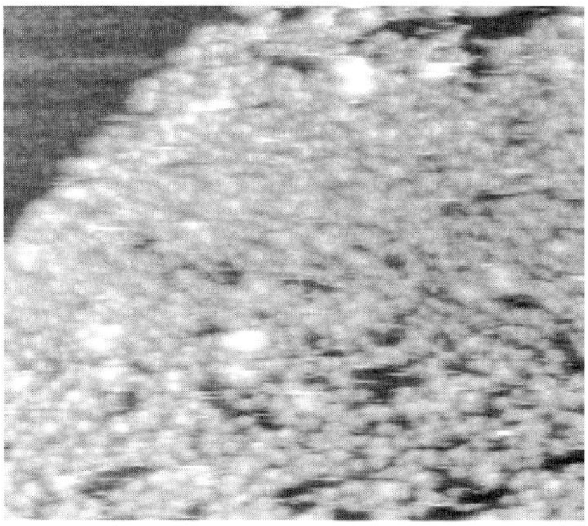

Figure 18

Figure 19 is a microdot of a protein deposited with an excess of buffer salts using a micro-dotting robot. Single-protein molecules cannot be resolved because of the excessive deposition of buffer salt crystals. This Figure indicates a non-appropriate deposition mixture, which should not be used for micro-dotting.

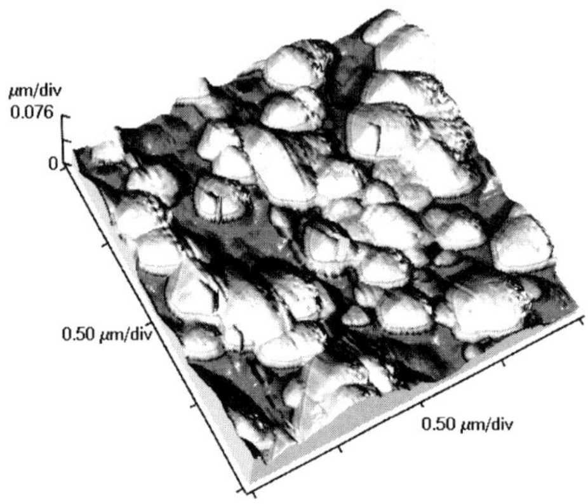

Figure 19

7.4 Lipids and vesicles

Vesicles are deposited at the surface of mica, which leads to fusion, coating, and collapse of the vesicles. Thus, a bilayer membrane is formed at the surface.

In order to avoid wiping away the lipid layer from the surface, the AFM image needs to be done in non-contact or tapping mode.

To illustrate the effect, compare Figures 20 and 21. Whereas Figure 20 was done in NC mode and no damage of the layer is observed, Figure 21 was "imprinted" via scanning in contact mode.

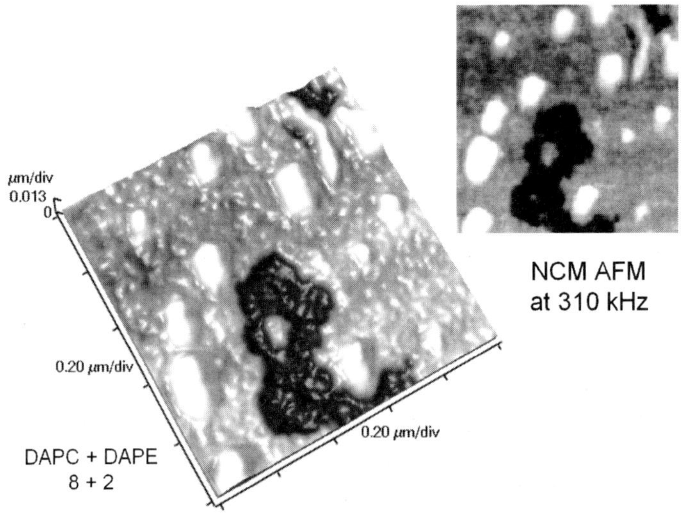

Figure 20

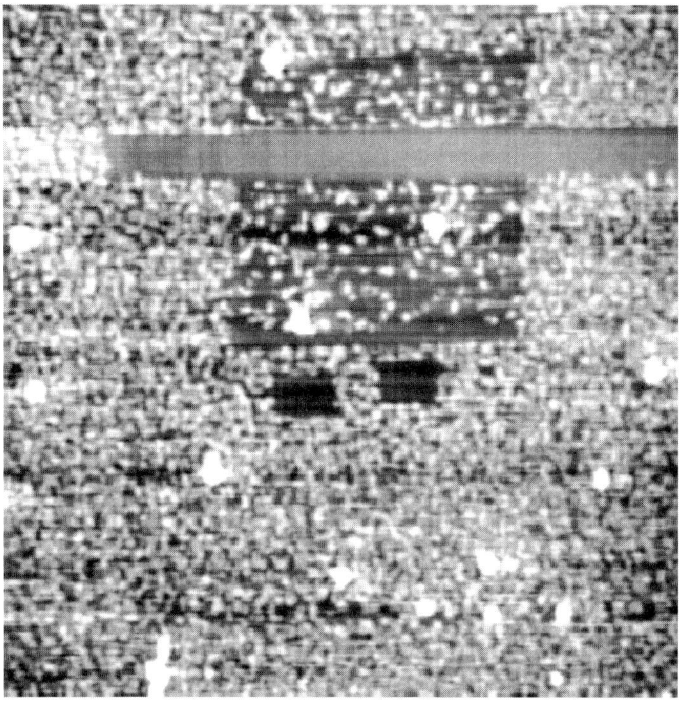

Figure 21

15 x 15 µm

By using the tip to adjust the force, all or some of the lipid is moved or only loosly attached vesicles are moved. Thus, in some areas of the surface all lipid molecules have been removed (black) and in some areas the lipid is concentrated (white). Areas scanned with the non-contact tapping mode show a surface with lipids and vesicles.

Figure 22 visualizes the difference of C mode, amplitude, and phase-imaging in non-contact mode.

Regions with low contrast in the C mode are more visible in the phase or NCM mode image because of the difference in signal processing.

Protocol 3 Imaging of lipids and vesicles

1. Dissolve 30 mg lecithine (PC) in 1 ml Ethanol [17].
2. Introduce 10 ml aqueous medium into a 25 ml flask and stir rapidly with a magnetic stirrer.
3. Fit a fine-gauge needle to a 1 ml glass syringe and draw up 750 µl of the lipid solution.
4. Position the tip of the needle just below the surface of stirred aqueous solution, and inject the organic solution as rapidly as possible into the medium. Liposomes will be formed immediately. The final concentration is approximately 2 mg of PC per ml.

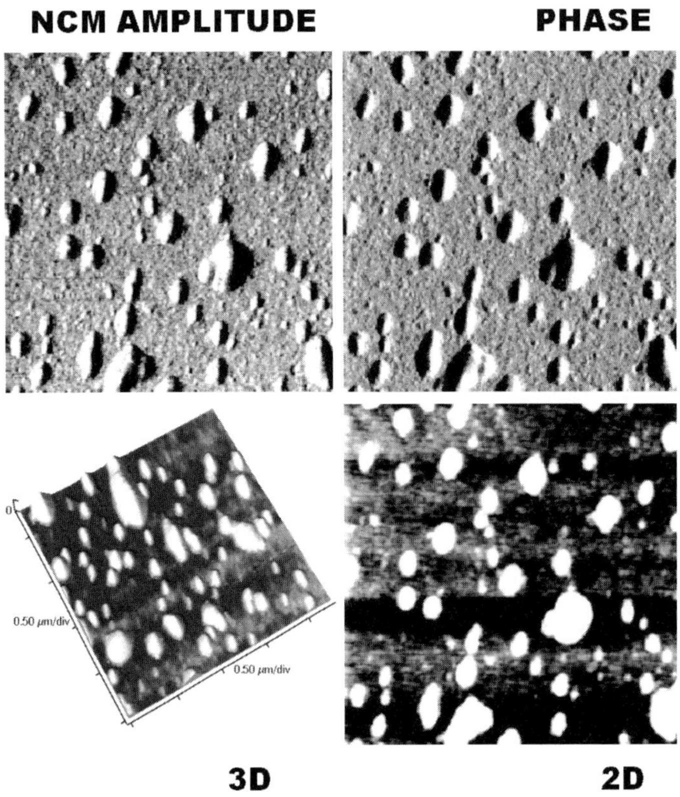

Figure 22

5. Cleave a fresh mica sheet.
6. Add 30 μl of the vesicle solution to its surface.
7. Leave for 10 min. to be adsorbed on the mica's surfaces.
8. Spin the mica with 2500 rpm and wash during spinning with deionized H_2O.
9. Dry by spinning.
10. Take AFM image.

References

1 Binnig G, Quate CF, Gerber C (1986). Atomic force microscope. *Phys Rev Lett* 56: 930–933

2 Drake B, Prater C B, Weisenhorn AL, et al. (1989) Imaging crystals, polymers and processes in water with the atomic force microscope. *Science* 243: 1586–1588

3 Hoh JH, Sosinsky GE, Revel J, Hansma PK (1993) Structure of the extracellular surface of the gap junction by atomic forcw microscopy. *Biophys J* 65: 149–163

4 Dammer U, Hegner M, Anselmetti D, et al. (1995). Specific antigen/antibody reactions measured by force microscopy. *Biophysical Journal* 70: 2437–2441

5 Lee GU, Kidwell DA, Colton RC (1994) Sensing Discrete Streptavidin-Biotin In-

teractions with Atomic Force Microscopy. *Langmuir* 10: 354–357

6 Coury JE, Anderson JR, McFail-Isom L, et al. (1997) Scanning Force Microscopy of Small Ligand-Nucleic Acid Complexes: Tris(o-phenanthroline)ruthenium(II) as a Test for a New Assay *J Am Chem Soc* 119: 3792–3796

7 Albrecht TR, Akamine S, Carver TE, Quate CF (1990) Microfabrication of cantilever styli for the atomic force microscope. *J Vac Sci Technol* A8(4): 3386–3396

8 Hansma HG, Bezanilla M, Laney DL, et al. (1995) Applications for Atomic Force Microscopy of DNA. *Biophys J* 68: 1672–1677

9 Binnig G, Quate C, Gerber Ch (1986) Atomic force microscope. *Phys Rev Lett* 56: 930–933

10 Daniel JM, Frank A, Schabert, et al. (1995) Imaging purple membranes in aqueous solutions at sub-nanometer resolution by atomic force microcopy. *Biophys J* 68: 1681–1686

11 Umemura K, Arakawa H, Ikai A (1993) High-resolution images of cell surface using a tapping-mode atomic force microscope *Jpn J Appl Phys* 32: 1711–1714.

12 Bar G, Brandsch R, Bruch M, et al. (2000) Importance of the indentation depth in tapping-mode atomic force microscopy study of compliant *Materials Surf Sci* 444, L11

13 Hansma PK (1994) Tapping mode atomic force microscopy in liquids *Applied Physics Letters Vol.* 64: 1738–40

14 Kim Y, Lieber CM (1992) Machining oxide thin films with an atomic force microscope: pattern and object formation on the nanometer scale. *Science* 257: 375

15 Raghavan D, Van Landingham M, Gu X, Nguyen T (2000) Characterization of heterogeneous regions in polymer systems using tapping mode and force mode atomic force microscopy. *Langmuir* 16: 9448–9459

16 Allison DP, Bottomley LA, Thundat T, et al. (1992) Immobilization of DNA for scanning probe microscopy *Proc Natl Acad Sci USA* 89(21): 10129–10133

17 New RRC (ed) (1990) Liposomes: A practical approach (The Practical Approach Series). IRL Press, Oxford, NY, Tokyo

9 Analysis in Complex Biological Fluids

Roland Palkovits, Christian Mayer and Thomas G. M. Schalkhammer

Contents

Methods and Tools in Biosciences and Medicine
Analytical Biotechnology, ed. by Thomas G.M. Schalkhammer
© 2002 Birkhäuser Verlag Basel/Switzerland

1 Introduction

Control of the composition of the culture media in which bacterial or mammalian cells are grown is important. Nutritional depletion, accumulation of side products, and pH may change significantly during the incubation period, thereby reducing cell viability as well as interfering with the production of the desired product.

This chapter is a summary of techniques and protocols to quantify most of the standard ingredients in complex biological media.

2 Materials (selected)

Glutamate dehydrogenase from bovine liver [EC 1, 4, 1, 3] [Sigma, 49390]
INT, p-iodonitrotetrazolium violet [Sigma, i8377]
Diaphorase [Sigma, D5540]
Aspartate aminotransferase (AST/GOT) [Sigma, 505]
Asparaginase [Sigma, A4887]
Glutaminase [Sigma, G8880]
Phosphatase, alkaline [Sigma, P4978]
BCIP/NBT solution, premixed [Sigma, B6404]

3 Analytes

3.1 Ions and metals

Ammonia (-ium)

Enzymatic methods

The determination of ammonia is of considerable value in fermentation control as well as clinical diagnosis. If any of the processes involved in ammonia's metabolic disposal is impaired, the level of ammonia, especially ammonium ion, will increase, leading to ammonia intoxication.

In eucaryontic cell fermentation, this induces rapid cell death in humans. It induces complications, including nausea, vomiting, ataxia, convulsions, lethargy, coma, and finally death. Mostly, patients with hepatic intoxication or reduced liver function are affected, as the liver is the site of the conversion of ammonium ion to urea.

Therefore, it is particularly important to monitor the concentration of both ammonia and carbon dioxide. Ammonia is determined via reductive amination using glutamate dehydrogenase (GLDH). A decrease of NADPH is measured photometrically at 340 nm.

$$\text{2-Oxoglutarate} + NH_3 + \text{NADPH} \xrightarrow{\textit{Glutamate dehydrogenase}} \text{Glutamate} + \text{NADP}^+$$

For this assay relatively minor amounts of glutamate dehydrogenase might be used, as there is only one enzymatic step involved.

Protocol 1 Ammonia

1. Freshly prepare before use NaH_2PO_4 buffer 50 mM, pH 7, with 2-oxoglutarate, 0.11 mM, and NADH, 0.7 mM.
2. Prepare a series of ammonia dilution in expected concentration range.
3. Mix 950 μl of buffer, 40 μl GlDH, and 10 μl of samples or standard ammonia dilution
4. Shake well and incubate for 10 min.
5. Read the absorbance at 340 nm against a reagent blank
6. Plot calibration graph from standard series and determine concentration of unknown sample.

Electrochemical methods

The NH_4^+-selective membrane is based on a mixture of the antibiotics Nonactin and Monactin, which are neutral carriers in a PVC matrix.

Bicarbonate

In fermentation broth as well as in body fluids (e. g., blood, plasma or serum) at about pH 8, CO_2 is present substantially as bicarbonate ion. In animals HCO_3 is carried via the blood to the lungs, which release it into the air. The kidneys filter some of the bicarbonate through the urine, but the lungs release most of it as carbon dioxide each time we exhale.

The level of bicarbonate may be too high or too low if the lungs are not functioning properly or in fermentation if carbon dioxide *versus* base supply is not balanced. Abnormal bicarbonate may be seen with certain metabolic problems.

Enzymatic methods

$$\text{(1) Phosphenolpyruvate} + HCO_3^- \xrightarrow{\textit{Phosphoenolpyruvate carboxylase}} \text{Oxalacetate} + H_2PO_4^-$$

$$\text{(2) Oxalacetate} + \text{NADH} \xrightarrow{\textit{Malatdehydrogenase}} \text{Malate} + \text{NAD}^+$$

Protocol 2 Bicarbonate

1. The specimen is first alkalinized to convert all carbon dioxide and carbonic acid to HCO^{3-}.
2. Phosphoenolpyruvate and HCO^{3-} reacts with phosphoenolpyruvate-carboxylase (1) (PEPC) and malat dehydrogenase (MDH) to Malate (2).
3. The decrease in absorbance of NADH at 340 nm is proportional to the total carbon dioxide content.

CO_2 – electrodes
A pCO_2-sensitive electrode determines the gaseous CO_2 after acidification. The rate of pH change of the buffer inside the membrane of the measuring electrode is taken as the measure of total CO_2 in the sample.

Calcium

Calcium is the most abundant mineral in animals, 99% of which is located in the bones and teeth. Calcium is needed to form teeth and bones and is also required for muscle contraction, blood clotting, and signal transmission in nerve cells. Its probably most well-known role is to prevent osteoporosis. Severe deficiencies of both calcium and vitamin D is called "rickets" in children and "osteomalacia" in adults.

Most laboratories measure calcium with photometric methods, but atomic absorption spectrometry (AAS) or ion-sensitive electrodes also have been used.

Photometric methods
Metal-complexing dyes such as O-Cresolphthalein or Arsenzo III form stable complexes with Ca^{2+} in alkaline or acidic solution.

O-Cresolphthalein forms a red complex with calcium in alkaline solution with an absorption peak between 570 and 580 nm. The sample is first acidified to release protein-bound and complexed calcium [1]. Organic base is added to buffer the reaction and to produce an alkaline pH (1).

Arsenazo III exhibits a high affinity for calcium, at pH 6, that is much higher than for magnesium [2]. The solution must be thoroughly buffered because binding of Ca^{2+} to Arsenazo III can be influenced by buffer and sodium concentrations (2).

$$\text{(1) Calcium + Cresolphthalein Complexone} \xrightarrow{\textit{Alkaline pH}} \text{Calcium-Cresolphthalein Complexone}$$

$$\text{(2) Calcium + Arsenazo III} \xrightarrow{\textit{Acid pH}} \text{Calcium-Arsenazo III Complex}$$

Ion-sensitive electrodes

Calcium ion-sensitive electrodes use liquid membranes containing the calcium-sensitive ionophore dissolved in an organic liquid trapped in a polymeric matrix (PVC).

Atomic absorption spectrometry

Total calcium is determined after diluting the sample with lanthanum chloride. The diluted specimen is aspirated into an air-acetylene flame, in which the ground-state calcium ions absorb light from a calcium hollow lamp. The absorption at 422.7 nm is directly proportional to the ground state calcium atoms in the flame.

Chloride

Chloride represents a major extra cellular anion, functions as an electrolyte balance associated with sodium, and preserves the electrolyte neutrality/maintenance of homeostasis.

Spectroscopy and ion-sensitive electrodes are the most common methods for determination of chloride.

Spectrophotometric methods

Chloride ions react with undissociated mercury thiocyanate to form undissociated mercuric chloride and free thiocyanate (1). The thiocanate ions react with Fe^{3+} to form a reddish-brown complex of ferric thiocyanate (2) with absorption maximum at 480 nm [3]. Addition of perchloric acid increases the color intensity.

$$\text{(1) } 2\,Cl^- + Hg(SCN)_2 \longrightarrow HgCl_2 + 2\,SCN^-$$

$$\text{(2) } SCN^- + Fe_3^+ \longrightarrow Fe(SCN)_3$$

Ion-sensitive electrodes

Solvent polymeric membranes that incorporate quaternary ammonium salt anion-exchangers (e.g., tri-*n*-octyl propyl ammonium chloride) are used for construction of Cl⁻sensitive electrodes [4].

Iron

At acid pH and in the presence of a reducing agent, transferrin-bound iron is released and reduced from Fe^{3+} to Fe^{2+}. It reacts with a specific chromogen (ferrozine or bathophenathroline), forming a colored complex. The color formed is proportional to the serum total iron concentration.

For determination of the serum unsaturated iron-binding capacity (UIBC), a standard amount of ferrous iron is added to serum at alkaline pH and is sequestered by transferrin, filling all available binding sites on the protein. The remaining unbound ferrous ions are then measured colorimetrically by use of ferrozine or bathophenathroline. The difference between the amount of unbound iron and the total amount of iron added is the UIBC.

The serum total iron-binding capacity (TIBC) is the sum of the serum total iron and the unsaturated iron-binding capacity. This is a measurement of the maximum concentration of iron that serum proteins can bind [5].

Magnesium

Today laboratories measure magnesium with photometric methods, but atomic absorption spectrometry (AAS) and ion-sensitive electrodes also have been used.

Photometric methods

A number of metallochromic indicators that change color on selectively binding magnesium have been used for measuring it in biological samples.

Magnesium forms a colored complex with Calmagite, the most commonly used indicator [6], in alkaline solution. The intensity of the color, measured at 520 nm, is proportional to the magnesium concentration measured at 530 to 550 nm. EGTA is added to reduce interference by calcium.

Xylidyl blue (Magon) forms a red complex with magnesium under alkaline conditions. This complex is measured at 520 nm. The intensity of the color is proportional to the magnesium concentration in the sample.

Atomic absorption spectrometry

Magnesium is determined after diluting the specimen with a lanthanum chloride solution to eliminate interference from anions. Then the sample is aspirated into an air-acetylene flame, in which the ground-state magnesium ions absorb light from a magnesium hollow lamp. The absorption at 285.2 nm is directly proportional to the magnesium atoms in the flame.

Ion-sensitive electrodes

These instruments use ionophores in a hydrophobic membrane for measuring Mg^{2+} concentration. But current ionophores have insufficient selectivity for Mg^{2+} over Ca^{2+}. Free calcium is simultaneously determined and used to correct free Mg^{2+} levels.

Phosphate

All common methods of phosphate determination are based on the reaction of phosphate ions with ammonium molybdate at acid pH to form a phosphomolybdate complex (1) [7]. The colorless phosphomolybdate complex can be measured either directly at 340 nm or after reduction by various reducing agents to produce molybdenum blue. Reduction of this compound produces a blue phosphomolybdenum complex measured at 620 to 700 nm (2) [8].

(1) Phosphate + Ammonium Molybdate + H^+ $\longrightarrow$ Ammonium Phosphomolybdate Complex

(2) AP-complex + Aminonaphtholsulfonic Acid $\longrightarrow$ blue complex

Sodium and potassium

Determination of sodium and potassium in body fluids is carried out with flame emission spectroscopy (FES), atomic absorption spectroscopy (AAS), spectrophotometrically, or electrochemically via ion-selective electrodes.

Flame emission spectrophotometry
Methods using flame photometry have been developed for both sodium and potassium, but today it is no longer a common laboratory method [9, 10].

Spectrophotometric methods
The spectral shift produced when either sodium or potassium binds to a macrocyclic chromphore is detected spectrophotometrically [11]. Macrocyclic ionophores (crown ethers, cryoptands, etc.) are molecules whose atoms are organized to form a specific structure into which metal ions fit and bind with high affinity. For example, ChromoLyte is a sodium-specific ionophore. Valinomycin in conjugation with a pH indicator is used to determine serum potassium concentration.

Ion-sensitive electrodes [12]
Analyzers contain sodium-sensitive electrodes with glass membranes. Potassium-sensitive electrodes contain liquid ion-exchange membranes that incorporate valinomycin. The measuring system is calibrated by introduction of calibrator solutions containing Na^+ and K^+.

3.2 Low-molecular-weight components

Cholesterol

Enzymatic methods [13] are the most commonly used methods for cholesterol measurements. Cholesteryl esters are hydrolyzed by an esterase to free cholesterol and fatty acids (1). The 3-OH group of cholesterol is then oxidized to a ketone by cholesterol oxidase (2). The resulting hydrogen peroxide is measured in a peroxidase-catalyzed reaction forming a quinoneimine dye (3).

$$\text{(1) Cholesterol Ester} \xrightarrow{\textit{Cholesterol esterase}} \text{Cholesterol + Fatty Acids}$$

$$\text{(2) Cholesterol} + O_2 \xrightarrow{\textit{Colesterol oxidase}} \text{Cholest-4-en-3-one} + H_2O_2$$

$$\text{(3) } 2\ H_2O_2 + \text{4-Aminoantipyridine + Phenol} \xrightarrow{\textit{Peroxidase}} \text{Quinone imine Dye} + 4\ H_2O$$

Cholesterol, high-density lipoprotein

High-density lipoprotein (HDL) cholesterol is measured after serum low-density (LDL) and very low-density (VLDL) lipoproteins are selectively precipitated by dextrane sulfate or phosphotungstic acid and removed by centrifugation.

In another method, the precipitant is complexed with magnetic particles. The supernatant contains cholesterol associated with the soluble HDL fraction and is measured by a standard method described as above [14].

In a direct method, an anti-human β-lipoprotein antibody binds to lipoproteins (LDL, VLDL, and chylomicrons). The antigen-antibody complexes formed block enzyme reactions when an enzymatic cholesterol reagent is added (see above). The blue color complex formed in that reaction is measured photometrically at 600 nm.

Glucose

Almost all commonly used techniques for glucose determination are enzymatic.

Hexokinase method [15]

ATP phosphorylates glucose in the presence of hexokinase and magnesium (1). The glucose-6-phosphate formed is oxidized by glucose-6-phosphate dehydrogenase in the presence of NAD^+ or $NADP^+$. The amount of NADH generated is directly proportional to the concentration of glucose in the sample and is measured by absorbance at 340 nm (2).

Alternatively, an oxidation/reduction system containing phenazine methosulfate (PMS) and iodonitrotetrazolium (INT) reacts with the NADH produced in the reaction. The reduction of INT forms a colored INT-formazan that is measured at 520 nm (3, 4).

$$\text{(1) Glucose + ATP} \xrightarrow{\textit{Hexokinase}} \text{Glucose-6-phosphate + ADP}$$

$$\text{(2) Glucose-6-phosphate} + NAD^+ \xrightarrow{\textit{G-6-PDH}} \text{6-Phosphogluconate + NADH} + H^+$$

$$\text{(3) NADH + PMS} \longrightarrow NAD^+ + \text{PMSH}$$

$$\text{(4) PMSH + INT} \longrightarrow \text{PMS + INT-Formazan}$$

Glucose oxidase method [16]

Glucose is oxidized to gluconic acid and hydrogen peroxide by the enzyme glucose oxidase (5). The hydrogen peroxide produced reacts with a chromogenic oxygen acceptor, such as 4-aminoantipyridine (6) or dianisidine (7), and results in a colored compound that can be measured.

$$\text{(5) Glucose} + 2\,H_2O + O_2 \xrightarrow{\textit{Glucose Oxidase}} \text{Gluconic Acid} + H_2O_2$$

$$\text{(6)}\ 2\,H_2O_2 + \text{4-Aminoantipyridine + p-Hydroxybenzene Sulfonate} \xrightarrow{\textit{Peroxidase}} \text{Quione imine Dye} + 4\,H_2O$$

$$\text{(7)}\ H_2O_2 + \text{o-Dianisidine} \xrightarrow{\textit{Peroxidase}} \text{Oxidized o-Diansidine} + H_2O$$

Glucose dehydrogenase method

β-D-glucose is oxidized to gluconolactone by glucose dehydrogenase (GDH). The amount of NADH generated is proportional to the glucose concentration and can be measured photometrically at 340 nm.

$$\text{(8) Glucose} + NAD^+ \xrightarrow{\textit{Glucose dehydrogenase}} \text{D-Gluconolactone + NADH} + H^+$$

Glutamate

L-Glutamic acid plays an important role in metabolism; it is, e. g., the precursor of other amino acids. L-Glutamic acid is part of many fermenter solutions and occurs in large quantities in most proteins.

$$\text{(1) L-Glutamate} + NAD^+ + H_2O \xrightarrow{\textit{Glutamate dehydrogenase}} \text{2-oxoglutarate + NADH} + NH^{4+}$$

$$\text{(2) NADH + H}^+ \text{ INT} \xrightarrow{\textit{Diaphorase}} \text{NAD}^+ + \text{formazan}$$

The formation of NADH in the first reaction is not favored; therefore, the produced NADH is oxidized, producing the colored formazan. An alternative flow injection system using immobilized Glutamate Dehydrogenase (GlDH) and aspartate aminotransferase is reported [17].

Protocol 3 L-Glutamic acid

1. Prepare buffer 350 mM imidazole / HCl pH 8.6, 1 g/l NAD^+, 0.1 g/l INT.
2. Prepare standard glutamate concentration series in 50 mM imidazole pH 7 buffer in the expected sample range.
3. Mix 960 µl buffer, 10 µl glutamate solution (standard or sample), 10 µl Diaphorase, and 10 µl GlDH.
4. Measure at 520 nm after 5 min.
5. Plot calibration from standard series and determine concentration of unknown sample.

Remarks and troubleshooting
The method is specific for L-glutamic acid, and the detection limit is 0.02 mg/l. High concentrations of ammonium ions or reducing substances disturb the reaction, as do very acid or basic samples, which influence the working pH. For high glutamate concentrations, the observed absorbance will be non-linear. In complex matrices it is useful to monitor the assay during the 10 min. of incubation. In the case of saturation, the enzyme or substrate needs to be reduced.

Glutamine

The major carbon and nitrogen sources in cultivation media for mammalian cells are glucose and L-glutamine. After conversion of glutamine to glutamate, a standard glutamate assay is used to quantify glutamine. An appropriate glutaminase exists, but the assay also can be run with the more cost-efficient asparaginase. An alternative flow injection system using the immobilized enzymes Glutamate Dehydrogenase (GlDH), aspartate aminotransferase, and asparaginase is reported [17].

$$\text{(1) L-Glutamine} + H_2O \xrightarrow{\textit{Glutaminase/Asparaginase}} \text{Glutamate} + NH_3$$

$$\text{(2) L-Glutamate} + \text{NAD}^+ + H_2O \xrightarrow{\textit{Glutamate dehydrogenase}} \text{2-oxoglutarate} + \text{NADH} + NH_4^+$$

$$\text{(3) NADH + H}^+ \text{ INT +} \xrightarrow{\textit{Diaphorase}} \text{NAD}^+ + \text{formazan}$$

Protocol 4 L-Glutamine

1. Prepare buffer 350 mM imidazole / HCl pH 8.7, 1 g/l NAD^+, 0.1 g/l INT.
2. Prepare buffer 50 mM imidazole / HCl pH 5.0.
3. Prepare standard glutamine concentration series in 50 mM imidazole pH 5 buffer in the expected sample range.
4. Mix 480 µl pH 5 buffer, 10 µl glutamate solution (standard or sample), and 10 µl asparaginase.
5. Incubate at room temperature for 5 min.
6. Add 480 µl pH 8.7 buffer, 10 µl diaphorase, and 10 µl GlDH.
7. Measure at 520 nm after 5 min.
8. Plot calibration graph from standard series and determine concentration of unknown sample.

Remarks and troubleshooting

The method is specific for L-glutamine; aspartate does not disturb because it is not oxidized by the GlDH. Clearly, the presence of glutamate will interfere with the assay, and the value for glutamate has to be determined with a separate experiment and subtracted from the "total" glutamine value. The detection limit is 0.05 mg/l. High concentrations of ammonium ions or reducing substances disturb the reaction, as do very acid or basic samples, which influence the working pH.

Lactate [18]

Lactate is oxidized to pyruvate and hydrogen peroxide by lactate oxidase (1). In the presence of the H_2O_2 generated, peroxidase catalyzes the oxidation of a chromogen to produce a colored dye with absorption maximum at 540 nm (2). The increase in absorbance at 540 nm is directly proportional to the lactate concentration in the specimen.

Lactate is oxidized to pyruvate by lactate dehydrogenase (LD) in the presence of NAD^+. The NADH formed in this reaction is measured spectrophotometrically at 340 nm (3). The equilibrium of the reaction normally lies on the lactate side. At pH 9 to 9.6, an excess of NAD^+, and use of tris-buffer, the equilibrium can be shifted to the right.

$$\text{(1) Lactate} + O_2 \xrightarrow{\textit{Lactate oxidase}} \text{Pyruvate} + H_2O_2$$

$$\text{(2) } 2\,H_2O_2 + \text{4-Aminoantipyridine} + \text{1,7-Dihydroxynaphthalene} \xrightarrow{\textit{Peroxidase}} \text{Red Dye}$$

$$\text{(3) Lactate} + NAD^+ \xrightarrow{\textit{Lactate dehydrogenase}} \text{Pyruvate} + NADH + H^+$$

Triglycerides

Triglycerides are measured enzymatically directly in biofluids such as plasma or serum. In all of the methods, the first step is the lipase-catalyzed hydrolysis of triglycerides to glycerol and fatty acids (1). Glycerol is then phosphorylated by glycerokinase (2). In the most used methods, glycerophosphate is oxidized to dihydroxyacetone and H_2O_2 (3), which is measured in a peroxidase-catalyzed reaction forming a quinone imine dye (4). The increase in absorbance, measured at 540 nm, is directly proportional to the triglyceride content in the sample. Free glycerol is assayed using a reagent without lipase. True triglyceride is calculated by subtracting the free glycerol concentration in the sample from the total triglycerides.

Alternatively, glycerophosphate can be measured in an NADH-producing reaction and the NADH measured spectrophotometrically at 340 nm (5) or in a diaphorase-catalyzed reaction forming a product measured at 500 nm (6).

Another method measures the ADP produced in reactions shown in (7) and (8). The decrease of NADH is measured photometrically at 340 nm.

$$\text{(1) Triglyceride} \xrightarrow{\textit{Lipase}} \text{Glycerol + Fatty Acids}$$

$$\text{(2) Glycerol + ATP} \xrightarrow{\textit{Glycerokinase}} \text{Glycerol-1-phosphate + ADP}$$

$$\text{(3) Glycerol-1-phosphate} + O_2 \xrightarrow{\textit{Glycerophosphate oxidase}} \text{Dihydroxyacetone Phosphate} + H_2O_2$$

$$\text{(4) } H_2O_2 + \text{4-Aminoantipyridine + Phenol} \xrightarrow{\textit{Peroxidase}} \text{Quinone imine Dye} + 2\ H_2O$$

$$\text{(5) Glycerol-1-Phosphate} + NAD^+ \xrightarrow{\textit{Glycerophosphate dehydrogenase}} \text{Dihydroxyacetone Phosphate} + NADH + H^+$$

$$\text{(6) NADH + Tetrazolium Dye} \xrightarrow{\textit{Diaphorase}} \text{Formazan} + NAD^+$$

$$\text{(7) ADP + Phospho-enol Pyruvate} \xrightarrow{\textit{Pyruvate kinase}} \text{ATP + Pyruvate}$$

$$\text{(8) Pyruvate} + NADH + H^+ \xrightarrow{\textit{Lactate dehydrogenase}} \text{Lactate} + NAD^+$$

Urea

Enzymatic methods
These methods are based on hydrolysis of urea with urease (1) followed by ammonium quantitation with glutamate dehydrogenase (GLDH) (2). A decrease of NADH concentration from the GLDH-reaction is measured photometerically at 340 nm in either an equilibrium or kinetic mode.

$$(1)\ \text{Urea} + 2\,H_2O \xrightarrow{\textit{Urease}} 2\,NH_4^+ + CO_3^{2-}$$

$$(2)\ NH_4^+ + \text{2-Oxoglutarate} + \text{NADH} + H^+ \xrightarrow{\textit{GLDH}} \text{Glutamate} + H_2O + NAD^+$$

Chemical methods
Most chemical methods for urea are based on the condensation of urea with diacetyl monoxime to form pink diazine (Fearon reaction), which absorbs strongly at 515 to 540 nm [19]. Ferric ions and thiosemicarbazide are added to the system to enhance the color.

$$\text{Urea} + \text{Diacetyl} \xrightarrow{H^+/\textit{heat}} \text{Diazine} + 2\,H_2O$$

Uric acid

Uric acid reacts with uricase at pH 9.4 to form allantoin. The absorbance of the reaction is measured before and after the addition of uricase. Since uric acid absorbs at 292 nm and the end products do not react; the decrease in absorbance is proportional to the uric acid concentration [20].

Many enzymatic assays for uric acid in serum involve a peroxidase system coupled with oxygen acceptors to produce a chromogen. The most common oxygen acceptor is a 4-aminophenazone together with a substituted phenol.

$$\text{Uric Acid} + 2\,H_2O + O_2 \xrightarrow{\textit{Uricase}} \text{Allantoin} + CO_2 + H_2O_2$$

3.3 Proteins

Total protein

Biuret method
This method is based on the presence of peptide bonds in all proteins. The copper ions in the Biuret reagent react with peptide bonds of serum proteins in alkaline medium to form a purple color with an absorbance maximum at 540 nm. The intensity of the color is proportional to the protein concentration [21].

Lowry method

A pyrogallol red-molybdate complex binds basic amino acid groups (Tyr, Trp) of protein molecules, which results in a shift in the reagent absorbance. The increase in absorbance at 600 nm is directly proportional to protein concentration in the sample. The reagent is intended for the quantitative determination of total protein in various fluids.

A modification of the Lowry [22] micro method uses Folin & Ciocalteu's phenol reagent to intensify the color of the purple-blue complex. The absorbance, which is read at 550 to 750 nm, is used to determine results from a standard curve.

Another modification of the Lowry method utilizes SDS for dissolving lipoproteins and offers excellent recovery of membrane proteins. An optional deoxycholate-trichloroacetic acid (TCA) precipitation technique permits removal of interfering substances such as tris, sucrose, EDTA, etc. The absorbance, read at 500 to 800 nm, is used to determine protein concentration.

Bradford method

Coomassie brilliant blue G reacts with proteins in an alcohol-acid medium to form a blue-colored protein dye complex. The absorbance is measured at 595 nm and is proportional to the protein concentration [23].

Protocol 5 Standard protein quantification

1. Prepare dye reagent by diluting 1 part Dye Concentrate with 4 parts distilled DDI water.
2. Filter through Whatman #1 filter (or equivalent) to remove particles.
3. The diluted reagent may be used for approximately 2 weeks when kept at room temperature.
4. Prepare three to five dilutions of a protein standard, which is representative of the protein solution to be tested.
5. The linear range of the assay for BSA is 0.2 to 0.9 mg/ml, whereas with IgG the linear range is 0.2 to 1.5 mg/ml.
6. Transfer 100 µl of each standard and sample solution into a clean, dry test tube.
7. Add 5.0 ml of diluted dye reagent to each tube and vortex.
8. Incubate at room temperature for at least 5 min.
9. The absorbance will increase over time. Thus, samples should incubate at room temperature for no more than 1 h.
10. Measure absorbance at 595 nm.

Protocol 6 Protein in microtiter plates

1. Prepare dye reagent as cited above.
2. The linear range of this microtiter plate assay is 0.05 mg/ml to approximately 0.5 mg/ml.
3. Protein solutions are normally assayed in duplicate or triplicate.
4. Transfer 10 µl of each standard and sample solution into separate microtiter plate wells.
5. Add 200 µl of diluted dye reagent to each well.
6. Mix the sample and reagent thoroughly using a microplate mixer.
7. Alternatively, use a multi-channel pipette to dispense the reagent. Depress the plunger repeatedly to mix the sample and reagent in the well. Replace with clean tips and add reagent to the next set of wells.
8. Incubate at room temperature for at least 5 min.
9. Absorbance will increase over time; samples should incubate at room temperature for no more than 1 h.
10. Measure absorbance at 595 nm.

Occasionally, a protein is assayed with exceptionally low color response to the Bradford reagent (e. g., gelatine). Nevertheless, quantitation of the protein is possible when the sample-to-dye ratio is changed. By using the sample-to-dye ratio of 800 µl sample + 200 µl dye reagent concentrate, a usable standard curve with reduced sensitivity is obtained.

If the absorbance is very low, the dye concentrate may be old. If it is over one year old, replace with a new bottle of reagent. Sometimes the sample may contain a substance that interferes with the reaction, such as detergents or sodium hydroxide. For low-molecular-weight proteins or peptides (< 5000 Daltons), this assay is not well suited.

Albumin

Albumin is the cheapest, well-defined single protein and represents the most abundant protein in blood (35 to 58 g/l, i. e., 55% to 65% of the total protein content). It has diverse functions, including the maintenance of the colloid osmotic pressure of the blood. Moreover, it is responsible for transportation of various ions, hormones, and drugs. Various chemical and immunological methods are available for the analysis of serum albumin. Mostly, methods are applied that are based on interaction with dyes, such as bromcresol green (BCG) and bromcresol purple (BCP). Bromcresol green binds quantitatively with human serum albumin at pH 4.2 to form an intensive blue-green complex that is determined spectrophotometrically at 628 nm. At pH 4.2, albumin acts as a cation to bind the anionic dye [24].

Bromcresol purple forms a stable complex with albumin at pH 5.2. The absorbance maximum is measured at 600 nm. The intensity of the color produced is directly proportional to the albumin concentration in the sample [25].

Protocol 7 Albumin

1. Transfer 3 ml BCG reagent into a spectrophotometer cell.
2. Add 20 µl of serum sample or standard.
3. Mix well and incubate for 2 min.
4. Read the absorbance at 640 nm against a reagent blank.
5. Plot calibration graph from standard series and determine concentration of unknown sample.

Hemoglobin [26]

Hemoglobin reacts with Drabkin's reagent, which contains potassium ferricyanide, potassium cyanide, and sodium bicarbonate. The Fe^{2+} of hemoglobin is oxidized to Fe^{3+} of methemoglobin by ferricyanide (1). Methemoglobin is converted into stable cyanomethemoglobin (2) by potassium cyanide (KCN). The absorbance of this derivative at 540 nm is proportional to the hemoglobin content.

For measurements of hemoglobin in complex fluids, 3,3′,5,5′-tetramethylbenzidine (TMB) is used as chromogen. Hemoglobin catalyzes the oxidation of TMB by hydrogen peroxide. The color formed in that reaction is proportional to the hemoglobin concentration in the test sample.

$$(1)\ HBFe^{2+} + Fe^{3+}(CN)_6^{3-} \longrightarrow HBFe^{3+} + Fe^{2+}(CN)_6^{4-}$$

$$(2)\ HBFe^{3+} + CN^- \longrightarrow HBFe^{3+}CN$$

3.4 Enzymes

Alanine aminotransferase

$$(1)\ \text{L-Alanine} + \text{2-Oxoglutarate} \xrightarrow{\textit{Alanine aminotransferase}} \text{Pyruvate} + \text{Glutamate}$$

$$(2)\ \text{Pyruvate} + \text{NADH} \xrightarrow{\textit{Lactate dehydrogenase}} \text{Lactate} + \text{NAD}^+$$

$$(3)\ \text{Pyruvate} + \text{2,4-Dinitrophenylhydrazine} \longrightarrow \text{Dinitrophenylhydrazone}$$

Protocol 8 Alanine aminotransferase (ALT)

1. Alanine aminotransferase (transaminase) catalyzes the reaction from L-alanine and 2-oxoglutarate, forming glutamate and pyruvate (1).
2. The pyruvate produced is reduced to lactate by lactate dehydrogenase (LDH) with simultaneous oxidation of NADH to NAD^+ that can be measured at 340 nm (2).
3. The decrease in NADH concentration is directly proportional to the ALT-activity.
4. The pyruvate that formed can react with 2,4-dinitrophenylhydrazine.
5. The resulting hydrazone of pyruvate is highly colored and its absorbance at 490–520 nm is proportional to ALT-activity (3).

Alkaline phosphatase [27]

Alkaline phosphatase is a broadly used enzyme that is applied for an easy, optically detectable interaction analysis. It catalyzes the hydrolysis of p-nitrophenyl phosphate (colorless) forming phosphate and free p-nitrophenol (yellow) under alkaline conditions. The rate of increase in absorbance at 405 nm that is due to the formation of p-nitrophenol is proportional to the AP activity.

$$\text{p-Nitrophenyl phosphate} \xrightarrow{\textit{Alkaline phosphatase}} \text{p-Nitrophenol} + P_i$$

Protocol 9 Alkaline phosphatase

1. Mix one part each of BCIP (5-bromo-4-chloro-3-indolyl-phosphate) concentrate and NBT (nitroblue tetrazolium) concentrate with 10 parts of Tris buffer solution in a glass container prior to use (i. e., 1 ml BCIP + 1 ml NBT + 10 ml Tris / HCl pH 8.2 Buffer). Alternatively, a premixed BCIP/NBT solution can be obtained from Sigma.
2. Prepare standard alkaline phosphatase concentrations in expected sample range.
3. Add 10 µl of samples or standard alkaline phosphatase dilution.
4. Mix well and incubate for 2 min.
5. Read the absorbance at 405 nm against a reagent blank.
6. Plot calibration curve from standard series and determine concentration of unknown sample.

Aspartate aminotransferase

$$\text{(1) L-Aspartate} + \text{2-Oxoglutarate} \xrightarrow{\textit{Aspartate aminotransferase}} \text{Oxalacetate} + \text{Glutamate}$$

$$\text{(2) Oxalacetate} + \text{NADH} \xrightarrow{\textit{Malate dehydrogenase}} \text{Malate} + NAD^+$$

$$\text{(3) Oxalacetate} + \text{2,4-Dinitrophenylhydrazine} \longrightarrow \text{Dinitrophenyl-hydrazone}$$

Protocol 10 Aspartate aminotransferase (AST)

1. Aspartate aminotransferase (transaminase) catalyzes the reaction from L-aspartate and 2-oxoglutarate, forming glutamate and oxalacetate (1).
2. The produced oxalacetate is reduced to malate by malate dehydrogenase (MDH).
3. Simultaneously, NADH is oxidized to NAD^+.
4. The reaction is measured at 340 nm (2).
5. The decrease in NADH concentration is directly proportional to the AST activity.
6. The pyruvate that is formed can react with 2,4-dinitrophenylhydrazine.
7. The resulting hydrazone of pyruvate is highly colored and its absorbance at 490–520 nm is proportional to AST activity (3).

Creatine kinase [28]

Creatine phosphate is converted to creatine with a phosphorylation of ADP to ATP by creatine kinase (CK) (1). The ATP produced is necessary for the hexokinase-catalyzed reaction of glucose to glucose-6-phosphate, which is converted to glucose-6-phosphate by glucose-6-phosphate dehydrogenase (G-6-PDH) (2). The rate of increase in absorbance at 340 nm that is due to the formation of NADH is directly proportional to the creatine kinase activity (3).

$$\text{(1) ADP + Phospho-creatine} \xrightarrow{\textit{Creatine kinase}} \text{ATP + Creatine}$$

$$\text{(2) ATP + Glucose} \xrightarrow{\textit{Hexokinase}} \text{ADP + Glucose-6-Phosphate}$$

$$\text{(3) Glucose-6-Phosphate + NAD}^+ \xrightarrow{\textit{G-6-PDH}} \text{6-Phosphogluconate + NADH}$$

Creatine kinase isoenzymes [29, 30]
CK isoenzymes are separated by electrophoresis, ion-exchange chromatography, or immunological methods.

Electrophoretic methods
The most commonly used technique is based on electrophoresis on agarose gels or cellulose acetate. The isoenzyme bands are visualized via incubation in a medium that promotes the following reactions:

$$\text{(1) ADP + Creatine Phosphate} \xrightarrow{\textit{Creatine kinase}} \text{Creatine + ATP}$$

$$\text{(2) ATP + Glucose} \xrightarrow{\textit{Hexokinase}} \text{ADP + Glucose-6-Phosphate}$$

$$\text{(3) Glucose-6-Phosphate + NAD}^+ \xrightarrow{\textit{Glucose-6-phosphate dehydrogenase}} \text{6-Phosphogluconate + NADH}$$

(4) NADH + Phenazine Methosulfate (oxid.) ⟶ NAD^+ + PMS (red.)

(5) PMS (red.) + Tetrazolium Dye ⟶ PMS (oxid.) + Formazan

The highly colored, insoluble formazan (NBT, MTT, or INT) localizes in electrophoresis zones of activity. The electrophoretic patterns may be visualized by photo-scanners.

Immunological methods

Immunological methods for measuring the CK iso-enzymes require specific antisera against M and B subunits. Immune-inhibition techniques are simpler and faster than immune-precipitation.

Serum is incubated with a CK-MB reagent containing antibody specific to the CK-M subunit, which completely inhibits the CK-M monomer. The activity of the CK-B that is not inhibited by the antibody is measured by the following reaction sequence:

(1) ADP + Creatine Phosphate ⟶ (*Creatine kinase*) Creatine + ATP

(2) ATP + Glucose ⟶ (*Hexokinase*) ADP + Glucose-6-Phosphate

(3) Glucose-6-Phosphate + NAD^+ ⟶ (*Glucose-6-phosphate dehydrogenase*) 6-Phosphogluconate + NADH

The rate of change in absorbance, that is measured at 340 nm, is directly proportional to the CK activity.

γ-Glutamyl transferase [31]

γ-Glutamyl transferase (γ-GT) catalyzes the transfer of glutamyl from γ-glutamyl-3-carboxy-4-nitroanilide to glycylglycine at pH 8.2. The increase in absorbance at 405 nm that is due to the p-nitroaniline formed in the reaction is measured spectrophotometrically.

γ-Glutamyl-3-Carboxy-4-Nitroanilide + Glycylglycine ⟶ (*γ-Glutamyl-Transferase*)

γ-Glutamylglycylglycine + 5-Amino-2-Nitrobenzoate

Lactate dehydrogenase [32]

Lactate dehydrogenase (LDH) catalyzes the conversion of l-lactate to pyruvate at alkaline pH 9.4. The rate of increase in absorbance at 340 nm that is due the formation of NADH is directly proportional to the LDH activity. In the presence of 2,4-dinitrophenylhydrazine, the pyruvate produces a highly colored phenyl hydrazone. The absorbance of that color formed is directly proportional to the LDH activity.

$$\text{Lactate} + NAD^+ \xrightarrow{\textit{Lactate dehydrogenase}} \text{Pyruvate} + NADH + H^+$$

Lipase

Titrimetric, turbidimetric, spectrophotometric, and fluorometric methods, and methods based on immunoassays, are commercially available for the measurement of lipases in biological samples. The most popular are spectrophotometric and titrimetric assays.

Spectrophotometric method [33]
Pancreatic lipase catalyzes the reaction from 1,2-diacylglycerol to 2-monoacylglycerol and fatty acids at pH 8.7 (1). In additional reaction steps (2)-(4), hydrogen peroxide is produced, which is reduced by peroxidase together with oxidation of a redox dye (5). The rate of increase in absorbance at 550 nm that is due to the formation of quinone diimine dye is directly proportional to the lipase activity in the specimen.

$$(1)\ \text{1,2-Diacylglycerol} + H_2O \xrightarrow{\textit{Pancreatic lipase}} \text{2-Monoacylglycerol} + \text{Fatty Acid}$$

$$(2)\ \text{2-Monoacylglycerol} + H_2O \xrightarrow{\textit{Monoglyceride lipase}} \text{Glycerol} + \text{Fatty Acid}$$

$$3)\ \text{Glycerol} + ATP \xrightarrow{\textit{Glycerol kinase}} \text{Glycerol-3-Phosphate} + ADP$$

$$\text{Glycerol-3-Phosphate} + O_2 \xrightarrow{\textit{Glycerol-3-Phosphate oxidase}} \text{Dihydroxyacetone Phosphate} + H_2O_2$$

$$(5)\ 2\ H_2O_2 + \text{4-Aminoantipyridine} + \text{TOOS}^* \xrightarrow{\textit{Peroxidase}} \text{Quinone Diimine Dye} + 2\ H_2O$$

*N-Ethyl-N-(2-hydroxy-3-sulfopropyl)-m-toluidine

Titrimetric method [34]
Lipase catalyzes the hydrolysis of fatty acids from an emulsion of olive oil triglycerides. The quantity of fatty acids formed is measured by titration with dilute sodium hydroxide solution. The quantity of alkali required to reach the indicator (thymolphthalein) endpoint is proportional to lipase activity.

$$\text{Olive Oil Triglycerides} \xrightarrow{\textit{Lipase}} \text{Fatty Acids} + \text{Diglycerides}$$

Immunoassay
There are some ELISA-based assays for lipase measurement based on the sandwich technique [35]. Specific anti-lipase antibodies, conjugated to peroxidase (PO) or alkaline phosphatase (AP), bind to various lipases from cell,

serum, or microorganisms. The measured PO/AP activity is proportional to the amount of lipase bound.

4 Standard values of selected parameters

Standard values of selected parameters in **blood plasma** and serum.

Substance	Value	Conc. unit
Alanine aminotransferase	*m* 10–40 *f* 7–35	U/l
Albumin	35–55	g/l
Alkaline phosphatase	*m* 38–94* *f* 28–111*	U/l
Ammonia-N (whole blood)	52–144	µmol/l
Asparte aminotransferase	8–20	U/l
Bicarbonate	21–25	mmol/l
Bilirubin, direct	< 6.8	µmol/l
Bilirubin, total	5–18.8	µmol/l
Calcium	2.2–2,7	mmol/l
Chloride	94–112	mmol/l
Cholesterol	3.35–6.70*	mmol/l
Creatine	*m* 11–22 *f* 13.3–24.3	µmol/l
Creatinine	*m* 23–61 *f* 23–92	µmol/l
Creatine kinase	*m* 62–106 *f* 44–88	U/l
C-reactive protein	0–9	mg/l
γ-glutamyl transferas)	*m* 2–30 *f* 1–24	U/l
Glucose	3.31–5.57*	mmol/l
Hemoglobin	*m* 131–174* *f 118–157**	g/l
Iron	*m* 16–25 *f* 14.3–21.6	µmol/l
Lactate	1–1.8	mmol/l
Lactate dehydrogenase	100–190*	U/l
Lipase	30–190	U/l
Magnesium	0.6–0.9	mmol/l
Phosphate	0.81–1.55	mmol/l
Potassium	4.1–5.6	mmol/l
Protein, Total	66–86	g/l
Sodium	135–149	mmol/l
Triglycerides	0.95–2.7*	mmol/l
Urea nitrogen	3.3–6.7	mmol/l
Uric acid	*m* 150–400 *f* 120–375	µmol/l

* age dependent m = male f = female

References

1 Connerty HV, Briggs AR (1966) Determination of serum calcium by means of orthocresolphthalein complexone. *Am J Clin Pathol* 45: 290–296

2 Bauer PJ (1981) Affinity and stoichiometry of calcium binding by arsenazo III. *Anal Biochem* 110: 61–72

3 Levinson SS (1976) Direct determination of serum chloride with a semi-automated discrete analyzer. *Clin Chem* 22: 273–274

4 Oesch U, Ammann D, Simon W (1986) Ion-selective membrane electrodes for clinical use. *Clin Chem* 32: 1448–1559

5 International Committee for Standardization in haematology (1978) The measurement of total and unsaturated iron binding capacity in serum. *Br J Haematol* 38: 281–290, 291–294

6 Liedtke RJ, Kroon G (1984) Automated calmagite compleximetric measurement of magnesium in serum, with sequential addition of EDTA to eliminate endogenous interference. *Clin Chem* 30: 1801–1804

7 Garber CC, Miller RC (1983) Revision of the 1963 semidine HCL standard method for inorganic phosphorous. *Clin Chem* 29: 184–188

8 Atkinson A, Gatenby AD, Lowe AG (1973) The determination of inorganic orthophosphate in biological systems. *Biochim Biophys Acta* 320: 195–204

9 Velapoldi RA, Paule RC, Schaffer R (1978) A reference method for the determination of sodium in serum. U. S. Department of Commerce, NBS special publication 260: Washington, DC, National Bureau of Standards

10 Velapoldi RA, Paule RC, Schaffer R (1978) A reference method for the determination of potassium in serum. U. S. Department of Commerce, NBS special publication 260-263: Washington, DC, National Bureau of Standards

11 Kumar A, Chapoteau E, Czech BP (1988) Chromogenic ionophore-based methods for spectrophotometric assay of sodium and potassium in serum and plasma. *Clin Chem* 34: 1709–1712

12 Maas AH, Siggaard-Andersen O, Weisberg HF (1985) Ionselective electrodes for sodium and potassium: A new problem of what is measured and what should be reported. *Clin Chem* 31: 482–485

13 Abell LL, Levy BB, Brodie BB (1952) A simplified method for the estimation of total cholesterol in serum and demonstration of its specifity. *J Biol Chem* 195: 357–366

14 National Cholesterol Education Program (1995) Recommendations on lipoprotein measurement: from the working group on lipoprotein measurement. NIH/NHLBINIH Publication No. 95-3044. Bethesda, MD, National Institutes of Health

15 Wright WR, Rainwater JC, Tolle LD (1971) Glucose assay systems: Evaluation of a colorimetric hexokinase procedure. *Clin Chem* 17: 1010–1015

16 Trinder P (1969) Determination of glucose in blood using glucose oxidase with an alternative oxygen acceptor. *Ann Clin Biochem* 6: 24–27

17 Mayer C, Frauer A., Schalkhammer T., Pittner F. (1999) Enzyme-Based Flow Injection Analysis System for Glutamine and Glutamate in Mammalian Cell Culture Media *Analytical Biochemistry* 268: 110–116

18 Livesley B, Atkinson L (1974) Accurate quantitative estimation of lactate in whole blood. *Clin Chem* 20: 1478

19 Sampson EJ, Baird MA (1979) Chemical inhibition used in a kinetic urease/glutamate dehydrogenase method for urea in serum. *Clin Chem* 25: 1721–1729

20 Duncan P, Gochman N, Cooper T (1982) A candidate reference method for uric acid in serum: I. Optimization and evaluation. *Clin Chem* 28: 284–290

21 Kingsley GR (1942)The direct biuret method for the determination of serum proteins as applied to photoelectric and

visual colorimetry. *J Lab Clin Med* 27: 840–845

22 Henry RJ, Cannon DC, Winkelman JW (eds) (1974) *Clinical chemistry: Principles and techniques.* 2nd ed. New York, Harper & Row, 424–428

23 Johnson A, Lott JA (1978) Standardization of the Coomassie brilliant blue method for cerebrospinal fluid proteins. *Clin Chim* 24: 1931–1933

24 Duggan J, Duggan PF (1982) Albumin by bromcresol green: A case of laboratory conservatism. *Clin Chem* 28: 1407–1408

25 Louderback A, Mealy EH, Taylor NA (1968) A new dye-binding technique using bromcresol purple for determination of albumin in serum. *Clin Chem* 14: 793–794

26 Van Kampen EJ, Zijlstra WG (1965) Determination of hemoglobin and its derivates. *Adv Clin Chem* 8: 141–187

27 Bowers GN, McComb RB (1966) A continuous spectrophotometric method for measuring the activity of serum alkaline phosphatase. *Clin Chem* 12: 70–89

28 Gerhardt W, Wulff K (1983) Creatine kinase. *In: Methods of Enzymatic Analysis. Vol. III, Enzymes 1: Oxidoreductases, Transferases.* 3rd ed. (H. U. Bergmeyer, J. Bergmeyer, M. Grassl (eds)) Weinheim, Verlag Chemie: 508–539

29 Apple FS, Preese LM (1994) Creatine kinase-MB: Detection of myocardial infarction and monitoring reperfusion. *J Clin Immunoassay* 17: 24–29

30 Brandt DR, Gates RC, Eng KK (1990) Quantifying the MB isoenzyme of creatine kinase with the Abbott "IMx" immunoassay analyzer. *Clin Chem* 36: 375–378

31 Rosalki SB, Tarlow D (1974) Optimized determination of γ-glutamyltransferase by reaction-rate analysis. *Clin Chem* 20: 1121–1124

32 McComb RB (1983) The measurement of lactate dehydrogenase. *In:* Clinical and Analytical Concepts in Enzymology (H. A. Homburger, Ed. Skokie, (eds), IL) College of American Pathologists: 157–171

33 Ventrucci M, Pezzili R, Garulli L, et al. (1994) Clinical evaluation of a new rapid assay for serum lipase determination. *Ital J Gastroenterol* 26:132–136

34 Tietz NW, Fiereck EA (1972) Measurement of lipase activity in serum. *In:* Standard Methods of Clinical Chemistry. (G. R. Cooper (ed)) New York, Academic Press, 1972, 19–31.

35 Grenner G, Deutsch G, Schmidtberger R, Dati F (1982) A highly sensitive enzyme immunoassay for the determination of pancreatic lipase. *J Clin Chem Clin Biochem* 20: 515–519

Guide to Solutions

Guide to Protocols

Index

UNIVERSITÄTSBIBLIOTHEK
BODENKULTUR
Wien
Peter-Jordanstraße 82
A-1190 Wien